SCIENTIFIC
COMPUTING

Springer

Berlin
Heidelberg
New York
Barcelona
Budapest
Hong Kong
London
Milan
Paris
Santa Clara
Tokyo

Springer
Singapore
Berlin
Heidelberg
New York
Barcelona
Budapest
Hong Kong
London
Milan
Paris
Santa Clara
Tokyo

PROCEEDINGS OF THE WORKSHOP ON

SCIENTIFIC COMPUTING

Hong Kong 10 –12 March, 1997

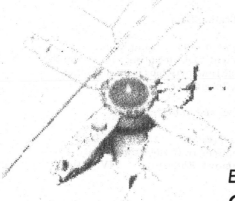

Editor-in-Chief
Gene Howard Golub

Managing Editor
Shiu Hong Lui

Editors
Franklin T. Luk
Robert James Plemmons

 Springer

Gene Howard Golub
Department of Computer Science
Stanford University
Stanford, CA 94305, U.S.A.

Shiu Hong Lui
Department of Mathematics
Hong Kong University of Science and Technology
Clear Water Bay, Kowloon, Hong Kong

Franklin T. Luk
Department of Computer Science
Rensselaer Polytechnic Institute
Troy, NY 12180, U.S.A.

Robert James Plemmons
Department of Mathematics and Computer Science
Wake Forest University
Winston-Salem, NC 27109, U.S.A.

Library of Congress Cataloging-in-Publication Data

Workshop on Scientific Computing (6th : 1997 : Hong Kong)
Proceedings of the Workshop on Scientific Computing : 10–12 March,
1997, Hong Kong / editor-in-chief, Gene Howard Golub ; managing editor,
Shiu Hong Lui ; editors, Franklin T. Luk, Robert James Plemmons.
          p.   cm.
     Includes bibliographical references and index.
     ISBN 9813083603
     1. Electronic data processing--Congresses.   2. Science--Data
processing--Congresses.   I. Golub, Gene H. (Gene Howard), 1932–
II. Luk, Franklin T.   III. Plemmons, Robert J.   IV. Title.
QA75.5.W626   1997
621.382'2'0285--dc21
                                                                  97-34830
                                                                  CIP

ISBN 981-3083-60-3

© Springer-Verlag Singapore Pte. Ltd. 1997
Printed in Singapore

Typesetting: Camera-ready by authors
5 4 3 2 1 0

Foreword

The Workshop on Scientific Computing 97 was held on March 10-12, 1997 on the campuses of the Hong Kong Baptist University and the Chinese University of Hong Kong. The papers in this volume are formal versions of the lectures; all contributed papers had been refereed.

The three-day workshop is the sixth in a series of workshops on scientific computing held in Hong Kong since 1990; the objectives are to promote research interest in scientific computing for local mathematicians and engineers, and to foster contacts and exchanges with experts from China and the rest of the world. The major themes of this conference were numerical linear algebra, signal processing, and image processing.

This meeting was particularly noteworthy because it celebrated Gene Golub's 65th birthday. Gene's colleagues and former students came to Hong Kong from all over the world to pay special tributes to his leadership and pioneering roles in scientific computation. It was a great honor for Hong Kong to host the conference.

I thank members of the Program Committee: Raymond Chan, Franklin Luk, and Robert Plemmons, who gathered together a superb cast of speakers; and the Local Organizing Committee: Kam-moon Liu, Shiu-hong Lui, Tsi-min Shih, Tao Tang, Chong Sze Tong, Wei-min Xue (Chairman), Siu-pang Yung, who worked tirelessly on organizational details. We acknowledge generous financial support from the Chinese University of Hong Kong, City University of Hong Kong, Hong Kong Baptist University, Hong Kong University of Science and Technology, Hong Kong Polytechnic University, University of Hong Kong, Institute of Mathematical Sciences (CUHK), Hong Kong Mathematical Society, Hong Kong Pei Hua Education Foundation Ltd, and the United States Army Research Office (Far East). We also thank all the participants for their contributions toward the success of this conference. Finally, we express our gratitude to Mr. Wai Wa Fung for his help in the preparation of this document.

May 1997

S. H. Lui
Managing Editor
WSC97

Table of Contents

Part I
Invited Papers

Tikhonov Regularization
for Large Scale Problems

Gene H. Golub[1]* and Urs von Matt[2]

[1] Department of Computer Science, Stanford University, Stanford, CA 94305,
golub@na-net.ornl.gov
[2] ISE Integrated Systems Engineering AG, Technopark Zürich, Technoparkstr. 1,
CH-8005 Zürich, Switzerland, vonmatt@ise.ch

Abstract. Tikhonov regularization is a powerful tool for the solution of
ill-posed linear systems and linear least squares problems. The choice of
the regularization parameter is a crucial step, and many methods have
been proposed for this purpose. However, efficient and reliable methods
for large scale problems are still missing.

In this paper approximation techniques based on the Lanczos algorithm
and the theory of Gauss quadrature are proposed to reduce the computational complexity for large scale problems. The new approach is
applied to 5 different heuristics: Morozov's discrepancy principle, the
Gfrerer/Raus-method, the quasi-optimality criterion, generalized cross-
validation, and the L-curve criterion. Numerical experiments are used to
determine the efficiency and robustness of the various methods.

1 Introduction

The solution of ill-posed linear systems and linear least squares problems is
a frequent task in numerical analysis. We refer the reader to [5,17,30] for an
overview of applications. In this paper we consider the (overdetermined) linear
system

$$b = Ax + e,$$

where A is an m-by-n matrix with $m \geq n$, b and e are vectors of size m, and x
is an n-vector. The matrix A and the vector b are given, and e is assumed to be
a random noise vector.

The direct solution of the least squares problem

$$\|Ax - b\|_2 = \min \tag{1}$$

may lead to a vector x that is severely contaminated with noise. Therefore
regularization is employed to get a more meaningful answer. In this paper we
focus on Tikhonov regularization. This approach improves the condition of the
problem by solving the linear least squares problem

$$\|Ax - b\|_2^2 + \alpha\|x\|_2^2 = \min \tag{2}$$

* The work of this author was in part supported by NSF under grant CCR-9505393.

instead of (1). The solution x_α satisfies the equation

$$(A^T A + \alpha I)x_\alpha = A^T b.$$

Equivalently x_α can be computed as the solution to the damped linear least squares problem

$$\left\| \begin{bmatrix} A \\ \sqrt{\alpha}I \end{bmatrix} x - \begin{bmatrix} b \\ 0 \end{bmatrix} \right\|_2 = \min. \tag{3}$$

In [21] P. C. Hansen and D. P. O'Leary show that, for a given $\alpha > 0$, x_α also solves the regularized total least squares problem

$$\min_{A_0, b_0, x} \| [A, b] - [A_0, b_0] \|_F,$$

$$\text{s.t.} \quad b_0 = A_0 x, \quad \|x\|_2 \leq \delta := \|(A^T A + \alpha I)^{-1} A^T b\|_2.$$

The choice of an appropriate regularization parameter α is crucial, and many methods have been proposed to compute α. In Section 2 we give an overview of some frequently used criteria. The computational core of these methods can be expressed in terms of certain matrix moments. Since the main focus of this paper is the calculation of regularization parameters for large scale problems we will present an iterative method to approximate matrix moments. The required theory of Gauss quadrature is reviewed in Section 3, and stable methods based on the Lanczos algorithm are presented in Sections 4 and 5. As our methods of computing the regularization parameter α are based on an iterative algorithm, the Lanczos algorithm, we must provide an appropriate termination criterion. This turns out to be a nontrivial subject, and it is discussed in Section 6. Finally we present numerical results in Section 7.

2   Regularization Criteria

The choice of an appropriate regularization parameter α is crucial, and many methods have been proposed for this purpose. We will present a number of popular methods which may also be applied to large scale problems by means of the approximation techniques of Sections 3–5. The first two methods assume that the norm of the noise vector e is known, whereas the last three methods make no such assumption.

2.1   Morozov's Discrepancy Principle

The value of α is chosen such that the norm of the residual $b - Ax_\alpha$ equals the norm of the error term [25,26]:

$$\|b - A(A^T A + \alpha I)^{-1} A^T b\|_2 = \|e\|_2.$$

For $\alpha > 0$ the identity

$$I - A(A^T A + \alpha I)^{-1} A^T = \alpha (A A^T + \alpha I)^{-1}$$

holds. Consequently the equation for α can also be written as

$$\phi_M(\alpha) := \alpha^2 b^T (A A^T + \alpha I)^{-2} b = \|e\|_2^2. \tag{4}$$

Because

$$\phi'_M(\alpha) = 2\alpha b^T A (A^T A + \alpha I)^{-3} A^T b$$

the function $\phi_M(\alpha)$ is strictly increasing for $\alpha > 0$, and the equation (4) has a unique positive solution.

2.2 The Gfrerer/Raus-Method

The Gfrerer/Raus-method [8,17] may be seen as an improved variant of the discrepancy principle. It determines α such that

$$\phi_{GR}(\alpha) := \alpha^3 b^T (A A^T + \alpha I)^{-3} b = \|e\|_2^2. \tag{5}$$

The first derivative of $\phi_{GR}(\alpha)$ is given by

$$\phi'_{GR}(\alpha) = 3\alpha^2 b^T A (A^T A + \alpha I)^{-4} A^T b.$$

Consequently $\phi_{GR}(\alpha)$ is a strictly increasing function for $\alpha > 0$, and the solution of (5) is unique.

2.3 The Quasi-Optimality Criterion

The quasi-optimality criterion (cf. [1,24], [26, Sect. 27], [31, Sect. II.6], and [32]) determines $\alpha > 0$ such that

$$\phi_Q(\alpha) := \alpha^2 b^T A (A^T A + \alpha I)^{-4} A^T b = \min.$$

Note that we have $\phi_Q(0) = 0$ such that the global minimizer of $\phi_Q(\alpha)$ is attained at $\alpha = 0$, provided that the matrix A has full rank. As M. Hanke and P. C. Hansen point out in [17, p. 283] this is an artificial feature of a finite-dimensional regularization problem. The function $\phi_Q(\alpha)$ usually has a large maximum in close vicinity of $\alpha = 0$, and the desired minimizer of $\phi_Q(\alpha)$ lies to the right of this maximum.

2.4 Generalized Cross-Validation

Generalized cross-validation (GCV) [12,35,36] determines the regularization parameter α as the global minimizer of

$$\phi_{GCV}(\alpha) := \frac{\|(A A^T + \alpha I)^{-1} b\|_2}{\operatorname{trace}\left((A A^T + \alpha I)^{-1}\right)}.$$

The trace term in the denominator of $\phi_{\text{GCV}}(\alpha)$ may be infeasible to evaluate for large matrices A. In [2,9,10,22] stochastic trace estimators are introduced to approximate the trace. We will use the approach proposed by M. F. Hutchinson [22]. Let U be the discrete random variable which takes the values $+1$ and -1 each with probability $1/2$, and let $\boldsymbol{u}$ be a vector of size m whose entries are independent samples from U. Then

$$\tilde{t}(\alpha) = \boldsymbol{u}^{\text{T}}(AA^{\text{T}} + \alpha I)^{-1}\boldsymbol{u} \tag{6}$$

is an unbiased estimator of

$$t(\alpha) = \text{trace}((AA^{\text{T}} + \alpha I)^{-1}).$$

Therefore we will only consider the minimization of the stochastic GCV function

$$\tilde{\phi}_{\text{GCV}}(\alpha) := \frac{\sqrt{\boldsymbol{b}^{\text{T}}(AA^{\text{T}} + \alpha I)^{-2}\boldsymbol{b}}}{\boldsymbol{u}^{\text{T}}(AA^{\text{T}} + \alpha I)^{-1}\boldsymbol{u}}$$

from now on. In [15] it is shown that the minimizers of ϕ_{GCV} and $\tilde{\phi}_{\text{GCV}}$ are equally well suited for the Tikhonov regularization in (2).

2.5   The L-Curve Criterion

The L-curve criterion is based on a plot of the solution norm $\|\boldsymbol{x}_\alpha\|_2$ versus the residual norm $\|\boldsymbol{b} - A\boldsymbol{x}_\alpha\|_2$ in a log-log scale [18,20]. The optimal regularization parameter α is characterized by a corner of this graph, i.e., the point on the L-curve with maximum curvature. Therefore the L-curve criterion maximizes the curvature

$$\phi_{\text{L}}(\alpha) := \frac{\rho'\eta'' - \rho''\eta'}{\left((\rho')^2 + (\eta')^2\right)^{3/2}} = \max,$$

where

$$\rho(\alpha) = \log\|\boldsymbol{b} - A(A^{\text{T}}A + \alpha I)^{-1}A^{\text{T}}\boldsymbol{b}\|_2 = \log\|\alpha(AA^{\text{T}} + \alpha I)^{-1}\boldsymbol{b}\|_2,$$
$$\eta(\alpha) = \log\|(A^{\text{T}}A + \alpha I)^{-1}A^{\text{T}}\boldsymbol{b}\|_2.$$

The prime denotes differentiation with respect to α.

Note that

$$\rho'(\alpha) = \frac{\boldsymbol{b}^{\text{T}}A(A^{\text{T}}A + \alpha I)^{-3}A^{\text{T}}\boldsymbol{b}}{\alpha\boldsymbol{b}^{\text{T}}(AA^{\text{T}} + \alpha I)^{-2}\boldsymbol{b}},$$

$$\eta'(\alpha) = -\frac{\boldsymbol{b}^{\text{T}}A(A^{\text{T}}A + \alpha I)^{-3}A^{\text{T}}\boldsymbol{b}}{\boldsymbol{b}^{\text{T}}A(A^{\text{T}}A + \alpha I)^{-2}A^{\text{T}}\boldsymbol{b}}.$$

The numerator of $\phi_{\text{L}}(\alpha)$ can be written as

$$\rho'\eta'' - \rho''\eta' = \rho'^2\left(\frac{\eta'}{\rho'}\right)' = -\eta'^2\left(\frac{\rho'}{\eta'}\right)'.$$

Therefore we have

$$\rho'\eta'' - \rho''\eta' = \left(\frac{b^{\mathrm{T}} A (A^{\mathrm{T}} A + \alpha I)^{-3} A^{\mathrm{T}} b}{\alpha b^{\mathrm{T}} (AA^{\mathrm{T}} + \alpha I)^{-2} b \cdot b^{\mathrm{T}} A (A^{\mathrm{T}} A + \alpha I)^{-2} A^{\mathrm{T}} b} \right)^2 .$$
$$(- b^{\mathrm{T}} (AA^{\mathrm{T}} + \alpha I)^{-2} b \cdot b^{\mathrm{T}} A (A^{\mathrm{T}} A + \alpha I)^{-2} A^{\mathrm{T}} b$$
$$+ 2\alpha b^{\mathrm{T}} (AA^{\mathrm{T}} + \alpha I)^{-3} b \cdot b^{\mathrm{T}} A (A^{\mathrm{T}} A + \alpha I)^{-2} A^{\mathrm{T}} b$$
$$- 2\alpha b^{\mathrm{T}} (AA^{\mathrm{T}} + \alpha I)^{-2} b \cdot b^{\mathrm{T}} A (A^{\mathrm{T}} A + \alpha I)^{-3} A^{\mathrm{T}} b).$$

2.6   Large Scale Problems

All of the regularization criteria discussed so far require the evaluation of matrix moments of the form

$$\mu_p(\alpha) = b^{\mathrm{T}} (AA^{\mathrm{T}} + \alpha I)^p b, \qquad (7)$$
$$\nu_p(\alpha) = c^{\mathrm{T}} (A^{\mathrm{T}} A + \alpha I)^p c. \qquad (8)$$

The exponent p is always a negative integer. If the matrix A is not too large, say $n \leq 1000$, then a direct method can be used to compute a regularization parameter α. For instance one can employ the singular value decomposition of A to evaluate the functions $\phi(\alpha)$ in the five heuristics above. This is the approach taken in the regularization tool box by P. C. Hansen [19].

As the size of the matrix A increases the direct evaluation of these moments becomes less and less feasible. We will therefore present an iterative method which will enable us to compute lower and upper bounds on these moments. The bounds will become tighter as the iteration index k increases. In the next section we will review the necessary theory from Gauss quadrature.

3   Gauss Quadrature

Let us consider the problem of approximating the quadratic form

$$s := g^{\mathrm{T}} \varphi(M) g, \qquad (9)$$

where φ is an analytic function and M is a symmetric n-by-n matrix. Let

$$M = U \Lambda U^{\mathrm{T}}$$

be the eigenvalue decomposition of M, with $\lambda_1 \leq \cdots \leq \lambda_n$ as the eigenvalues. We assume that a and b denote lower and upper bounds on the spectrum of M, i.e., $a \leq \lambda_1$ and $\lambda_n \leq b$. The expression (9) can be rewritten as

$$s = h^{\mathrm{T}} \varphi(\Lambda) h = \sum_{i=1}^{n} \varphi(\lambda_i) h_i^2 = \int_a^b \varphi(\lambda) \, d\omega(\lambda), \qquad (10)$$

where the vector h is given by

$$h := U^T g.$$

In (10) we also expressed the finite sum as a Stieltjes integral where the measure $\omega(x)$ is a staircase function with steps of size h_i^2 at the eigenvalues λ_i. This representation of s as an integral suggests to use numerical integration to approximate it. Because the moments

$$\mu_k = \int_a^b \lambda^k \, d\omega(\lambda) = g^T M^k g$$

are easy to compute for $k \geq 0$ we will focus on Gauss quadrature:

$$s = \int_a^b \varphi(\lambda) \, d\omega(\lambda) \approx \sum_{i=1}^k \varphi(x_i) w_i. \tag{11}$$

The quantities $x_1 < \cdots < x_k$ denote the abscissas or nodes of the quadrature rule, and the w_i's are the corresponding weights. The degree of the quadrature rule is given by the index k.

It is sometimes useful to prescribe an abscissa $x_1 = a$ or $x_k = b$. One obtains a Gauss-Radau quadrature rule if $m = 1$ abscissa is prescribed. If both endpoints a and b are prescribed, i.e., $m = 2$, one obtains a Gauss-Lobatto quadrature rule.

There is an extensive literature on Gauss quadrature (cf. [3,7,23,28,34]). In [16] it is shown how the nodes x_i and the weights w_i can be computed by means of an eigenvalue decomposition. In [11] this approach is extended to Gauss-Radau and Gauss-Lobatto quadrature rules. In this section we give a short overview of the relevant theory.

3.1 Orthogonal Polynomials

A sequence of orthogonal polynomials can be associated with a weight function $\omega(\lambda)$. For the sake of simplicity we may assume that all the eigenvalues λ_i of M are distinct and that all the entries in the vector h are nonzero. Then there exists the sequence $\{p_i\}_{i=0}^{n-1}$, where p_i is a polynomial of degree i, such that

$$\int_a^b p_i(\lambda) p_j(\lambda) \, d\omega(\lambda) = \begin{cases} 1, & \text{if } i = j, \\ 0, & \text{if } i \neq j. \end{cases}$$

These polynomials satisfy the three-term recurrence relationship

$$\lambda \begin{bmatrix} p_0(\lambda) \\ \vdots \\ \vdots \\ p_{k-1}(\lambda) \end{bmatrix} = \begin{bmatrix} \alpha_1 & \beta_1 & & \\ \beta_1 & \ddots & \ddots & \\ & \ddots & \alpha_{k-1} & \beta_{k-1} \\ & & \beta_{k-1} & \alpha_k \end{bmatrix} \begin{bmatrix} p_0(\lambda) \\ \vdots \\ \vdots \\ p_{k-1}(\lambda) \end{bmatrix} + \begin{bmatrix} 0 \\ \vdots \\ 0 \\ \beta_k p_k(\lambda) \end{bmatrix}, \tag{12}$$

Algorithm 1. Lanczos Algorithm.

$q_{-1} := 0$
$\beta_0 := 0$
$q_0 := g/\|g\|_2$
$\alpha_1 := q_0^T M q_0$
for $k := 2$ **to** n **do**
$\quad r_{k-1} := (M - \alpha_{k-1}I)q_{k-2} - \beta_{k-2}q_{k-3}$
$\quad \beta_{k-1} := \|r_{k-1}\|_2$
$\quad q_{k-1} := r_{k-1}/\beta_{k-1}$
$\quad \alpha_k := q_{k-1}^T M q_{k-1}$
end

or,

$$\lambda p_k(\lambda) = T_k p_k(\lambda) + \beta_k p_k(\lambda) e_k$$

for $k = 1, \ldots, n$. The coefficients α_i and β_i can be computed by means of the Lanczos algorithm 1 (cf. [14, Chapt. 9]). An arbitrary nonzero value can be chosen for β_n. The zeros of $p_k(\lambda)$ are also the eigenvalues of T_k.

In exact arithmetic Algorithm 1 computes the orthogonal n-by-n matrix

$$Q_n = \begin{bmatrix} q_0 \cdots q_{n-1} \end{bmatrix}$$

and the tridiagonal matrix T_n such that

$$M = Q_n T_n Q_n^T.$$

If we execute only $k < n$ iterations we get Q_k, the first k columns of Q_n, and T_k, the k-by-k leading principal submatrix of T_n. It is easy to see that these quantities satisfy the equation

$$MQ_k = Q_k T_k + \beta_k q_k e_k^T. \tag{13}$$

3.2 Construction of Gauss Quadrature Rules

There are $2k - m$ degrees of freedom when determining a Gauss quadrature rule (11) of degree k with m prescribed abscissas. Thus we can determine the nodes x_i and the corresponding weights w_i such that all polynomials up to degree $2k - m - 1$ are integrated exactly. As V. I. Krylov shows in [23, p. 161] an equivalent requirement is to find a polynomial $\tilde{p}_k(\lambda)$ of degree k, whose zeros are the nodes $x_1 < \cdots < x_k$, and the weights $w_1, \ldots, w_k$, such that

1. the equation

$$\int_a^b \tilde{p}_k(\lambda)p(\lambda) \, d\omega(\lambda) = 0 \tag{14}$$

holds for all polynomials p of degree $d < k - m$, and,

2. the equation

$$\int_a^b p(\lambda)\, d\omega(\lambda) = \sum_{i=1}^k p(x_i) w_i \qquad (15)$$

holds for all polynomials p of degree $d < k$.

3.3   Abscissas

First we show how the polynomial $\tilde{p}_k(\lambda)$ can be determined such that condition (14) holds. In the case of a Gauss quadrature rule, where no abscissas are prescribed, we can choose $\tilde{p}_k = p_k$, the kth orthogonal polynomial with respect to the weight function $\omega(\lambda)$. This ensures that $\tilde{p}_k$ is orthogonal to all polynomials of degree $d < k$.

Next we consider a Gauss-Radau quadrature rule where we prescribe $x_1 = a$. We choose the following representation for $\tilde{p}_k$:

$$\tilde{p}_k = \beta_k p_k + (\alpha_k - \tilde{\alpha}_k) p_{k-1}.$$

This ensures that $\tilde{p}_k$ is orthogonal to all polynomials of degree $d < k - 1$. It is not hard to see that $\tilde{p}_k$ also satisfies the equation

$$\lambda \begin{bmatrix} p_0(\lambda) \\ \vdots \\ \vdots \\ p_{k-1}(\lambda) \end{bmatrix} = \begin{bmatrix} \alpha_1 & \beta_1 & & \\ \beta_1 & \ddots & \ddots & \\ & \ddots & \alpha_{k-1} & \beta_{k-1} \\ & & \beta_{k-1} & \tilde{\alpha}_k \end{bmatrix} \begin{bmatrix} p_0(\lambda) \\ \vdots \\ \vdots \\ p_{k-1}(\lambda) \end{bmatrix} + \begin{bmatrix} 0 \\ \vdots \\ 0 \\ \tilde{p}_k(\lambda) \end{bmatrix},$$

or,

$$\lambda \boldsymbol{p}_k(\lambda) = \tilde{T}_k \boldsymbol{p}_k(\lambda) + \tilde{p}_k(\lambda) \boldsymbol{e}_k.$$

Thus the zeros of $\tilde{p}_k$ are just the eigenvalues of $\tilde{T}_k$, and if we set

$$\tilde{\alpha}_k = a + \beta_{k-1}^2 \boldsymbol{e}_{k-1}^{\mathrm{T}} (T_{k-1} - aI)^{-1} \boldsymbol{e}_{k-1},$$

we ensure that $\tilde{T}_k$ has the eigenvalue $x_1 = a$.

If we want to compute a Gauss-Radau quadrature rule with the prescribed abscissa $x_k = b$ we would just set

$$\tilde{\alpha}_k = b + \beta_{k-1}^2 \boldsymbol{e}_{k-1}^{\mathrm{T}} (T_{k-1} - bI)^{-1} \boldsymbol{e}_{k-1}.$$

Finally we want to determine the polynomial $\tilde{p}_k$ for a Gauss-Lobatto quadrature rule, where $x_1 = a$ and $x_k = b$ are prescribed. We may write $\tilde{p}_k$ as

$$\tilde{p}_k = \frac{\beta_k \beta_{k-1}}{\tilde{\beta}_{k-1}} p_k + (\alpha_k - \tilde{\alpha}_k) \frac{\beta_{k-1}}{\tilde{\beta}_{k-1}} p_{k-1} + \frac{\beta_{k-1}^2 - \tilde{\beta}_{k-1}^2}{\tilde{\beta}_{k-1}} p_{k-2},$$

where the parameters $\tilde{\alpha}_k$ and $\tilde{\beta}_{k-1}$ are to be determined such that $\tilde{p}_k(a) = \tilde{p}_k(b) = 0$. Obviously $\tilde{p}_k$ is orthogonal to all polynomials of degree $d < k - 2$. It can also be verified that $\tilde{p}_k$ satisfies the equation

$$
\lambda \begin{bmatrix} p_0(\lambda) \\ \vdots \\ \vdots \\ \tilde{p}_{k-1}(\lambda) \end{bmatrix} = \begin{bmatrix} \alpha_1 \ \beta_1 & & \\ \beta_1 & \ddots & \ddots \\ & \ddots & \alpha_{k-1} \ \tilde{\beta}_{k-1} \\ & & \tilde{\beta}_{k-1} \ \tilde{\alpha}_k \end{bmatrix} \begin{bmatrix} p_0(\lambda) \\ \vdots \\ \vdots \\ \tilde{p}_{k-1}(\lambda) \end{bmatrix} + \begin{bmatrix} 0 \\ \vdots \\ 0 \\ \tilde{p}_k(\lambda) \end{bmatrix},
$$

or,

$$
\lambda \tilde{\boldsymbol{p}}_k(\lambda) = \tilde{T}_k \tilde{\boldsymbol{p}}_k(\lambda) + \tilde{p}_k(\lambda) e_k.
$$

By setting

$$
\tilde{\alpha}_k = \frac{b y_{k-1} - a z_{k-1}}{y_{k-1} - z_{k-1}},
$$

$$
\tilde{\beta}_{k-1} = \sqrt{\frac{b - a}{y_{k-1} - z_{k-1}}},
$$

where

$$
y_{k-1} = e_{k-1}^{T} (T_{k-1} - aI)^{-1} e_{k-1},
$$

$$
z_{k-1} = e_{k-1}^{T} (T_{k-1} - bI)^{-1} e_{k-1},
$$

we can assign the two eigenvalues a and b to the matrix $\tilde{T}_k$. Since the eigenvalues of $\tilde{T}_k$ are the zeros of $\tilde{p}_k$ we have the desired property $\tilde{p}_k(a) = \tilde{p}_k(b) = 0$.

3.4   Weights

The weights $\{w_i\}_{i=1}^k$ must be determined in such a way that condition (15) is satisfied. We will only discuss the case of a Gauss quadrature rule. The cases of a Gauss-Radau or a Gauss-Lobatto quadrature rule are handled similarly: just replace the matrix T_k by $\tilde{T}_k$.

Condition (15) is equivalent to the requirement that the orthogonal polynomials $\{p_i\}_{i=0}^{k-1}$ are integrated exactly:

$$
\int_a^b p_0 \, d\omega(\lambda) = \sum_{j=1}^k p_0 w_j = \frac{1}{p_0} = \sqrt{\int_a^b d\omega(\lambda)} = \sqrt{\mu_0},
$$

$$
\int_a^b p_i(\lambda) \, d\omega(\lambda) = \sum_{j=1}^k p_i(x_j) w_j = 0, \qquad i = 1, \dots, k - 1,
$$

or, in matrix terms:

$$
Pw = \sqrt{\mu_0} e_1. \tag{16}
$$

Note that the abscissas $x_1 < \cdots < x_k$ are the zeros of p_k, which are also the eigenvalues of T_k. This is obvious from the three-term recurrence relationship (12). Therefore the columns of P are the eigenvectors of T_k, and they are orthogonal to each other:

$$PX = T_k P.$$

Here the quantity X denotes the diagonal eigenvalue matrix $\mathrm{diag}(x_i)$. However the columns of P are not normalized. Let

$$T_k = QXQ^\mathrm{T}$$

denote the eigenvalue decomposition of T_k with $Q^\mathrm{T} Q = I$. Then we have

$$Q = PD,$$

where $D = \mathrm{diag}(q_{1i}/p_0)$. Consequently the solution of the linear system (16) is given by

$$w = \sqrt{\mu_0} D Q^\mathrm{T} e_1,$$

or,

$$w_i = \mu_0 q_{1i}^2.$$

3.5   Evaluation of Gauss Quadrature Rules

In order to approximate the quadratic form (9) by a quadrature rule of degree k we must execute k iterations of the Lanczos Algorithm 1. The value of the Gauss quadrature rule can then be evaluated as follows:

$$G = \sum_{i=1}^{k} \varphi(x_i) w_i = \mu_0 \sum_{i=1}^{k} \varphi(x_i) q_{1i}^2$$
$$= \|g\|_2^2 e_1^\mathrm{T} Q \varphi(X) Q^\mathrm{T} e_1 = \|g\|_2^2 e_1^\mathrm{T} \varphi(T_k) e_1.$$

A Gauss-Radau quadrature rule R and a Gauss-Lobatto quadrature rule L can be expressed in the same way. We only need to replace the matrix T_k by $\tilde{T}_k$ as discussed in Section 3.3.

3.6   Quadrature Error

In [23, pp. 162–163] and [27, p. 134] explicit expressions for the quadrature error

$$E[\varphi] := \int_a^b \varphi(\lambda) \, d\omega(\lambda) - \sum_{i=1}^{k} \varphi(x_i) w_i$$

are derived. In the case of a Gauss quadrature rule, there exists a ξ_G such that

$$E[\varphi] = \frac{\varphi^{(2k)}(\xi_G)}{(2k)!} \int_a^b \prod_{i=1}^k (x - x_i)^2 \, d\omega(x), \qquad a \leq \xi_G \leq b,$$

and for a Gauss-Radau quadrature rule with $x_1 = a$ we have

$$E[\varphi] = \frac{\varphi^{(2k-1)}(\xi_R)}{(2k-1)!} \int_a^b (x - a) \prod_{i=2}^k (x - x_i)^2 \, d\omega(x), \qquad a \leq \xi_R \leq b.$$

Similarly the error for a Gauss-Lobatto quadrature rule is given by

$$E[\varphi] = \frac{\varphi^{(2k-2)}(\xi_L)}{(2k-2)!} \int_a^b (x - a)(x - b) \prod_{i=2}^{k-1} (x - x_i)^2 \, d\omega(x), \qquad a \leq \xi_L \leq b.$$

3.7 Matrix Moments

As we have seen in Section 2.6 we are particularly interested in approximating the matrix moments $\mu_p(\alpha)$ and $\nu_p(\alpha)$. These moments can be expressed as a quadratic form (9) with

$$\varphi(\lambda) = (\lambda + \alpha)^p.$$

The key derivatives of φ are given as follows:

$$\varphi^{(2k)}(\lambda) = p(p-1) \cdots (p - 2k + 1)(\lambda + \alpha)^{p-2k},$$
$$\varphi^{(2k-1)}(\lambda) = p(p-1) \cdots (p - 2k + 2)(\lambda + \alpha)^{p-2k+1},$$
$$\varphi^{(2k-2)}(\lambda) = p(p-1) \cdots (p - 2k + 3)(\lambda + \alpha)^{p-2k+2}.$$

The exponent p is always negative. We may also assume that $\alpha > 0$ and $\lambda \geq 0$ since both AA^T and $A^T A$ are symmetric positive semidefinite matrices. Consequently we know the sign of the derivatives of φ:

$$\varphi^{(2k)}(\lambda) \geq 0,$$
$$\varphi^{(2k-1)}(\lambda) \leq 0,$$
$$\varphi^{(2k-2)}(\lambda) \geq 0.$$

Therefore we can use a Gauss or a Gauss-Radau ($x_k = b$) quadrature rule to compute a lower bound on the matrix moments of Section 2.6. Conversely we get upper bounds from a Gauss-Radau ($x_1 = a$) or a Gauss-Lobatto quadrature rule. In the remainder of this paper we will only use a Gauss quadrature rule to obtain a lower bound, and a Gauss-Radau quadrature rule with $x_1 = a = 0$ for the upper bound.

Algorithm 2. Lanczos Bidiagonalization I.

$q_0 := b/\|b\|_2$
$s_0 := A^T q_0$
$\gamma_1 := \|s_0\|_2$
$w_0 := s_0/\gamma_1$
for $k := 2$ **to** n **do**
    $r_{k-1} := A w_{k-2} - \gamma_{k-1} q_{k-2}$
    $\delta_{k-1} := \|r_{k-1}\|_2$
    $q_{k-1} := r_{k-1}/\delta_{k-1}$
    $s_{k-1} := A^T q_{k-1} - \delta_{k-1} w_{k-2}$
    $\gamma_k := \|s_{k-1}\|_2$
    $w_{k-1} := s_{k-1}/\gamma_k$
end

4   Lanczos Bidiagonalization I

In this section we consider the numerical aspects of approximating the matrix moment $\mu_p(\alpha)$ in (7). The discussion in Section 3 seems to suggest that we apply Algorithm 1 to the matrix $M = AA^T$. However, we may lose accuracy if we compute AA^T explicitly. This problem can be avoided by using the Lanczos bidiagonalization of Algorithm 2 (cf. [13]). If we execute k iterations of this algorithm in exact arithmetic we get the orthogonal m-by-k matrix

$$Q_k = [q_0 \cdots q_{k-1}]$$

and the orthogonal n-by-k matrix

$$W_k = [w_0 \cdots w_{k-1}].$$

The matrices Q_k and W_k are connected by the two equations

$$AW_k = Q_k B_k + \delta_k q_k e_k^T, \tag{17}$$
$$A^T Q_k = W_k B_k^T, \tag{18}$$

where B_k denotes the k-by-k lower bidiagonal matrix

$$B_k = \begin{bmatrix} \gamma_1 & & & \\ \delta_1 & \ddots & & \\ & \ddots & \ddots & \\ & & \delta_{k-1} & \gamma_k \end{bmatrix}.$$

By combining (17) and (18) we get

$$AA^T Q_k = Q_k B_k B_k^T + \gamma_k \delta_k q_k e_k^T.$$

It is now straightforward to identify $B_k B_k^T$ with the matrix T_k in (13), and $\gamma_k \delta_k$ is equal to β_k in (13). Thus the Gauss quadrature rule for the approximation of (7) can be written as

$$G_p(\alpha) = \|b\|_2^2 e_1^T (B_k B_k^T + \alpha I)^p e_1.$$

To compute a Gauss-Radau quadrature rule we prescribe the abscissa $x_1 = a = 0$. This is an obvious choice since AA^T is always positive semidefinite. Thus we have to compute the entry $\tilde{\alpha}_k$ in $\tilde{T}_k$ such that $\tilde{T}_k$ becomes singular. It is not hard to see that we can express $\tilde{T}_k$ by

$$\tilde{T}_k = \tilde{B}_k \tilde{B}_k^T,$$

where we obtain $\tilde{B}_k$ from B_k by setting γ_k to zero. Consequently, the Gauss-Radau quadrature rule is given by

$$R_p(\alpha) = \|b\|_2^2 e_1^T (\tilde{B}_k \tilde{B}_k^T + \alpha I)^p e_1.$$

As discussed in Section 3.7 the Gauss and Gauss-Radau quadrature rules are actually lower and upper bounds on the matrix moment $\mu_p(\alpha)$:

$$G_p(\alpha) \le \mu_p(\alpha) \le R_p(\alpha).$$

Note that

$$\lim_{\alpha \downarrow 0} R_p(\alpha) = \infty$$

in general for $p < 0$. On the other hand $G_p(\alpha)$ remains bounded for $\alpha \downarrow 0$ if B_k is nonsingular.

We do not want to clutter the notation more than necessary, and we omit the degree k of the quadrature rule from the functions G_p and R_p. The reader should bear in mind, however, that these functions will approximate $\mu_p(\alpha)$ more and more accurately as k increases.

In order to evaluate $G_p(\alpha)$ and $R_p(\alpha)$ we need to solve linear systems of the form

$$(B_k B_k^T + \alpha I)x = y.$$

It is important to note that for $\alpha > 0$ the solution x also satisfies the equivalent linear least squares problem

$$\left\| \begin{bmatrix} B_k^T \\ \sqrt{\alpha} I \end{bmatrix} x - \begin{bmatrix} 0 \\ y/\sqrt{\alpha} \end{bmatrix} \right\|_2 = \min.$$

This least squares problem can be solved efficiently by a sequence of Givens transformations. The details of this approach are discussed in [4, pp. 140–141], [6, pp. 45–52], and [34, pp. 103–108].

Algorithm 3. Lanczos Bidiagonalization II.

$q_0 := c/\|c\|_2$
$r_0 := Aq_0$
$\gamma_1 := \|r_0\|_2$
$p_0 := r_0/\gamma_1$
for $k := 2$ **to** n **do**
    $s_{k-1} := A^T p_{k-2} - \gamma_{k-1} q_{k-2}$
    $\delta_{k-1} := \|s_{k-1}\|_2$
    $q_{k-1} := s_{k-1}/\delta_{k-1}$
    $r_{k-1} := Aq_{k-1} - \delta_{k-1} p_{k-2}$
    $\gamma_k := \|r_{k-1}\|_2$
    $p_{k-1} := r_{k-1}/\gamma_k$
end

5   Lanczos Bidiagonalization II

We will now address the approximation of the matrix moment $\nu_p(\alpha)$ in (8).
The Lanczos bidiagonalization of Algorithm 3 operates on the matrix A instead
of $A^T A$ (cf. [13]). After k iterations in exact arithmetic we obtain the orthogonal
m-by-k matrix

$$P_k = [p_0 \; \cdots \; p_{k-1}]$$

and the orthogonal n-by-k matrix

$$Q_k = [q_0 \; \cdots \; q_{k-1}] .$$

If we define the k-by-k upper bidiagonal matrix

$$B_k = \begin{bmatrix} \gamma_1 & \delta_1 & & \\ & \ddots & \ddots & \\ & & \ddots & \delta_{k-1} \\ & & & \gamma_k \end{bmatrix},$$

then P_k and Q_k satisfy the equations

$$AQ_k = P_k B_k, \tag{19}$$
$$A^T P_k = Q_k B_k^T + \delta_k q_k e_k^T. \tag{20}$$

From (19) and (20) we conclude that

$$A^T A Q_k = Q_k B_k^T B_k + \gamma_k \delta_k q_k e_k^T.$$

Obviously the matrix $B_k^T B_k$ is the same as T_k in (13), and $\gamma_k \delta_k$ corresponds to β_k
in (13). We obtain the following Gauss quadrature rule for the approximation
of (8):

$$G_p(\alpha) = \|c\|_2^2 e_1^T (B_k^T B_k + \alpha I)^p e_1.$$

Again, we can also use a Gauss-Radau quadrature rule to approximate $\nu_p(\alpha)$, and we will prescribe the abscissa $x_1 = a = 0$. Similarly as in Section 4 we can write the matrix $\tilde{T}_k$ as

$$\tilde{T}_k = \tilde{B}_k^T \tilde{B}_k,$$

where we obtain $\tilde{B}_k$ from B_k by setting γ_k to zero. Consequently, the Gauss-Radau quadrature rule is given by

$$R_p(\alpha) = \|c\|_2^2 e_1^T (\tilde{B}_k^T \tilde{B}_k + \alpha I)^p e_1.$$

We are now able to bound $\nu_p(\alpha)$ from below and above:

$$G_p(\alpha) \leq \nu_p(\alpha) \leq R_p(\alpha).$$

In general the function $R_p(\alpha)$ is unbounded for $p < 0$ as α goes to zero:

$$\lim_{\alpha \downarrow 0} R_p(\alpha) = \infty.$$

If B_k is nonsingular, however, $G_p(\alpha)$ remains bounded for $\alpha \downarrow 0$. Both bounds become tighter as we execute more and more iterations in Algorithm 3.

The evaluation of $G_p(\alpha)$ and $R_p(\alpha)$ involves the solution of linear systems of the form

$$(B_k^T B_k + \alpha I)x = y.$$

For $\alpha > 0$ we prefer to compute x by means of a linear least squares problem as discussed in the previous section.

6   Termination Criteria

The theory of Sections 3–5 allows us to compute lower and upper bounds on the functions $\phi(\alpha)$ of Section 2. If we execute k steps of Algorithm 2 (Lanczos bidiagonalization I) with the initial vector b we get the following bounds on the moment $\mu_p(\alpha)$:

$$\|b\|_2^2 e_1^T (B_k B_k^T + \alpha I)^p e_1 \leq b^T (AA^T + \alpha I)^p b \leq \|b\|_2^2 e_1^T (\tilde{B}_k \tilde{B}_k^T + \alpha I)^p e_1.$$

Note that the same Lanczos algorithm can also be used to compute bounds on $\tilde{t}(\alpha)$ defined in (6). We just have to use the random vector u as the initial vector.

Similarly we run Algorithm 3 (Lanczos bidiagonalization II) with the initial vector $c = A^T b$. Thus we get the following bounds on $\nu_p(\alpha)$:

$$\|c\|_2^2 e_1^T (B_k^T B_k + \alpha I)^p e_1 \leq c^T (A^T A + \alpha I)^p c \leq \|c\|_2^2 e_1^T (\tilde{B}_k^T \tilde{B}_k + \alpha I)^p e_1.$$

It is straightforward to compute lower and upper bounds on the various functions $\phi(\alpha)$ defined in Section 2:

$$\mathfrak{L}_k(\alpha) \leq \phi(\alpha) \leq \mathfrak{U}_k(\alpha).$$

These bounds get tighter as the number k of Lanczos iterations increases. We will now discuss the choice of appropriate stopping criteria.

Only certain values of the regularization parameter α are meaningful in the least squares problem (3). We will only consider α's in the range

$$\alpha_{\min} \leq \alpha \leq \alpha_{\max},$$

where

$$\alpha_{\min} = \|A\|_2^2 \, \varepsilon^2,$$
$$\alpha_{\max} = \|A\|_2^2,$$

where ε denotes the unit roundoff of the computer. If $\alpha < \alpha_{\min}$ the matrix $\sqrt{\alpha} I$ in (3) is numerically zero compared to A. On the other hand, if $\alpha > \alpha_{\max}$, the damping matrix $\sqrt{\alpha} I$ becomes larger than the data matrix A.

As a simple guard against premature termination we require that at least

$$k_{\min} := \lceil 3 \log \min(m, n) \rceil$$

Lanczos iterations are executed. Thus the following stopping criteria are only checked for $k \geq k_{\min}$.

In the case of Morozov's discrepancy principle and the Gfrerer/Raus-method, $\phi(\alpha)$ is a strictly increasing function for $\alpha > 0$. Therefore, there is a unique regularization parameter α_* such that $\phi(\alpha) = \|e\|_2^2$. The solution of the two equations

$$\mathfrak{U}_k(\alpha) = \|e\|_2^2,$$
$$\mathfrak{L}_k(\alpha) = \|e\|_2^2,$$

yields the bounds α_l and α_u, respectively, such that

$$\alpha_l \leq \alpha_* \leq \alpha_u.$$

Consequently we may stop the iteration as soon as the bounds on α_* are tight enough. We use the criterion

$$0.99 \, \alpha_u \leq \alpha_l, \tag{21}$$

which allows us to determine α_* to two digits of accuracy.

The criterion (21) works well if $\phi'(\alpha_*)$ is sufficiently large, say, $\phi'(\alpha_*) \geq 1$. In the case of $\phi'(\alpha_*) \approx 0$ we need to relax the criterion (21). We only require that $\phi(\alpha)$ is approximated to two digits of accuracy in the range $\alpha_l \leq \alpha \leq \alpha_u$. In our implementation we use the condition

$$0.99 \, \mathfrak{U}_k(\alpha_l) \leq \|e\|_2^2 \leq 1.01 \, \mathfrak{L}_k(\alpha_l). \tag{22}$$

The Lanczos iteration is stopped as soon as either (21) or (22) is satisfied.

In the case of the quasi-optimality criterion, GCV, and the L-curve criterion we need to compute a global minimizer or maximizer of $\phi(\alpha)$. Unfortunately, $\phi(\alpha)$ is usually neither convex nor concave, and thus no simple bounds on the global minimizer/maximizer α_* are available.

But the bounds on $\phi(\alpha)$ can still be used to compute approximations to α_*. We observe that the bounds $\mathfrak{L}_k(\alpha)$ and $\mathfrak{U}_k(\alpha)$ are very tight for large α's. As $\alpha \downarrow 0$, these bounds become increasingly loose. It is usually infeasible to execute the Lanczos algorithm until $\phi(\alpha)$ is approximated well over the whole interval $[\alpha_{\min}, \alpha_{\max}]$. In our implementation we only iterate until the largest local minimizer or maximizer of $\phi(\alpha)$ has been identified. Obviously this strategy fails if the global solution α_* is smaller than the largest local minimizer/maximizer. In Section 7 we will discuss the robustness of the three regularization criteria with respect to this type of premature termination.

We will now discuss the approximation of the largest local minimizer of $\phi(\alpha)$ for the quasi-optimality criterion and GCV. The largest local maximizer of $\phi_L(\alpha)$ can be computed analogously.

First we compute a global minimizer α_u of the upper bound $\mathfrak{U}_k(\alpha)$. As a next step we would like to determine $\alpha_1 < \alpha_u < \alpha_2$ such that

$$\mathfrak{L}_k(\alpha_1) > \mathfrak{U}_k(\alpha_u) < \mathfrak{L}_k(\alpha_2). \tag{23}$$

This would mean that the interval $[\alpha_1, \alpha_2]$ contains a local minimizer of $\phi(\alpha)$. Unfortunately, condition (23) may need a lot of iterations to be satisfied in practice. Therefore we only try to find an $\alpha_1 < \alpha_u$ such that

$$\mathfrak{L}_k(\alpha_1) > \mathfrak{L}_k(\alpha_u).$$

To account for rounding errors in the calculation of the bounds we require that

$$\begin{aligned} \alpha_1 &\leq (1 - \sqrt{\varepsilon})\,\alpha_u, \\ \mathfrak{L}_k(\alpha_1) &\geq (1 + \sqrt{\varepsilon})\,\mathfrak{L}_k(\alpha_u), \end{aligned} \tag{24}$$

where ε denotes the unit roundoff of the computer.

7   Numerical Results

We implemented our methods to compute regularization parameters for large scale problems in MATLAB [29]. Our experiments were conducted on a Sun Ultra workstation with a unit roundoff of $\varepsilon = 2^{-52} \approx 2.2204 \cdot 10^{-16}$. The software is available from the second author upon request.

We used the functions "baart," "phillips," and "shaw" in Per Christian Hansen's regularization tool box [19] to generate test matrices. These matrices represent discretizations of Fredholm integral equations, a frequent source of ill-posed linear systems. However, they are dense, and we will only solve them for small values of n.

Table 1. Test Cases.

	Test Matrix	m	n	σ_1	σ_n	relerr$_e$
1	large	2000	1000	1.000	$1.690 \cdot 10^{-87}$	10^{-1}
2	large	2000	1000	1.000	$1.690 \cdot 10^{-87}$	10^{-2}
3	large	2000	1000	1.000	$1.690 \cdot 10^{-87}$	10^{-3}
4	large	20000	10000	1.000	$3.147 \cdot 10^{-869}$	10^{-1}
5	large	20000	10000	1.000	$3.147 \cdot 10^{-869}$	10^{-2}
6	large	20000	10000	1.000	$3.147 \cdot 10^{-869}$	10^{-3}
7	baart	100	100	3.229	$1.581 \cdot 10^{-18}$	10^{-1}
8	baart	100	100	3.229	$1.581 \cdot 10^{-18}$	10^{-3}
9	baart	100	100	3.229	$1.581 \cdot 10^{-18}$	10^{-5}
10	baart	100	100	3.229	$1.581 \cdot 10^{-18}$	10^{-7}
11	phillips	200	200	5.803	$1.372 \cdot 10^{-7}$	10^{-1}
12	phillips	200	200	5.803	$1.372 \cdot 10^{-7}$	10^{-2}
13	phillips	200	200	5.803	$1.372 \cdot 10^{-7}$	10^{-3}
14	phillips	200	200	5.803	$1.372 \cdot 10^{-7}$	10^{-4}
15	shaw	100	100	2.993	$5.486 \cdot 10^{-19}$	10^{-1}
16	shaw	100	100	2.993	$5.486 \cdot 10^{-19}$	10^{-3}
17	shaw	100	100	2.993	$5.486 \cdot 10^{-19}$	10^{-5}
18	shaw	100	100	2.993	$5.486 \cdot 10^{-19}$	10^{-7}

In order to obtain large scale test problems, which can also be solved explicitly, we use the m-by-n matrix A defined by its singular value decomposition

$$A = U\Sigma V^{\mathrm{T}},$$

where

$$U = I - 2\frac{uu^{\mathrm{T}}}{\|u\|_2^2},$$

$$V = I - 2\frac{vv^{\mathrm{T}}}{\|v\|_2^2},$$

and the entries of u and v are random numbers. Σ is a diagonal matrix with the singular values $\sigma_i = e^{-0.2(i-1)}$. We will call this the "large" test problem.

We use a random vector x_0 as the true solution of (1) and compute the right-hand side b according to

$$b = Ax_0 + e,$$

where e is a random vector with

$$\|e\|_2 = \mathrm{relerr}_e \, \|Ax_0\|_2.$$

We used the 18 problems from Table 1 to test our regularization algorithms. The columns σ_1 and σ_n show the values of the largest and the smallest singular values of A, respectively. Note that σ_1 and σ_n were computed by using

Table 2. Average Value of the Exact Regularization Parameter α_*.

	Morozov	Gfrerer/Raus	Quasi-Optimality	GCV	L-Curve
1	$2.998 \cdot 10^{-4}$	$5.420 \cdot 10^{-4}$	$9.538 \cdot 10^{-4}$	$3.319 \cdot 10^{-5}$	$9.380 \cdot 10^{-4}$
2	$2.303 \cdot 10^{-6}$	$4.737 \cdot 10^{-6}$	$2.452 \cdot 10^{-4}$	$1.800 \cdot 10^{-7}$	$7.384 \cdot 10^{-6}$
3	$6.278 \cdot 10^{-9}$	$1.409 \cdot 10^{-8}$	$1.630 \cdot 10^{-5}$	$4.630 \cdot 10^{-10}$	$2.136 \cdot 10^{-8}$
4	$8.784 \cdot 10^{-6}$	$1.674 \cdot 10^{-5}$	$5.333 \cdot 10^{-5}$	$8.359 \cdot 10^{-7}$	$1.326 \cdot 10^{-4}$
5	$4.553 \cdot 10^{-7}$	$1.036 \cdot 10^{-6}$	$7.441 \cdot 10^{-7}$	$3.786 \cdot 10^{-8}$	$8.779 \cdot 10^{-6}$
6	$3.885 \cdot 10^{-9}$	$8.053 \cdot 10^{-9}$	$8.719 \cdot 10^{-6}$	$1.809 \cdot 10^{-10}$	$1.756 \cdot 10^{-8}$
7	$5.929 \cdot 10^{-4}$	$1.367 \cdot 10^{-3}$	$4.394 \cdot 10^{-4}$	$1.362 \cdot 10^{-4}$	$5.894 \cdot 10^{-4}$
8	$3.869 \cdot 10^{-6}$	$7.385 \cdot 10^{-6}$	$1.191 \cdot 10^{-6}$	$3.138 \cdot 10^{-7}$	$4.887 \cdot 10^{-7}$
9	$3.062 \cdot 10^{-9}$	$9.005 \cdot 10^{-9}$	$2.037 \cdot 10^{-9}$	$2.809 \cdot 10^{-10}$	$2.614 \cdot 10^{-10}$
10	$1.227 \cdot 10^{-14}$	$2.979 \cdot 10^{-14}$	$3.921 \cdot 10^{-12}$	$9.020 \cdot 10^{-16}$	$4.760 \cdot 10^{-15}$
11	$1.398 \cdot 10^{-2}$	$2.064 \cdot 10^{-2}$	$3.593 \cdot 10^{-33}$	$1.224 \cdot 10^{-3}$	$7.050 \cdot 10^{-2}$
12	$1.871 \cdot 10^{-4}$	$3.756 \cdot 10^{-4}$	$1.181 \cdot 10^{-32}$	$3.103 \cdot 10^{-5}$	$7.388 \cdot 10^{-4}$
13	$3.131 \cdot 10^{-6}$	$6.626 \cdot 10^{-6}$	$2.562 \cdot 10^{-32}$	$3.663 \cdot 10^{-7}$	$5.803 \cdot 10^{-6}$
14	$4.941 \cdot 10^{-8}$	$8.631 \cdot 10^{-8}$	$2.562 \cdot 10^{-32}$	$8.465 \cdot 10^{-9}$	$9.112 \cdot 10^{-9}$
15	$7.259 \cdot 10^{-3}$	$9.898 \cdot 10^{-3}$	$1.885 \cdot 10^{-2}$	$3.049 \cdot 10^{-4}$	$1.307 \cdot 10^{-2}$
16	$2.077 \cdot 10^{-6}$	$3.283 \cdot 10^{-6}$	$1.660 \cdot 10^{-2}$	$8.562 \cdot 10^{-8}$	$4.283 \cdot 10^{-7}$
17	$5.511 \cdot 10^{-9}$	$2.597 \cdot 10^{-8}$	$1.876 \cdot 10^{-8}$	$2.604 \cdot 10^{-10}$	$1.503 \cdot 10^{-10}$
18	$7.917 \cdot 10^{-14}$	$5.636 \cdot 10^{-14}$	$2.008 \cdot 10^{-8}$	$4.422 \cdot 10^{-16}$	$1.927 \cdot 10^{-14}$

MATLAB's svd command. If $\sigma_n \approx \varepsilon \sigma_1$, the matrix A is numerically singular and the value of σ_n may carry no significant digits. However, for the large test matrix, the extreme singular values are exact. Note that the test cases 11–14 are only moderately ill-conditioned, whereas all the other test cases are severely ill-conditioned.

We solved each test problem 10 times with a given regularization technique. Each time we used a different random perturbation e of the right-hand side. This gives us a better way to assess the average properties of each regularization method.

Table 2 shows the average values for the exact regularization parameters α_* produced by each regularization technique. The value of α_* tends to 0 as the norm of the perturbation becomes smaller and smaller. For the test cases 11–14 the quasi-optimality criterion fails to compute a regularization parameter. As already pointed out in Section 2.3 this anomaly is caused by the properties of the finite-dimensional regularization problem. The function $\phi_Q(\alpha)$ has a large maximum around $\alpha \approx 10^{-14}$, and the desired minimizer α_* lies to the right of this maximum. Unfortunately, $\phi_Q(\alpha)$ tends to 0 as $\alpha \downarrow 0$, and our black box minimizer determined $\alpha = \alpha_{\min}$ as the global minimizer of $\phi_Q(\alpha)$. Because of this drawback the quasi-optimality criterion is not as robust as the other regularization methods.

Table 3. Average Relative Error relerr$_\alpha$ of the Regularization Parameter α.

	Morozov	Gfrerer/Raus	Quasi-Optimality	GCV	L-Curve
1	$1.274 \cdot 10^{-1}$	$1.104 \cdot 10^{-1}$	$7.718 \cdot 10^{-2}$	$3.744 \cdot 10^{-2}$	$4.186 \cdot 10^{-2}$
2	$1.765 \cdot 10^{-1}$	$1.337 \cdot 10^{-1}$	$1.818 \cdot 10^{-1}$	$8.277 \cdot 10^{-2}$	$4.069 \cdot 10^{-2}$
3	$1.542 \cdot 10^{-1}$	$1.411 \cdot 10^{-1}$	$6.483 \cdot 10^{+1}$	$9.226 \cdot 10^{-2}$	$5.027 \cdot 10^{+4}$
4	$5.045 \cdot 10^{-1}$	$4.552 \cdot 10^{-1}$	$8.546 \cdot 10^{+0}$	$8.670 \cdot 10^{-2}$	$1.349 \cdot 10^{-15}$
5	$5.972 \cdot 10^{-1}$	$3.591 \cdot 10^{-1}$	$4.657 \cdot 10^{+2}$	$7.426 \cdot 10^{-2}$	$4.728 \cdot 10^{-2}$
6	$5.047 \cdot 10^{-1}$	$5.455 \cdot 10^{-1}$	$2.563 \cdot 10^{+4}$	$5.223 \cdot 10^{-2}$	$9.272 \cdot 10^{+3}$
7	$3.909 \cdot 10^{-13}$	$7.513 \cdot 10^{-14}$	$1.309 \cdot 10^{-8}$	$4.186 \cdot 10^{+5}$	$3.995 \cdot 10^{-8}$
8	$1.759 \cdot 10^{-12}$	$2.482 \cdot 10^{-12}$	$4.026 \cdot 10^{-7}$	$2.567 \cdot 10^{+19}$	$2.047 \cdot 10^{-8}$
9	$2.226 \cdot 10^{-9}$	$2.629 \cdot 10^{-9}$	$2.052 \cdot 10^{-5}$	$1.278 \cdot 10^{+10}$	$9.772 \cdot 10^{-3}$
10	$7.754 \cdot 10^{-3}$	$1.084 \cdot 10^{-2}$	$3.735 \cdot 10^{-2}$	$1.691 \cdot 10^{+13}$	$3.987 \cdot 10^{+12}$
11	$2.564 \cdot 10^{-2}$	$3.739 \cdot 10^{-5}$	$2.028 \cdot 10^{+31}$	$8.344 \cdot 10^{-1}$	$1.713 \cdot 10^{-8}$
12	$3.352 \cdot 10^{-2}$	$1.962 \cdot 10^{-2}$	$2.028 \cdot 10^{+31}$	$7.778 \cdot 10^{-1}$	$2.699 \cdot 10^{-2}$
13	$2.398 \cdot 10^{-2}$	$2.479 \cdot 10^{-2}$	$2.316 \cdot 10^{+30}$	$1.228 \cdot 10^{+0}$	$6.415 \cdot 10^{+1}$
14	$1.185 \cdot 10^{-2}$	$1.405 \cdot 10^{-2}$	$1.717 \cdot 10^{+30}$	$5.042 \cdot 10^{-1}$	$7.539 \cdot 10^{+5}$
15	$7.889 \cdot 10^{-14}$	$7.845 \cdot 10^{-14}$	$2.024 \cdot 10^{-16}$	$1.145 \cdot 10^{+25}$	$1.510 \cdot 10^{-8}$
16	$1.180 \cdot 10^{-2}$	$1.587 \cdot 10^{-2}$	$2.344 \cdot 10^{+5}$	$8.841 \cdot 10^{+3}$	$2.890 \cdot 10^{-2}$
17	$2.500 \cdot 10^{-2}$	$2.762 \cdot 10^{-2}$	$7.222 \cdot 10^{+3}$	$5.539 \cdot 10^{+9}$	$7.629 \cdot 10^{+7}$
18	$1.613 \cdot 10^{-2}$	$6.770 \cdot 10^{-2}$	$4.124 \cdot 10^{+6}$	$3.262 \cdot 10^{+15}$	$9.405 \cdot 10^{+11}$

Let α be an approximation to α_* computed by one of our approximation methods. Then we define the relative error

$$\text{relerr}_\alpha := \frac{|\alpha - \alpha_*|}{\min(\alpha, \alpha_*)}.$$

Table 3 presents the average relative errors for all the test cases.

Our approximation techniques work extremely well for Morozov's discrepancy principle and the Gfrerer/Raus-method. Because $\phi_M(\alpha)$ and $\phi_{GR}(\alpha)$ are strictly increasing for $\alpha > 0$ it is easy to give lower and upper bounds on α_*.

In the cases of the quasi-optimality criterion, GCV, and the L-curve criterion, however, a global minimizer or maximizer of $\phi(\alpha)$ must be determined. As discussed in Section 6 we terminate as soon as the largest local minimizer or maximizer has been identified. This explains the large value of relerr$_\alpha$ for some of the test cases.

On the other hand it should be noted that even regularization parameters α of vastly different sizes can lead to acceptable solutions x_α. This was also observed by J. M. Varah [33, p. 175]. We call a solution x_α acceptable if the norm of the corresponding residual $\|Ax_\alpha - b\|_2$ is a good estimate for the norm of the perturbation $\|e\|_2$ of the right-hand side. Therefore, we also compute the relative error of the residual, i.e.,

$$\text{relerr}_{\text{res}} := \left| \frac{\|Ax_\alpha - b\|_2 - \|e\|_2}{\|e\|_2} \right|.$$

Table 4. Average Relative Error relerr$_{res}$ of the Residual Norm $\|\alpha(AA^T + \alpha I)^{-1}b\|_2$.

	Morozov	Gfrerer/Raus	Quasi-Optimality	GCV	L-Curve
1	$5.447 \cdot 10^{-4}$	$6.839 \cdot 10^{-3}$	$1.782 \cdot 10^{-2}$	$6.650 \cdot 10^{-3}$	$1.493 \cdot 10^{-2}$
2	$1.558 \cdot 10^{-3}$	$1.207 \cdot 10^{-2}$	$5.944 \cdot 10^{-1}$	$1.003 \cdot 10^{-2}$	$2.042 \cdot 10^{-2}$
3	$1.735 \cdot 10^{-3}$	$1.297 \cdot 10^{-2}$	$5.350 \cdot 10^{+1}$	$1.385 \cdot 10^{-2}$	$5.350 \cdot 10^{+1}$
4	$4.794 \cdot 10^{-4}$	$1.672 \cdot 10^{-3}$	$9.107 \cdot 10^{-3}$	$1.154 \cdot 10^{-3}$	$1.253 \cdot 10^{-2}$
5	$5.789 \cdot 10^{-4}$	$2.054 \cdot 10^{-3}$	$6.362 \cdot 10^{-1}$	$1.094 \cdot 10^{-3}$	$2.357 \cdot 10^{-2}$
6	$8.176 \cdot 10^{-4}$	$4.494 \cdot 10^{-3}$	$7.972 \cdot 10^{+0}$	$1.405 \cdot 10^{-3}$	$1.139 \cdot 10^{+1}$
7	$2.417 \cdot 10^{-15}$	$1.220 \cdot 10^{-2}$	$1.510 \cdot 10^{-2}$	$2.285 \cdot 10^{-2}$	$1.789 \cdot 10^{-2}$
8	$5.657 \cdot 10^{-14}$	$4.798 \cdot 10^{-2}$	$1.557 \cdot 10^{-2}$	$2.539 \cdot 10^{-2}$	$1.785 \cdot 10^{-2}$
9	$1.931 \cdot 10^{-11}$	$8.806 \cdot 10^{-2}$	$1.792 \cdot 10^{-2}$	$2.520 \cdot 10^{-2}$	$3.329 \cdot 10^{-2}$
10	$2.535 \cdot 10^{-4}$	$8.193 \cdot 10^{-2}$	$1.564 \cdot 10^{+0}$	$2.632 \cdot 10^{-2}$	$1.023 \cdot 10^{+6}$
11	$1.409 \cdot 10^{-3}$	$2.039 \cdot 10^{-2}$	$8.802 \cdot 10^{-2}$	$3.733 \cdot 10^{-2}$	$8.982 \cdot 10^{-2}$
12	$1.921 \cdot 10^{-3}$	$4.432 \cdot 10^{-2}$	$5.018 \cdot 10^{+0}$	$7.698 \cdot 10^{-2}$	$1.248 \cdot 10^{-1}$
13	$2.079 \cdot 10^{-3}$	$6.512 \cdot 10^{-2}$	$1.691 \cdot 10^{+1}$	$1.347 \cdot 10^{-1}$	$1.095 \cdot 10^{+0}$
14	$2.361 \cdot 10^{-3}$	$1.582 \cdot 10^{-1}$	$1.559 \cdot 10^{+2}$	$2.138 \cdot 10^{-1}$	$9.481 \cdot 10^{+1}$
15	$4.828 \cdot 10^{-4}$	$9.240 \cdot 10^{-3}$	$5.335 \cdot 10^{-2}$	$3.824 \cdot 10^{-2}$	$4.141 \cdot 10^{-2}$
16	$4.672 \cdot 10^{-4}$	$3.422 \cdot 10^{-2}$	$2.473 \cdot 10^{+1}$	$4.660 \cdot 10^{-2}$	$4.511 \cdot 10^{-2}$
17	$1.089 \cdot 10^{-3}$	$3.127 \cdot 10^{-1}$	$7.315 \cdot 10^{+1}$	$6.580 \cdot 10^{-2}$	$1.990 \cdot 10^{+3}$
18	$8.937 \cdot 10^{-4}$	$4.159 \cdot 10^{-2}$	$4.121 \cdot 10^{+5}$	$7.254 \cdot 10^{-2}$	$2.987 \cdot 10^{+5}$

Table 4 presents the average relative error of the residual norm for each test case.

Morozov's discrepancy principle and the Gfrerer/Raus-method give excellent estimates for $\|e\|_2$. This could be expected because we can approximate the optimal regularization parameter α_* very well for these methods. Furthermore, GCV also provides amazingly accurate estimates for $\|e\|_2$. This is particularly remarkable as the value of relerr$_\alpha$ can be quite large. On the other hand, the quasi-optimality and the L-curve criteria fail to compute acceptable solutions x_α in many cases. These methods suffer from the fact that $\phi_Q(\alpha)$ and $\phi_L(\alpha)$ have many local minimizers/maximizers that are significantly larger than α_*. Therefore these methods are not very robust in connection with our approximation techniques.

The computational cost of our regularization techniques is proportional to the number k of Lanczos iterations. In Table 5 we give an overview of the average number of required iterations. Morozov's discrepancy principle and the Gfrerer/Raus-method require roughly the same number of iterations to converge. GCV needs more iterations but it is always able to provide good estimates for $\|e\|_2$. On the other hand, the quasi-optimality and the L-curve criteria terminate prematurely in many cases.

Table 5. Average Number of Lanczos Iterations k.

	Morozov	Gfrerer/Raus	Quasi-Optimality	GCV	L-Curve
1	22.5	22.4	22.8	53.2	22.4
2	76.5	72.4	29.6	177.3	84.6
3	346.6	316.7	22.0	854.4	22.0
4	47.5	47.2	60.2	141.2	31.7
5	104.0	98.2	35.6	283.2	82.3
6	346.0	340.9	43.8	1450.1	29.0
7	15.0	15.0	15.0	16.2	15.0
8	15.0	15.0	15.0	15.0	15.0
9	15.0	15.0	15.0	15.0	15.1
10	22.1	21.7	17.0	28.2	15.0
11	17.0	17.0	17.0	21.3	17.0
12	27.2	25.5	17.0	52.8	25.2
13	84.7	74.4	17.0	176.7	70.1
14	350.9	329.8	17.0	726.0	17.0
15	15.0	15.0	15.0	15.1	15.0
16	15.7	15.5	15.0	23.7	16.6
17	24.8	22.4	15.0	46.4	15.0
18	70.7	80.2	15.0	151.8	15.0

8   Conclusions

We presented iterative methods to compute regularization parameters for ill-posed linear systems and linear least squares problems. Our approach is based on the Lanczos algorithm and the theory of Gauss quadrature. We are able to compute lower and upper bounds on the functions $\phi(\alpha)$ that arise in Morozov's discrepancy principle, the Gfrerer/Raus-method, the quasi-optimality criterion, GCV, and the L-curve criterion. These bounds are used to compute approximations to the optimal regularization parameter α_*.

When we apply our approximation techniques to the 5 regularization methods we obtain the following results: If the norm of the perturbation e of the right-hand side b is known, Morozov's discrepancy principle and the Gfrerer/Raus-method are reliable methods to compute a regularization parameter α. They converge equally fast and provide similar values for α. It can be argued that the Gfrerer/Raus-method is actually an improved variant of Morozov's discrepancy principle (cf [17, p. 280]).

If no accurate estimate of $\|e\|_2$ is available then GCV works much more reliably than the quasi-optimality or the L-curve criterion. The quasi-optimality and the L-curve criteria are likely to terminate prematurely and return a regularization parameter α which is significantly too large. GCV may require more Lanczos iterations but it provides reliable answers in return.

References

1. A. B. BAKUSHINSKII, *Remarks on choosing a regularization parameter using the quasi-optimality and ratio criterion*, U.S.S.R. Computational Mathematics and Mathematical Physics, 24/4 (1984), pp. 181–182.

2. S. BERNARDSON, P. MCCARTY AND C. THRON, *Monte Carlo methods for estimating linear combinations of inverse matrix entries in lattice QCD*, Computer Physics Communications, 78 (1994), pp. 256–264.

3. P. J. DAVIS AND P. RABINOWITZ, *Methods of Numerical Integration*, Academic Press, Orlando, 1984.

4. L. ELDÉN, *Algorithms for the regularization of ill-conditioned least squares problems*, BIT, 17 (1977), pp. 134–145.

5. HEINZ W. ENGL, *Regularization methods for the stable solution of inverse problems*, Surveys on Mathematics for Industry, 3 (1993), pp. 71–143.

6. WALTER GANDER, *On the Linear Least Squares Problem with a Quadratic Constraint*, Habilitationsschrift ETH Zürich, STAN-CS-78-697, Stanford University, 1978.

7. WALTER GAUTSCHI, *Orthogonal polynomials: applications and computation*, Acta Numerica, to appear.

8. HELMUT GFRERER, *An A Posteriori Parameter Choice for Ordinary and Iterated Tikhonov Regularization of Ill-Posed Problems Leading to Optimal Convergence Rates*, Math. Comp., 49 (1987), pp. 507–522.

9. D. A. GIRARD, *A Fast 'Monte-Carlo Cross-Validation' Procedure for Large Least Squares Problems with Noisy Data*, Numer. Math., 56 (1989), pp. 1–23.

10. D. A. GIRARD, *Asymptotic optimality of the fast randomized versions of GCV and C_L in ridge regression and regularization*, The Annals of Statistics, 19 (1991), pp. 1950–1963.

11. G. H. GOLUB, *Some Modified Matrix Eigenvalue Problems*, SIAM Review, 15 (1973), pp. 318–334.

12. G. H. GOLUB, M. HEATH AND G. WAHBA, *Generalized Cross-Validation as a Method for Choosing a Good Ridge Parameter*, Technometrics, 21 (1979), pp. 215–223.

13. G. H. GOLUB AND W. KAHAN, *Calculating the singular values and pseudo-inverse of a matrix*, SIAM J. Numer. Anal., 2 (1965), pp. 205–224.

14. G. H. GOLUB AND C. F. VAN LOAN, *Matrix Computations*, Second Edition, The Johns Hopkins University Press, Baltimore, 1989.

15. GENE H. GOLUB AND URS VON MATT, *Generalized Cross-Validation for Large Scale Problems*, Journal of Computational and Graphical Statistics, to appear, `ftp://ftp.cscs.ch/pub/CSCS/techreports/1996/TR-96-28.ps.gz`.

16. G. H. GOLUB AND J. H. WELSCH, *Calculation of Gauss Quadrature Rules*, Math. Comp., 23 (1969), pp. 221–230.

17. MARTIN HANKE AND PER CHRISTIAN HANSEN, *Regularization methods for large-scale problems*, Surveys on Mathematics for Industry, 3 (1993), pp. 253–315.

18. PER CHRISTIAN HANSEN, *Analysis of Discrete Ill-Posed Problems by Means of the L-curve*, SIAM Review, 34 (1992), pp. 561–580.

19. PER CHRISTIAN HANSEN, *REGULARIZATION TOOLS: A Matlab package for analysis and solution of discrete ill-posed problems*, Numerical Algorithms, 6 (1994), pp. 1–35.

20. PER CHRISTIAN HANSEN AND DIANNE P. O'LEARY, *The use of the L-curve in the regularization of discrete ill-posed problems*, SIAM J. Sci. Comput., 14 (1993), pp. 1487–1503.

21. PER CHRISTIAN HANSEN AND DIANNE P. O'LEARY, *Regularization Algorithms Based on Total Least Squares*, Department of Computer Science, University of Maryland, Technical Report CS-TR-3684, September 1996, ftp://ftp.cs.umd.edu/pub/papers/papers/3684/3684.ps.Z.
22. M. F. HUTCHINSON, *A stochastic estimator of the trace of the influence matrix for Laplacian smoothing splines*, Communications in Statistics, Simulation and Computation, 19 (1990), pp. 433–450.
23. V. I. KRYLOV, *Approximate Calculation of Integrals*, translated by A. H. Stroud, Macmillan, New York, 1962.
24. A. S. LEONOV, *On the choice of regularization parameters by means of the quasi-optimality and ratio criteria*, Soviet Math. Dokl., 19 (1978), pp. 537–540.
25. V. A. MOROZOV, *On the solution of functional equations by the method of regularization*, Soviet Math. Dokl., 7 (1966), pp. 414–417.
26. V. A. MOROZOV, *Methods for Solving Incorrectly Posed Problems*, Springer, New York, 1984.
27. J. STOER, *Einführung in die Numerische Mathematik I*, Springer-Verlag, 1983.
28. A. H. STROUD AND D. SECREST, *Gaussian Quadrature Formulas*, Prentice-Hall, Englewood Cliffs, New Jersey, 1966.
29. THE MATHWORKS INC., *MATLAB, High-Performance Numeric Computation and Visualization Software*, Natick, Massachusetts, 1992.
30. ANDREI N. TIKHONOV, *Ill-Posed Problems in Natural Sciences*, Proceedings of the International Conference, TVP Science Publishers, Moscow, 1992.
31. ANDREY N. TIKHONOV AND VASILIY Y. ARSENIN, *Solutions of Ill-Posed Problems*, John Wiley & Sons, New York, 1977.
32. A. N. TIKHONOV AND V. B. GLASKO, *Use of the regularization method in non-linear problems*, U.S.S.R. Computational Mathematics and Mathematical Physics, 5/3 (1965), pp. 93–107.
33. J. M. VARAH, *Pitfalls in the numerical solution of linear ill-posed problems*, SIAM J. Sci. Stat. Comput., 4 (1983), pp. 164–176.
34. U. VON MATT, *Large Constrained Quadratic Problems*, Verlag der Fachvereine, Zürich, 1993, http://vdf.ethz.ch/vdf/info/2023.html.
35. GRACE WAHBA, *Practical approximate solutions to linear operator equations when the data are noisy*, SIAM J. Numer. Anal., 14 (1977), pp. 651–667.
36. GRACE WAHBA, *Spline Models for Observational Data*, CBMS-NSF, Regional Conference Series in Applied Mathematics, Vol. 59, SIAM, Philadelphia, 1990.

Vandermonde Factorization of a Hankel Matrix[*]

Daniel L. Boley[1], Franklin T. Luk[2] and David Vandevoorde[2]

[1] University of Minnesota, Minneapolis MN 55455, USA
[2] Rensselaer Polytechnic Institute, Troy NY 12180, USA

Abstract. We show that an arbitrary Hankel matrix of a finite rank admits a Vandermonde decomposition: $H = V^T D V$, where V is a confluent Vandermonde matrix and D is a block diagonal matrix. This result was first derived by Vandevoorde; our contribution here is a presentation that uses only linear algebra, specifically, the Jordan canonical form. We discuss the choices for computing this decomposition in only $O(n^2)$ operations, and we illustrate how to employ the decomposition as a fast way to analyze a noisy signal.

1 Introduction

Let $\{h_k\}_{k=1}^{\infty}$ denote a complex-valued signal, and let H represent the associated infinite Hankel matrix whose (i, j)-element is defined by $H_{ij} = h_{i+j-1}$:

$$H = \begin{pmatrix} h_1 & h_2 & h_3 & h_4 & \cdots \\ h_2 & h_3 & h_4 & h_5 & \cdots \\ h_3 & h_4 & h_5 & h_6 & \cdots \\ h_4 & h_5 & h_6 & h_7 & \cdots \\ & & & & \ddots \end{pmatrix} \tag{1}$$

This matrix is symmetric (not Hermitian if complex): $H^T = H$. Throughout this paper, the notation M^T denotes the transpose of M and not the conjugate transpose. Suppose that the underlying signal is a sum of r exponentials, i.e., for $k = 1, 2, \ldots$,

$$h_k = \sum_{i=1}^{r} \lambda_i^k d_i, \tag{2}$$

where the λ_i's are distinct complex numbers. Then the Hankel matrix H will have rank r. In this case, the Hankel matrix admits the factorization:

$$H = V^T D V,$$

where D is diagonal and V is Vandermonde:

$$D \triangleq \mathrm{diag}(d_1, d_2, \ldots, d_r)$$

[*] This research was partially supported by NSF Grant CCR-9405380

and

$$V \triangleq \begin{pmatrix} 1 & \lambda_1 & \lambda_1^2 & \lambda_1^3 & \cdots \\ 1 & \lambda_2 & \lambda_2^2 & \lambda_2^3 & \cdots \\ 1 & \lambda_3 & \lambda_3^2 & \lambda_3^3 & \cdots \\ \vdots & \vdots & \vdots & \vdots & \ddots \\ 1 & \lambda_r & \lambda_r^2 & \lambda_r^3 & \cdots \end{pmatrix}.$$

We stress that a diagonal decomposition is possible only if the λ_j's are distinct.

In this paper, we consider the general case where the λ_j's are multiple. The factorization must be generalized so that the matrix D becomes block diagonal and the matrix V takes on a confluent Vandermonde structure. The theory was first developed in Vandevoorde's Ph.D. thesis [11]. The next section is devoted to a derivation of this generalized Vandermonde decomposition based entirely on concepts from linear algebra. In Section 3 we sketch how the decomposition can be computed quickly, viz., using $O(n^2)$ operations and $O(n)$ space. We conclude in Section 4 with an example illustrating to use the method to analyze a noisy signal.

2   Derivation via Jordan Canonical Form

Assume that the matrix H of (1) has rank r. By a theorem of Gantmacher [5, vol. 2, p. 207], the signal satisfies a recurrence relation of length r:

$$h_k = a_{r-1}h_{k-1} + a_{r-2}h_{k-2} + \cdots + a_0 h_{k-r}, \tag{3}$$

which generates the entire signal once the r initial values $\{h_1, h_2, \ldots, h_r\}$ are fixed. The recurrence (3) is a difference equation which can be used to solve for the a_i's after the next r values $\{h_{r+1}, h_{r+2}, \ldots, h_{2r}\}$ become known.

Let C denote the companion matrix corresponding to the polynomial:

$$p(\lambda) \triangleq \lambda^r - a_{r-1}\lambda^{r-1} - \cdots - a_1\lambda - a_0; \tag{4}$$

that is,

$$C \triangleq \begin{pmatrix} 0 & 1 & & & & \\ & 0 & 1 & & & \\ & & \ddots & \ddots & & \\ & & & \ddots & \ddots & \\ & & & & 0 & 1 \\ a_0 & a_1 & a_2 & \cdots & a_{r-2} & a_{r-1} \end{pmatrix}. \tag{5}$$

We show that the first r rows of H can be regarded as a Krylov sequence generated by C. Let

$$\mathbf{h}_k \triangleq \begin{pmatrix} h_k \\ h_{k+1} \\ \vdots \\ h_{k+r-1} \end{pmatrix} \tag{6}$$

denote the first r entries in the k-th column of H. The first r rows of H can be written as

$$H_{1:r,1:\infty} = (\mathbf{h}_1 \quad \mathbf{h}_2 \quad \mathbf{h}_3 \quad \cdots) = (\mathbf{h}_1 \quad C\mathbf{h}_1 \quad C^2\mathbf{h}_1 \quad \cdots). \tag{7}$$

Suppose $\lambda_1, \lambda_2, \ldots, \lambda_s$ denote the roots of the polynomial $p(\lambda)$ with respective multiplicities $m_1, m_2, \ldots, m_s$ so that

$$m_1 + m_2 + \cdots + m_s = r.$$

Denote a Jordan canonical decomposition of C by

$$C = PJP^{-1},$$

with J in the canonical form:

$$J = \begin{pmatrix} J_{m_1}(\lambda_1) & & & \\ & J_{m_2}(\lambda_2) & & \\ & & \ddots & \\ & & & J_{m_s}(\lambda_s) \end{pmatrix}_{r \times r},$$

where

$$J_{m_i}(\lambda) = \begin{pmatrix} \lambda_i & 1 & & & \\ & \lambda_i & 1 & & \\ & & \ddots & \ddots & \\ & & & \lambda_i & 1 \\ & & & & \lambda_i \end{pmatrix}_{m_i \times m_i}, \tag{8}$$

for $i = 1, 2, \ldots, s$. The companion matrix C is guaranteed to be nonderogatory, i.e., it has one Jordan block per distinct eigenvalue. The transformation P is not unique, but having fixed the order for the eigenvalues, any alternative transformation $\tilde{P}$ also yielding the same Jordan canonical form must be related to P by

$$\tilde{P}Q = P,$$

where

$$Q = \begin{pmatrix} Q_{m_1} & & & \\ & Q_{m_2} & & \\ & & \ddots & \\ & & & Q_{m_s} \end{pmatrix}_{r \times r}, \tag{9}$$

with each Q_{m_i} as an $m_i \times m_i$ nonsingular and upper triangular matrix whose diagonal entries are all the same. This is because the columns of any $\tilde{P}$ are Jordan chains, and there are only limited kinds of transformations to the Jordan chains that will yield the same Jordan form [5, vol. 1, p. 172]. Note that Q is block diagonal with blocks conforming to the Jordan blocks.

For the purposes of computation, or for fixing ideas, it is often convenient to choose a specific P. One such choice is that of a confluent Vandermonde matrix [6, p. 188]:

$$P \triangleq \begin{pmatrix} \mathbf{p}^T \\ \mathbf{p}^T J \\ \vdots \\ \mathbf{p}^T J^{r-1} \end{pmatrix}_{r \times r}, \tag{10}$$

where

$$\mathbf{p}^T = \begin{pmatrix} \mathbf{e}_1^{[m_1]T} & \mathbf{e}_1^{[m_2]T} & \cdots & \mathbf{e}_1^{[m_s]T} \end{pmatrix}_{1 \times r};$$

so $\mathbf{p}$ is partitioned conformally with the Jordan canonical form J and each partition $\mathbf{e}_1^{[m_i]T}$ represents a first unit-coordinate vector:

$$\mathbf{e}_1^{[m_i]T} = \begin{pmatrix} 1 & 0 & \cdots & 0 \end{pmatrix}_{1 \times m_i}.$$

It is easy to verify (for any choice of a starting vector $\mathbf{p}$) that

$$CP = \begin{pmatrix} \mathbf{p}^T J \\ \mathbf{p}^T J^2 \\ \vdots \\ \mathbf{p}^T J^{r-1} \\ \mathbf{p}^T \cdot (a_{r-1} J^{r-1} + \cdots + a_1 J + a_0 I) \end{pmatrix} = \begin{pmatrix} \mathbf{p}^T J \\ \mathbf{p}^T J^2 \\ \vdots \\ \mathbf{p}^T J^{r-1} \\ \mathbf{p}^T J^r \end{pmatrix} = PJ,$$

where we have used the the characteristic equation for J:

$$p(J) = J^r - a_{r-1} J^{r-1} - \cdots - a_1 J - a_0 I = 0.$$

Hence $C = PJP^{-1}$, supporting our choice of the special form of P in (10).

Now, define a generalized Vandermonde matrix V by

$$V \triangleq \begin{pmatrix} \mathbf{v} & J\mathbf{v} & J^2\mathbf{v} & \cdots \end{pmatrix},$$

where

$$\mathbf{v} \triangleq P^{-1}\mathbf{h}_1.$$

The relation (7) can be written as

$$H_{1:r,1:\infty} = PV. \tag{11}$$

Let V_r denote the first r columns of V. We will express the Jordan decomposition of C in terms of V_r. Forming the product $V_r C^T$ columnwise, we get

$$\begin{aligned} V_r C^T &= \begin{pmatrix} \mathbf{v} & J\mathbf{v} & \cdots & J^{r-2}\mathbf{v} & J^{r-1}\mathbf{v} \end{pmatrix} C^T \\ &= \begin{pmatrix} J\mathbf{v} & J^2\mathbf{v} & \cdots & J^{r-1}\mathbf{v} & (a_0 + a_1 J + \cdots + a_{r-1} J^{r-1})\mathbf{v} \end{pmatrix} \\ &= \begin{pmatrix} J\mathbf{v} & J^2\mathbf{v} & \cdots & J^{r-1}\mathbf{v} & J^r\mathbf{v} \end{pmatrix} \\ &= JV_r. \end{aligned} \tag{12}$$

Since the first r columns of (11) are independent by assumption, the matrix V_r must be nonsingular. So we obtain the decomposition:

$$C^T = V_r^{-1}JV_r,$$

or equivalently,

$$C = V_r^T J^T V_r^{-T}.$$

We will use this result to express the transformation P in terms of V_r. Define the block diagonal "flip" matrix as follows:

$$F \triangleq \begin{pmatrix} F_{m_1} & & & \\ & F_{m_2} & & \\ & & \ddots & \\ & & & F_{m_s} \end{pmatrix}_{r \times r}, \tag{13}$$

where

$$F_{m_i} \triangleq \begin{pmatrix} & & & 1 \\ & 0 & 1 & \\ & \cdot^{\cdot^{\cdot}} & 0 & \\ 1 & & & \\ 1 & & & \end{pmatrix}_{m_i \times m_i},$$

for $i = 1, 2, \ldots, s$, partitioned conformally with the Jordan block $J_{m_i}(\lambda_i)$. This matrix is involutory $(F^2 = I)$ and symmetric $(F^T = F)$. When applied to the Jordan matrix J, the matrix F has the effect of transposing it:

$$FJF = J^T.$$

We can thereby write the Jordan decomposition of C as

$$C = V_r^T J^T V_r^{-T} = V_r^T FJFV_r^{-T}. \tag{14}$$

Setting

$$P = V_r^T FQ$$

for some matrix Q of the form (9), we get the decomposition of the leading $r \times r$ part of the Hankel matrix as

$$H_r = PV_r = V_r^T FQV_r.$$

Given that H is symmetric and V_r is nonsingular, we obtain

$$V_r^T FQV_r = H = H^T = V_r^T (FQ)^T V_r.$$

Hence the matrix D, defined by

$$D \triangleq FQ,$$

must be symmetric; furthermore, it is block diagonal with blocks conforming to the Jordan blocks. Since each diagonal block of Q is upper triangular, we derive the following form for D:

$$D = \begin{pmatrix} D_1 & & & \\ & D_2 & & \\ & & \ddots & \\ & & & D_s \end{pmatrix}_{r \times r}, \tag{15}$$

where

$$D_i = \begin{pmatrix} * & * & * & * & d_i \\ * & * & * & d_i & \\ * & * & \cdot^{\cdot^{\cdot}} & & \\ * & d_i & & & 0 \\ d_i & & & & \end{pmatrix}_{m_i \times m_i}, $$

for $i = 1, 2, \ldots, s$; each block D_i is symmetric and upper anti-triangular, with a constant value along the main antidiagonal. Combining these formulas, we obtain both a relation between P and V_r:

$$P = V_r^T D,$$

and a symmetric decomposition for the leading $r \times r$ Hankel matrix:

$$H_r = PV_r = V_r^T DV_r = PD^{-1}P^T. \tag{16}$$

From (12) we get

$$V = (V_r \quad J^r V_r \quad J^{2r} V_r \quad \cdots) = (V_r \quad V_r(C^T)^r \quad V_r(C^T)^{2r} \quad \cdots).$$

Since $CH_r = H_r C^T$, we obtain

$$H = \begin{pmatrix} I_r \\ C^r \\ C^{2r} \\ \vdots \end{pmatrix} \cdot H_r \cdot (I_r \quad (C^T)^r \quad (C^T)^{2r} \quad \cdots) = V^T DV, \tag{17}$$

i.e., a factorization for the entire infinite Hankel matrix.

A further analysis of the structure of the matrices in (16) yields the fact that the diagonal blocks of D^{-1} and D have Hankel structure. For D^{-1}, we start with the identity

$$(PJP^{-1})(PD^{-1}P^T) = CH = HC^T = (PD^{-1}P^T)(P^{-T}J^T P^T),$$

which simplifies to $JD^{-1} = D^{-1}J^T$. The matrix D^{-1} is block diagonal with blocks conforming to those of J. So the i-th block of this last relation is

$$J_{m_i}(\lambda_i)D_i^{-1} = D_i^{-1}J_{m_i}^T(\lambda_i),$$

for $i = 1, \ldots, s$. Subtracting $\lambda_i I$ from both sides yields

$$Z_i D_i^{-1} = D_i^{-1} Z_i^T,$$

where Z has the form of an upshift matrix of appropriate size:

$$Z_i = J_{m_i}(\lambda_i) - \lambda_i I = \begin{pmatrix} 0 & 1 & & & \\ & 0 & 1 & & \\ & & \ddots & \ddots & \\ & & & 0 & 1 \\ & & & & 0 \end{pmatrix}_{m_i \times m_i}. \tag{18}$$

Hence D_i^{-1} must have a Hankel structure. For D, we carry out a similar argument using $V_r^T F$ instead of P and D instead of D^{-1}, and apply the factorization (14) to get a Hankel structure for the diagonal blocks D_i.

We now show how we may modify the Vandermonde matrix V_r so that its first column consists of all ones. Suppose that no component of $\mathbf{v}$ is zero. Define a new diagonal matrix D_v by

$$D_v \triangleq \operatorname{diag}(v_1, v_2, \ldots, v_r),$$

and a new generalized Vandermonde matrix W_r by

$$W_r \triangleq D_v^{-1} V_r.$$

Then

$$D_v^{-1} \mathbf{v} = \mathbf{e} \triangleq (1 \quad 1 \quad \cdots \quad 1)^T,$$

and we can re-write (16) as

$$H_r = W_r^T (D_v D D_v) W_r,$$

where the product $D_v D D_v$ is a block diagonal matrix with blocks conforming to J. The matrix W_r has the following structure:

$$\begin{aligned} W_r &= (\mathbf{e} \quad D_v^{-1} J D_v \mathbf{e} \quad D_v^{-1} J^2 D_v \mathbf{e} \quad \cdots \quad D_v^{-1} J^{r-1} D_v \mathbf{e}) \\ &= (\mathbf{e} \quad \hat{J} \mathbf{e} \quad \hat{J}^2 \mathbf{e} \quad \cdots \quad \hat{J}^{r-1} \mathbf{e}), \end{aligned} \tag{19}$$

with

$$\hat{J} \triangleq D_v^{-1} J D_v.$$

We observe that the matrix $\hat{J}$ possesses a sort of generalized Jordan structure.

We conclude this section with three notes.

First, consider the special case that all the eigenvalues of C are simple. In this case, the Jordan matrix $J = \hat{J}$ is diagonal, F is the identity, $D = Q$ is "scalar" diagonal, and

$$P = V_r^T D.$$

The decomposition (16) simplifies to

$$H_r = V_r^T (D) V_r = W_r^T (D D_v^2) W_r, \tag{20}$$

where the parts enclosed in parentheses are diagonal matrices. The last expression on the right applies if $\mathbf{v}$ has no zero component, which is guaranteed in this case by the nonsingularity of V_r. In fact, the matrix W_r of (19) for this case has the usual Vandermonde structure. The above holds for any choice of P, but if we fix P as given in (10), we see that $P = W_r^T$.

Second, the factorization (16) can be used to factor any nonsingular $r \times r$ Hankel matrix H_r. This matrix is filled by the entries $h_1, h_2, \ldots, h_{2r-1}$; hence to fix the polynomial (4) and carry out the rest of the development above, we must choose some value for h_{2r}. With such a choice, the rest of the development above goes through unchanged.

Third, we address the issue of factoring a singular $n \times n$ Hankel matrix H_n. One way to do this is to embed this $n \times n$ matrix inside an infinite Hankel matrix H_∞ of a finite rank r by extending it with infinitely many zeros. We could use a different choice for the extension to avoid a nilpotent C, and one open issue is how to choose the extension to minimize the resulting rank. Factor the infinite Hankel matrix as

$$H_\infty = V^T D V,$$

where D is $r \times r$ and V is $r \times \infty$, and extract the first n rows and columns of this decomposition to get

$$H_n = V_n^T D V_n,$$

where V_n is the $r \times n$ matrix consisting of the first n columns of V. Note that it could be that $r < n$ or $r > n$. The former case occurs, for example, if the rank of H_n is $r < n$ and the leading $r \times r$ part of H_n happens to be nonsingular.

3   Algorithms (Generic case)

We briefly discuss choices of algorithms that can be used to compute the Vandermonde decomposition. Many of the individual pieces to the algorithms are off-the-shelf methods; some are quite experimental and some have received very little attention in the literature. The methods we present are based on the use of the Lanczos algorithm or its derivatives. Most details can be found in [11].

We begin with an outline of the basic steps:

1. Compute the "modes" generating the Hankel matrix, viz., the roots of the polynomial $p(\lambda)$ of (4).
2. Compute the "diagonal" matrix D:

$$D = V^{-T} H V^{-1},$$

   where V is the Vandermonde matrix generated by the eigenvalues in step 1. The diagonal structure of D follows from the theory developed in the previous section.
3. Optionally scale the columns of V to unit norm, scaling the entries in D appropriately.

For each of these steps there are choices for the algorithm to use. For step 1, we could solve for the coefficients in the recurrence (3) by simply plugging the values for h_i $(i = 1, \ldots, 2n - 1)$, yielding a special set of n equations in n unknowns originally proposed by Prony [10] and popularized by Yule [14] and Walker [12]. Then we must find all the roots of the polynomial $p(\lambda)$.

Vandevoorde [11] proposed an alternative for step 1. We use a variant of a nonsymmetric Lanczos process to generate a tridiagonal matrix T whose eigenvalues match those of C. Then we compute the eigenvalues of T. It turns out that both these steps (generating T and computing its eigenvalues) can be performed very efficiently. Space does not permit a full description of the process, but we can give a hint on the basic ideas used. We have already seen that the $n \times n$ Hankel matrix H_n can be thought of as a Krylov sequence generated by C and h_1 (cf. (7)). If we let H_{2n} be the $2n \times 2n$ Hankel matrix obtained by extending the "signal" $\{h_k\}$ with all zeros, then H_{2n} can similarly be thought of as the Krylov sequence generated by the $2n \times 2n$ upshift matrix Z of the form (18).

The nonsymmetric Lanczos process in [11] is equivalent to a procedure that bi-orthogonalizes the Krylov sequence we just mentioned against another sequence (called the "left" Krylov sequence). If the coefficients computed by the enforcement of the bi-orthogonalization conditions involve only the first n entries of each vector, the Lanczos coefficients generated by C and by H_{2n} will be identical. This can happen, for example, by making the left Krylov sequence upper triangular, so that the right Krylov sequence will be bi-orthogonalized to lower triangular. This can be arranged by starting the left Krylov sequence with the coordinate unit vector e_1. As a result, each new vector is generated by shifting up the previous vector and then orthogonalizing it against the two previous vectors. This takes linear time and linear space, and there are at most $O(n)$ steps so that the total time is $O(n^2)$. The resulting algorithm is described in detail in [2]. There is also a symmetric variant originally proposed in [9] that can generate a symmetrized tridiagonal matrix directly from this non-symmetric recursion. This symmetric variant has similar costs.

Once the tridiagonal matrix has been generated, the task is to find its eigenvalues. There are two variants of the QR-type algorithm that can be applied here. One is the complex symmetric QR algorithm proposed in [4], for which the matrix T must be symmetrized. Even when T is real, if the signs of the corresponding superdiagonal and subdiagonal entries of T are opposite, then the symmetrized matrix will be complex. The resulting QR algorithm is a direct analog of the ordinary Hermitian QR method, but it uses complex orthogonal rotations and complex symmetric matrices instead of unitary rotations and Hermitian matrices, respectively. Another option is to use the LR algorithm [13], which is based on the LU factorization without pivoting to preserve the tridiagonal structure. The LR algorithm can break down, but if a random shift is applied when zero pivot occurs during the LU factorization, the process can still exhibit very rapid convergence. If T is real, an implicit double-shift LR algorithm can in principle be carried out in real arithmetic [13]. Both algorithms require linear time for each iteration in a manner very similar to the Hermitian analog, and the number of iterations is generally $O(n)$ in a manner very similar to the QR algorithm usually employed. The relative merits between these alternative algorithms have not been studied in detail.

The other major task is finding the diagonal matrix D in step 2. Because of the structure of V, the diagonal entries of D appear in the first column of the product DV. But $DV = V^{-T}H$. Hence this first column is the solution $\mathbf{d}$ to the Vandermonde system:

$$V^T\mathbf{d} = \mathbf{h}_1,$$

where $\mathbf{h}_1$ is the first column of H. This can be solved with a fast $O(n)$ Vandermonde solver [1], where to maintain stability Higham [7, p. 438] recommends arranging the eigenvalues with a so-called Leja ordering.

4   Analysis of a Signal

Consider a signal $\{h_k\}$ which suffers from the presence of noise. How can we recover the principal modes that generate the signal? A popular method by Kung [8] based on the singular value decomposition (SVD) is known to be an effective method for this purpose, but it suffers from the need to carry out both an SVD and a matrix eigensolution, each costing $O(n^3)$ operations. A second popular approach is to form the Hankel matrix generated by the signal, and then proceed to find a nearby Hankel matrix of a lower rank [3]. The Vandermonde decomposition of this nearby low-rank Hankel matrix yields the parameters in (2). The method of [3] iterates until it converges to a nearby Hankel matrix. Unfortunately, this method requires the repeated use of the SVD and hence costs up to $O(n^3)$ operations per iteration.

We indicated in Section 3 how the Vandermonde decomposition can be computed quickly. An obvious way to obtain a nearby Hankel matrix of a lower rank is to set to zero all the diagonal entries in D that are smaller than a certain tolerance. Although this crude method does not always yield the best approximation, a judicious combination of this approach with other criteria can yield a good result. We conclude this paper with an illustration of one such approach in the next paragraph.

Start with a signal generated by five modes, shown by circles on the complex plane in Figure 1, to which has been added white noise with a signal-to-noise ratio of 3.55dB. Form the 128×128 Hankel matrix H and compute its Vandermonde decomposition $H = V^T DV$. Figure 2 shows the absolute values of the diagonal entries of D in descending order. It turns out that selecting the modes corresponding to the five largest values of D does not yield satisfactory results, but we can almost recover the correct modes by the following simple procedure. Choose the modes corresponding to the largest entries in D (also called weights), specifically those that are within 10% of the largest entry (in absolute value); in this case fourteen modes remained. Then choose a subset of these fourteen using a second criterion based on the Discrete Fourier Transform (DFT) of the signal. The DFT of the original signal is shown by the dotted line in Figure 3. As most of the modes lie relatively close to the unit circle, their argument (angle on the

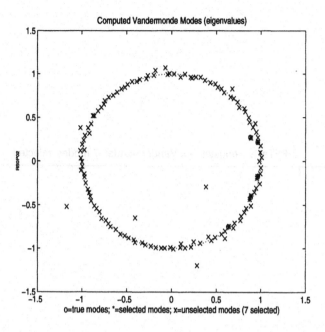

Fig. 1. Original modes (o) and those computed from the tridiagonal matrix discussed in in Section 3 (* & x).

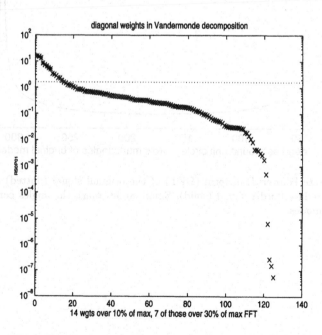

Fig. 2. Diagonal entries from Vandermonde decomposition

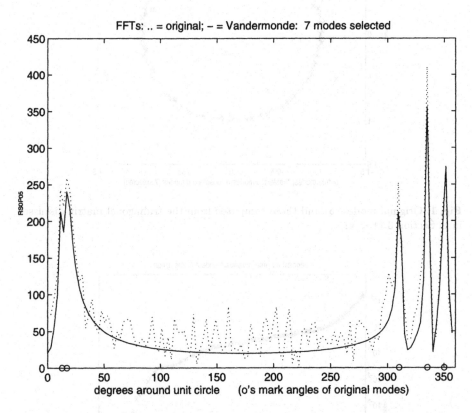

Fig. 3. Discrete Fourier Transform (DFT) of the original signal (dotted) and the reconstructed reduced-order signal (solid). Small circles mark the angles corresponding the original modes.

complex plane) maps to the horizontal axis of Figure 3. In fact, we have marked the angles corresponding to the five original "unknown" modes by means of circles along the x-axis. This leads to our second criterion, viz., select those modes for which the DFT is larger than a certain threshold (in this case 30%) of the largest value in the DFT (in absolute value). This selection criterion is applied only to those modes that survived the first selection process. In this example, out of the fourteen modes only seven survived the second selection process. These final seven modes are marked by *'s in Figure 1, and the resulting DFT using these seven modes is shown by the solid line in Figure 3. We remark that one can still distinguish the two close peaks in this DFT corresponding to the two very close original modes. We should emphasize that the choice of criteria requires further study. Indeed, a more sophisticated selection criterion is presented in [11].

References

1. Å. Björck and V. Pereyra. Solution of Vandermonde system of equations. *Math. Comp.*, 24:893–903, 1970.
2. D. L. Boley, T. J. Lee, and F. T. Luk. The Lanczos algorithm and Hankel matrix factorization. *Lin. Alg. & Appl.*, 172:109–133, 1992.
3. J. Cadzow. Signal enhancement – a composite property mapping algorithm. *IEEE Trans. Acoust., Speech., Sig. Proc.*, 36:49–62, 1988.
4. J. K. Cullum and R. A. Willoughby. A QL procedure for computing the eigenvalues of complex symmetric tridiagonal matrices. *SIAM J. Matrix Anal. Appl.*, 17(1):83–109, 1996.
5. F. R. Gantmacher. *Theory of Matrices*. Chelsea, New York, 1959.
6. G. H. Golub and C. F. Van Loan. *Matrix Computations*. Johns Hopkins Univ. Press, 3rd edition, 1996.
7. N. J. Higham. *Accuracy and Stability of Numerical Algorithms*. SIAM, Philadelphia, 1996.
8. S. Y. Kung. A new identification and model reduction algorithm via singular value decompositions. In *Proc. 12-th Asilomar Conf. Circ., Syst. Comp.*, pages 705–714, November 1984.
9. J. L. Phillips. The triangular decomposition of Hankel matrices. *Math. Comp.*, 25(115):599–602, 1971.
10. R. Prony. Essai expérimental et analytique sur les lois de la dilatabilité et sur celles de la force expansive de la vapeur de l'eau et de la vapeur de l'alkool, à différentes températures. *J. de l'École Polytechnique*, 1:24–76, 1795.
11. D. Vandevoorde. *A Fast Exponential Decomposition Algorithm and its Applications to Structured Matrices*. PhD thesis, Computer Science Department, Rensselaer Polytechnic Institute, 1996.
12. G. Walker. On periodicity in series of related terms. *Proc. Roy. Soc. London Ser. A*, 131A:518–532, 1931.
13. D. Watkins and L. Elsner. Chasing algorithms for the eigenvalue problem. *SIAM J. Matr. Anal.*, 12:374–384, 1991.
14. G. Yule. On a method of investigating periodicities in disturbed series, with special reference to Wolfer's sunspot numbers. *Trans. Roy. Soc. London Ser. A*, 226A:267–298, 1927.

Numerical Stability of Some Fast Algorithms for Structured Matrices

Richard P. Brent

Computer Sciences Laboratory
Australian National University
Canberra, ACT 0200, Australia

Richard.Brent@anu.edu.au

Abstract. We consider the numerical stability/instability of fast algorithms for solving systems of linear equations or linear least squares problems with a low displacement-rank structure. For example, the matrices involved may be Toeplitz or Hankel. In particular, we consider algorithms which incorporate pivoting without destroying the structure, such as the Gohberg-Kailath-Olshevsky (GKO) algorithm, and describe some recent results on the stability of these algorithms. We also compare these results with the corresponding stability results for algorithms based on the semi-normal equations and for the well known algorithms of Schur/Bareiss and Levinson.

1 Introduction

It is well known that systems of n linear equations with a low displacement rank (e.g. Toeplitz or Hankel matrices) can be solved in $O(n^2)$ arithmetic operations[1]. For positive definite Toeplitz matrices the first $O(n^2)$ algorithms were introduced by Kolmogorov [30], Wiener [43] and Levinson [31]. These algorithms are related to recursions of Szegö [40] for polynomials orthogonal on the unit circle. Another class of $O(n^2)$ algorithms, e.g. the Bareiss algorithm [2], are related to Schur's algorithm for finding the continued fraction representation of a holomorphic function in the unit disk [34]. This class can be generalized to cover unsymmetric matrices and more general "low displacement rank" matrices [28]. In this paper we consider the numerical stability of some of these algorithms. A more detailed survey is given in [29].

In the following, R denotes a structured matrix, T is a Toeplitz or Toeplitz-type matrix, P is a permutation matrix, L is lower triangular, U is upper triangular, and Q is orthogonal. In error bounds, $O_n(\varepsilon)$ means $O(\varepsilon f(n))$, where $f(n)$ is a polynomial in n.

---

[1] Asymptotically faster algorithms exist [1,8], but are not considered here.

2   Classes of Structured Matrices

Structured matrices R satisfy a *Sylvester equation* which has the form

$$\nabla_{\{A_f, A_b\}}(R) = A_f R - R A_b = \Phi \Psi \,, \tag{1}$$

where A_f and A_b have some simple structure (usually banded, with 3 or fewer full diagonals), Φ and Ψ are $n \times \alpha$ and $\alpha \times n$ respectively, and α is some fixed integer. The pair of matrices (Φ, Ψ) is called the $\{A_f, A_b\}$-*generator* of R.

α is called the $\{A_f, A_b\}$-*displacement rank* of R. We are interested in cases where α is small (say at most 4).

Cauchy matrices

Particular choices of A_f and A_b lead to definitions of basic classes of matrices. Thus, for a Cauchy matrix

$$C(\mathbf{t}, \mathbf{s}) = \left[\frac{1}{t_i - s_j} \right]_{ij} \,,$$

we have

$$A_f = D_t = \mathrm{diag}(t_1, t_2, \ldots, t_n) \,,$$
$$A_b = D_s = \mathrm{diag}(s_1, s_2, \ldots, s_n)$$

and

$$\Phi^T = \Psi = [1, 1, \ldots, 1] \,.$$

More general matrices, where Φ and Ψ are any rank-α matrices, are called *Cauchy-type*.

Toeplitz matrices

For a Toeplitz matrix $T = [t_{ij}] = [a_{i-j}]$, we take

$$A_f = Z_1 = \begin{bmatrix} 0 & 0 & \cdots & 0 & 1 \\ 1 & 0 & & & 0 \\ 0 & 1 & & & \vdots \\ \vdots & & \ddots & & \vdots \\ 0 & \cdots & 0 & 1 & 0 \end{bmatrix}, \quad A_b = Z_{-1} = \begin{bmatrix} 0 & 0 & \cdots & 0 & -1 \\ 1 & 0 & & & 0 \\ 0 & 1 & & & \vdots \\ \vdots & & \ddots & & \vdots \\ 0 & \cdots & 0 & 1 & 0 \end{bmatrix},$$

$$\Phi = \begin{bmatrix} 1 & 0 & \cdots & 0 \\ a_0 & a_{1-n} + a_1 & \cdots & a_{-1} + a_{n-1} \end{bmatrix}^T,$$

and

$$\Psi = \begin{bmatrix} a_{n-1} - a_{-1} & \cdots & a_1 - a_{1-n} & a_0 \\ 0 & \cdots & \cdots & 0 \end{bmatrix}.$$

We can generalize to *Toeplitz-type* matrices by taking Φ and Ψ to be general rank-α matrices.

3   Structured Gaussian Elimination

Let an input matrix, R_1, have the partitioning

$$R_1 = \begin{bmatrix} d_1 & \mathbf{w}_1^T \\ \mathbf{y}_1 & \tilde{R}_1 \end{bmatrix}.$$

The first step of normal Gaussian elimination is to premultiply R_1 by

$$\begin{bmatrix} 1 & \mathbf{0}^T \\ -\mathbf{y}_1/d_1 & I \end{bmatrix},$$

which reduces it to

$$\begin{bmatrix} d_1 & \mathbf{w}_1^T \\ 0 & R_2 \end{bmatrix},$$

where

$$R_2 = \tilde{R}_1 - \mathbf{y}_1 \mathbf{w}_1^T / d_1$$

is the *Schur complement* of d_1 in R_1. At this stage, R_1 has the factorization

$$R_1 = \begin{bmatrix} 1 & \mathbf{0}^T \\ \mathbf{y}_1/d_1 & I \end{bmatrix} \begin{bmatrix} d_1 & \mathbf{w}_1^T \\ 0 & R_2 \end{bmatrix}.$$

One can proceed recursively with the Schur complement R_2, eventually obtaining a factorization $R_1 = LU$.

The key to *structured* Gaussian elimination is the fact that the displacement structure is preserved under Schur complementation, and that the generators for the Schur complement of R_{k+1} can be computed from the generators of R_k in $O(k)$ operations.

Row and/or column interchanges destroy the structure of matrices such as Toeplitz matrices. However, if A_f is diagonal (which is the case for Cauchy and Vandermonde type matrices), then *the structure is preserved under row permutations*.

This observation leads to the *GKO-Cauchy* algorithm [21] for fast factorization of Cauchy-type matrices with partial pivoting, and many recent variations on the theme by Boros, Gohberg, Ming Gu, Heinig, Kailath, Olshevsky, M. Stewart, *et al*: see [7,21,23,26,35].

The GKO-Toeplitz algorithm

Heinig [26] showed that, if T is a Toeplitz-type matrix, then

$$R = FTD^{-1}F^*$$

is a Cauchy-type matrix, where

$$F = \frac{1}{\sqrt{n}} [e^{2\pi i(k-1)(j-1)/n}]_{1 \le k, j \le n}$$

is the Discrete Fourier Transform matrix,

$$D = \text{diag}(1, e^{\pi i/n}, \dots, e^{\pi i(n-1)/n}),$$

and the generators of T and R are simply related.

The transformation $T \leftrightarrow R$ is perfectly stable because F and D are unitary. Note that R is (in general) complex even if T is real.

Heinig's observation was exploited by Gohberg, Kailath and Olshevsky [21]: R can be factorized as $R = P^T LU$ using GKO-Cauchy. Thus, from the factorization

$$T = F^* P^T LU F D \,,$$

a linear system involving T can be solved in $O(n^2)$ operations. The full procedure of conversion to Cauchy form, factorization, and solution requires $O(n^2)$ (complex) operations.

Other structured matrices, such as Hankel, Toeplitz-plus-Hankel, Vandermonde, Chebyshev-Vandermonde, etc, can be converted to Cauchy-type matrices in a similar way.

Error Analysis

Because GKO-Cauchy and GKO-Toeplitz involve partial pivoting, we might guess that their stability would be similar to that of Gaussian elimination with partial pivoting. Unfortunately, there is a flaw in this reasoning. During GKO-Cauchy the *generators* have to be transformed, and the partial pivoting does not ensure that the transformed generators are small.

Sweet and Brent [39] show that significant generator growth can occur if all the elements of $\Phi\Psi$ are small compared to those of $|\Phi||\Psi|$. This can not happen for ordinary Cauchy matrices because $\Phi^{(k)}$ and $\Psi^{(k)}$ have only one column and one row respectively. However, it can happen for higher displacement-rank Cauchy-type matrices, even if the original matrix is well-conditioned.

The Toeplitz Case

In the Toeplitz case there is an extra constraint on the selection of Φ and Ψ, but it is still possible to give examples where the normalized solution error grows like κ^2 and the normalized residual grows like κ, where κ is the condition number of the Toeplitz matrix. Thus, the GKO-Toeplitz algorithm is (at best) weakly stable[2].

It is easy to think of modified algorithms which avoid the examples given by Sweet and Brent, but it is difficult to prove that they are stable in all cases. Stability depends on the worst case, which may be rare and hard to find by random sampling.

The problem with the original GKO algorithm is growth in the generators. Ming Gu suggested exploiting the fact that the generators are not unique. Recall

---

[2] For definitions of stability and weak stability, see [5,9,10].

the Sylvester equation (1). Clearly we can replace Φ by ΦM and Ψ by $M^{-1}\Psi$, where M is any invertible $\alpha \times \alpha$ matrix, because this does not change the product $\Phi\Psi$. Similarly at later stages of the GKO algorithm.

Ming Gu [23] proposes taking M to orthogonalize the columns of Φ (that is, at each stage we do an orthogonal factorization of the generators). Michael Stewart [35] proposes a (cheaper) LU factorization of the generators. In both cases, clever pivoting schemes give error bounds analogous to those for Gaussian elimination with partial pivoting.

Gu and Stewart's error bounds

The error bounds obtained by Ming Gu and Michael Stewart involve a factor K^n, where K depends on the ratio of the largest to smallest modulus elements in the Cauchy matrix

$$\left[\frac{1}{t_i - s_j}\right]_{ij}.$$

Although this is unsatisfactory, it is similar to the factor 2^{n-1} in the error bound for Gaussian elimination with partial pivoting.

Michael Stewart [35] gives some interesting numerical results which indicate that his scheme works well, but more numerical experience is necessary before a definite conclusion can be reached.

In practice, we can use an $O(n^2)$ algorithm such as Michael Stewart's, check the residual, and resort to iterative refinement or a stable $O(n^3)$ algorithm in the (rare) cases that it is necessary.

4   Positive Definite Structured Matrices

An important class of algorithms, typified by the algorithm of Bareiss [2], find an LU factorization of a Toeplitz matrix T, and (in the symmetric case) are related to the classical algorithm of Schur [20,34].

It is interesting to consider the numerical properties of these algorithms and compare with the numerical properties of the Levinson algorithm (which essentially finds an LU factorization of T^{-1}).

The Bareiss algorithm for positive definite matrices

Bojanczyk, Brent, de Hoog and Sweet[3] [6,37] have shown that the numerical properties of the Bareiss algorithm are similar to those of Gaussian elimination (*without* pivoting). Thus, the algorithm is stable for positive definite symmetric Toeplitz matrices.

The Levinson algorithm can be shown to be weakly stable for bounded n, and numerical results by Varah [42], BBHS and others suggest that this is all

---

[3] Abbreviated BBHS.

that we can expect. Thus, the Bareiss algorithm is (generally) better numerically than the Levinson algorithm.

Cybenko [15] showed that if certain quantities called "reflection coefficients" are positive then the Levinson-Durbin algorithm for solving the Yule-Walker equations (a positive-definite system with special right-hand side) is stable. However, "random" positive-definite Toeplitz matrices do not usually satisfy Cybenko's condition.

The generalized Schur algorithm

The Schur algorithm can be generalized to factor a large variety of structured matrices – see Kailath *et al* [27,28]. For example, the generalized Schur algorithm applies to block Toeplitz matrices, Toeplitz block matrices, and to matrices of the form $T^T T$, where T is rectangular Toeplitz.

It is natural to ask if the stability results of BBHS (which are for the classical Schur/Bareiss algorithm) extend to the generalized Schur algorithm. This was considered by M. Stewart and Van Dooren [36] and by Chandrasekharan and Sayed [12]. (The results were obtained independently by the two pairs of authors, and the "generalized Schur algorithm" considered in each case is slightly different – for details see [29].)

The conclusion is that the generalized Schur algorithm is stable for positive definite symmetric (or Hermitian) matrices, provided that the hyperbolic transformations in the algorithm are implemented correctly. In contrast, BBHS showed that stability of the classical Schur/Bareiss algorithm is not so dependent on details of the implementation.

5 Fast Orthogonal Factorization

In an attempt to achieve stability without pivoting, and to solve $m \times n$ least squares problems $(m \geq n)$, it is natural to consider algorithms for computing an orthogonal factorization

$$T = QU$$

of T. The first such $O(n^2)$ algorithm[4] was introduced by Sweet [37]. Unfortunately, Sweet's algorithm is unstable: it depends on the condition of a submatrix of T – see Luk and Qiao [32].

Other $O(n^2)$ algorithms for computing the matrices Q and U or U^{-1} were given by Bojanczyk, Brent and de Hoog[5] [4], Chun *et al* [14], Cybenko [16], and Qiao [33], but none of them has been shown to be stable, and in several cases examples show that they are unstable.

Unlike the classical $O(n^3)$ Givens or Householder algorithms, the $O(n^2)$ algorithms do not form Q in a numerically stable manner as a product of matrices which are (close to) orthogonal.

---

[4] More precisely, $O(mn)$. For simplicity, in the time bounds we assume $m = O(n)$.
[5] Abbreviated BBH.

For example, the algorithms of Bojanczyk, Brent and de Hoog [4] and Chun *et al* [14] depend on Cholesky downdating, and numerical experiments show that they do not give a Q which is close to orthogonal.

The generalized Schur algorithm, applied to $T^T T$, computes the upper triangular matrix U but not the orthogonal matrix Q.

Use of the semi-normal equations

It can be shown that, provided the Cholesky downdates are implemented in a certain way (analogous to the condition for the stability of the generalized Schur algorithm), the BBH algorithm computes U in a weakly stable manner [5]. In fact, the computed upper triangular matrix $\tilde{U}$ is about as good as can be obtained by performing a Cholesky factorization of $T^T T$, so

$$\|T^T T - \tilde{U}^T \tilde{U}\|/\|T^T T\| = O_m(\varepsilon) \ .$$

Thus, by solving

$$\tilde{U}^T \tilde{U} x = T^T b$$

(the so-called *semi-normal* equations) we have a *weakly stable* algorithm for the solution of general Toeplitz systems $Tx = b$ in $O(n^2)$ operations. The solution can be improved by iterative refinement if desired. Note that the computation of Q is avoided, and the algorithm is applicable to full-rank Toeplitz least squares problems.

Computing Q stably

It is difficult to give a satisfactory $O(n^2)$ algorithm for the computation of Q in the factorization

$$T = QU \tag{2}$$

Chandrasekharan and Sayed give a stable algorithm to compute the factorization

$$T = LQU \tag{3}$$

where L is lower triangular. Their algorithm can be used to solve linear equations, but not least squares problems, because T has to be square, and in any case the matrix Q in (3) is different from the matrix Q in (2). Because their algorithm involves embedding the $n \times n$ matrix T in a $2n \times 2n$ matrix

$$\begin{bmatrix} T^T T & T^T \\ T & 0 \end{bmatrix},$$

the constant factors in the operation count are large: $59n^2 + O(n \log n)$, which should be compared to $8n^2 + O(n \log n)$ for BBH and the semi-normal equations.

References

1. G. S. Ammar and W. B. Gragg, "Superfast solution of real positive definite Toeplitz systems", *SIMAX* 9 (1988), 61–76.
2. E. H. Bareiss, "Numerical solution of linear equations with Toeplitz and vector Toeplitz matrices", *Numer. Math.* 13 (1969), 404–424.
3. A. W. Bojanczyk, R. P. Brent, P. Van Dooren and F. R. de Hoog, "A note on downdating the Cholesky factorization", *SISSC* 8 (1987), 210–220.
4. A. W. Bojanczyk, R. P. Brent and F. R. de Hoog, "QR factorization of Toeplitz matrices", *Numer. Math.* 49 (1986), 81–94.
5. A. W. Bojanczyk, R. P. Brent and F. R. de Hoog, "Stability analysis of a general Toeplitz systems solver", *Numerical Algorithms* 10 (1995), 225–244.
6. A. W. Bojanczyk, R. P. Brent, F. R. de Hoog and D. R. Sweet, "On the stability of the Bareiss and related Toeplitz factorization algorithms", *SIAM J. Matrix Anal. Appl.* 16 (1995), 40–57.
7. T. Boros, T. Kailath and V. Olshevsky, "Fast algorithms for solving Cauchy linear systems", preprint, 1995.
8. R. P. Brent, F. G. Gustavson and D. Y. Y. Yun, "Fast solution of Toeplitz systems of equations and computation of Padé approximants", *J. Algorithms* 1 (1980), 259–295.
9. J. R. Bunch, "Stability of methods for solving Toeplitz systems of equations", *SISSC* 6 (1985), 349–364.
10. J. R. Bunch, "The weak and strong stability of algorithms in numerical linear algebra", *Linear Alg. Appl.* 88/89 (1987), 49–66.
11. J. R. Bunch, "Matrix properties of the Levinson and Schur algorithms", *J. Numerical Linear Algebra with Applications* 1 (1992), 183–198.
12. S. Chandrasekaran and A. H. Sayed, *Stabilizing the generalized Schur algorithm*, *SIMAX* 17 (1996), 950–983.
13. S. Chandrasekaran and A. H. Sayed, *A fast stable solver for nonsymmetric Toeplitz and quasi-Toeplitz systems of linear equations*, preprint, Jan. 1996.
14. J. Chun, T. Kailath and H. Lev-Ari, "Fast parallel algorithms for QR and triangular factorization", *SIAM J. Sci. Stat. Computing* 8 (1987), 899–913.
15. G. Cybenko, "The numerical stability of the Levinson-Durbin algorithm for Toeplitz systems of equations", *SIAM J. Sci. Stat. Computing* 1 (1980), 303–319.
16. G. Cybenko, "Fast Toeplitz orthogonalization using inner products", *SIAM J. Sci. Stat. Computing* 8 (1987), 734–740.
17. J.-M. Delosme and I. C. F. Ipsen, "Parallel solution of symmetric positive definite systems with hyperbolic rotations", *Linear Algebra Appl.* 77 (1986), 75–111.
18. P. Delsarte, Y. V. Genin and Y. G. Kamp, "A generalisation of the Levinson algorithm for Hermitian Toeplitz matrices with any rank profile", *IEEE Trans. Acoustics, Speech and Signal Processing* ASSP–33 (1985), 964–971.
19. J. Durbin, "The fitting of time-series models", *Rev. Int. Stat. Inst.* 28 (1959), 229–249.
20. I. Gohberg (editor), *I. Schur Methods in Operator Theory and Signal Processing* (Operator Theory: Advances and Applications, Volume 18), Birkhäuser Verlag, Basel, 1986.
21. I. Gohberg, T. Kailath and V. Olshevsky, "Gaussian elimination with partial pivoting for matrices with displacement structure", *Math. Comp.* 64 (1995), 1557–1576.
22. G. H. Golub and C. Van Loan, *Matrix Computations*, second edition, Johns Hopkins Press, Baltimore, Maryland, 1989.

23. Ming Gu, *Stable and efficient algorithms for structured systems of linear equations*, Tech. Report LBL-37690, Lawrence Berkeley Laboratory, Aug. 1995.

24. Ming Gu, *New fast algorithms for structured least squares problems*, Tech. Report LBL-37878, Lawrence Berkeley Laboratory, Nov. 1995.

25. P. C. Hansen and H. Gesmar, "Fast orthogonal decomposition of rank deficient Toeplitz matrices", *Numerical Algorithms* 4 (1993), 151–166.

26. G. Heinig, "Inversion of generalized Cauchy matrices and other classes of structured matrices", *Linear Algebra for Signal Processing, IMA Volumes in Mathematics and its Applications, Vol. 69*, Springer, 1994, 95–114.

27. T. Kailath and J. Chun, "Generalized displacement structure for block-Toeplitz, Toeplitz-block, and Toeplitz-derived matrices", *SIAM J. Matrix Anal. Appl.* 15 (1994), 114–128.

28. T. Kailath and A. H. Sayed, "Displacement structure: theory and applications", *SIAM Review* 37 (1995), 297–386.

29. T. Kailath and A. H. Sayed (editors), *Fast Reliable Algorithms for Matrices with Structure*, to appear.

30. A. N. Kolmogorov, "Interpolation and extrapolation of stationary random sequences", *Izvestia Akad. Nauk SSSR* 5 (1941), 3–11 (in Russian). German summary, *ibid* 11–14.

31. N. Levinson, "The Wiener RMS (Root-Mean-Square) error criterion in filter design and prediction", *J. Math. Phys.* 25 (1947), 261–278.

32. F. T. Luk and S. Qiao, "A fast but unstable orthogonal triangularization technique for Toeplitz matrices", *Linear Algebra Appl.* 88/89 (1987), 495–506.

33. S. Qiao, "Hybrid algorithm for fast Toeplitz orthogonalization", *Numer. Math.* 53 (1988), 351–366.

34. I. Schur, "Über Potenzreihen, die im Innern des Einheitskreises beschränkt sind", *J. fur die Reine und Angewandte Mathematik* 147 (1917), 205–232. English translation in [20], 31–59.

35. M. Stewart, "Stable pivoting for the fast factorization of Cauchy-like matrices", preprint, Jan. 13, 1997.

36. M. Stewart and P. Van Dooren, "Stability issues in the factorization of structured matrices", preprint, Aug. 1996. To appear in *SIMAX*.

37. D. R. Sweet, *Numerical Methods for Toeplitz Matrices*, PhD thesis, University of Adelaide, 1982.

38. D. R. Sweet, "Fast Toeplitz orthogonalization", *Numer. Math.* 43 (1984), 1–21.

39. D. R. Sweet and R. P. Brent, "Error analysis of a fast partial pivoting method for structured matrices", *Proceedings SPIE, Volume 2563, Advanced Signal Processing Algorithms* SPIE, Bellingham, Washington, 1995, 266–280.

40. G. Szegö, *Orthogonal Polynomials*, AMS Colloquium publ. XXIII, AMS, Providence, Rhode Island, 1939.

41. W. F. Trench, "An algorithm for the inversion of finite Toeplitz matrices", *J. SIAM (SIAM J. Appl. Math.)* 12 (1964), 515–522.

42. J. M. Varah, "The prolate matrix", *Linear Algebra Appl.* 187 (1993), 269–278.

43. N. Wiener, *Extrapolation, Interpolation and Smoothing of Stationary Time Series, with Engineering Applications*, Technology Press and Wiley, New York, 1949 (originally published in 1941 as a Technical Report).

On inhomogeneous eigenproblems and pseudospectra of matrices

Da Yong Cai and Xu Gong Lu

Department of Applied Mathematics
Tsinghua University
Beijing 100084

Abstract. The inhomogeneous eigenvalue(s) of matrices is the complex parameter(s) λ with which the equation

$$Ax - \lambda x = b$$

has solutions with $\|x\|_2 = 1$ where A is a $n \times n$ matrix, b a nonzero vector. This problem is resulted from many practical applications. It is also of theoretical interests.

In this paper, the geometry and topology of inhomogeneous eigenvalues are considered. A Gerschgorin-like theorem is given. Also an iterative algorithm is proposed of which the convergent speed is linear.

The relationship between inhomogeneous eigenvalue and the pseudospectra of matrices is explored. The later is getting more and more important in numerical linear algebra.

Throughout this paper only real matrices are considered. It is easy to generalize our results to complex cases.

1 Introduction

In 1985, G. Söderlind and R. Mattheij [1] considered the asymptotic estimation of the solution of linear differential systems.

$$\dot{U} = L(t) \cdot U + g, \tag{1.1}$$

where $L(t)$ is a $n \times n$ matrix. By introducing a new independent variable vector y and the generalized Liaponov transformation, this problem has an algebraic form

$$\mu y = Ay - b, \tag{1.2}$$

where $A \in R^{n \times n}$, $b \in R^n$ are given, $\mu \in R$, $y \in R^n$ are the expected solution(s).

In order to attain a well-defined problem as expected in engineering an extra condition

$$\|y\|_2 = 1 \tag{1.3}$$

has to be imposed.

In [2], a least square problem with a quadratic constraint is rewritten in the form of (1.2) and (1.3). In this case, μ denotes Lagrange multiplier.

More recently, L.Trefethen described the pseudospectra of matrices and its application in iterative solutions of linear systems. One of the definition for pseudospectra can be formulated as

$$\Lambda_\varepsilon(A) \equiv \{z \in \mathbb{C} | z = \sigma(A + E), \quad \|E\| \le \varepsilon\}. \tag{1.4}$$

Our computed results show that the pseudospectra of a matrix is tightly related to the geometry of the solution of (1.2) and (1.3).

The above facts motivated us to explore the geometry and topology of solutions of

$$Ax - \mu x = b \ne 0$$
$$\|x\|_2 = 1 \tag{1.5}$$

This problem is referred to be an inhomogeneous eigenvalue problem in literatures. In section 2, several theorems are given, which indicate the geometry and topology of the inhomogeneous eigenvalue set. Two algorithms and their convergence analysis are proposed in section 3.

Section 4 gives several numerical examples to show the relationship between the inhomogeneous eigenvalue set and the pseudospectra of matrices.

2   Geometry and topology of the inhomogeneous eigenvalue

Let us rewrite the inhomogeneous eigenvalue problem as below

$$Ax - \lambda x = b, \quad (b \ne 0),$$
$$(x, x) = 1, \tag{2.1}$$

where $A \in C^{n \times n}$ and $b \in C^n$ are given.

For the sake of simplicity and clearness of our description, only the real case is considered, i.e. $A \in R^{n \times n}$ and $b \in R^n$. It is easy to extend the result to complex case and $b \in C^{n \times m} \quad (m > 1)$.
In the following
$\| \cdot \|$ denotes the F-norm of a matrix or a vector,
$\sigma(A \backslash b)$ is the solution set of (2.1), and
$\sigma(A)$ is the spectrum of A.

Theorem 1. *Let A and b be a $n \times n$ matrix and a vector in R^n respectively. There exists $r(\theta) \ge 0$ such that $\lambda + r(\theta)e^{i\theta} \subset \sigma(A \backslash b)$ for any $\theta \in (0, 2\pi]$ and $\lambda \in \sigma(A)$. In the viewpoint of geometry, it means that $\sigma(A \backslash b)$ contains the components, each of which is a closed curve around an eigenvalue of A.*

Example 2.

$$A = \begin{pmatrix} 1 & 0 \\ 0 & 2 \end{pmatrix}, \quad b = t \begin{pmatrix} 1 \\ 1 \end{pmatrix},$$

where t is a scalar.

Obviously, we have

$$\sigma(A) = \{1, 2\}, \quad \text{and}$$
$$\sigma(A \backslash b) = \{\lambda \in \mathbb{C} |\ t^2 (|1 - \lambda|^2 + |2 - \lambda|^2) = |1 - \lambda|^2 |2 - \lambda|^2\}.$$

It is easy to verify that the equation

$$t^2 (|1 - \lambda|^2 + |2 - \lambda|^2) = |1 - \lambda|^2 |2 - \lambda|^2 \tag{2.2}$$

1. has only 2 real roots if $|t| > t^* > 0$.
2. has 4 real roots, one of which is multiple, if $|t| = t^*$.
3. otherwise, all roots of (2.2) are real and distinct.

By theorem 1, the topology of $\sigma(A \backslash b)$ should be

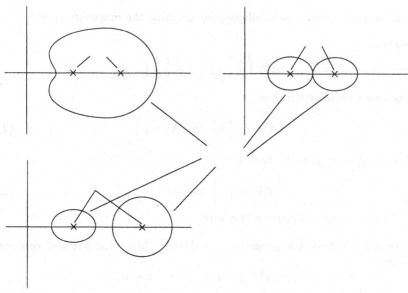

Figure 1. Topology of $\sigma(A \backslash b)$.

Remark 3. :

1. $\sigma(A)$ and $\sigma(A \backslash b)$ seems like independent each other, in a sense.
2. The topology of $\sigma(A \backslash b)$ could be very complicated while $\|b\|$ is getting smaller and smaller. It could be split into $l \le n$ components.
3. t^* is a bifurcation parameter of $\sigma(A \backslash b)$. It is very important for numerical computation of $\sigma(A \backslash b)$.

Theorem 4 (Gerschgorin-like). :
Let A be a $n \times n$ matrix and, b a vector.

i)

$$\sigma(A\backslash b) \subset \bigcup_{\mu \in \sigma(A)} \left\{ z \in C \,\middle|\, |z - \mu| \leq x(\|b\|) \right\}, \qquad (2.3)$$

where

$$X(r) = r + \max \left\{ (\Delta(A)(n-1)r)^{1/2}, (\Delta(A)^{1-1/n}(n-1)r)^{1/2} \right\}, \qquad (2.4)$$

$$\Delta(A) = \left(\|A\| - \sum_{\mu \in \sigma(A)} |\mu|^2 \right)^{1/2}. \qquad (2.5)$$

ii)

$$\sigma(A\backslash b) \subset \bigcup_{i=1}^{n} \left\{ z \in C \,\middle|\, |z - a_{ii}| \leq \sum_{i \neq j} |a_{i,j}| + \sqrt{n}\|b\| \right\}. \qquad (2.6)$$

The example given here is adopted for checking the inequality (2.6).

Example 5.

$$A = \begin{pmatrix} 1 & 0 \\ 0 & 1 \end{pmatrix}, \quad b = \begin{pmatrix} 1 \\ 0 \end{pmatrix}.$$

The exact pseudospectra set is

$$\sigma(A\backslash b) = \left\{ \lambda \,\middle|\, |\lambda - 1| = 1 \right\}. \qquad (2.7)$$

The estimation given by (2.6) is

$$\sigma(A\backslash b) = \left\{ \lambda \,\middle|\, |\lambda - 1| = 1 \right\}. \qquad (2.8)$$

It shows that the estimation is sharp!

Theorem 6. *(refined the geometry of $\sigma(A\backslash b)$). There exist bounded connected open sets*

$$D_j \subset C \quad j = 1, 2, \ldots, \ell, \quad \ell \leq n,$$

such that

1) $D_i \cap D_j = 0 \quad i \neq j$,
2) $D_j \cap \sigma(A) \neq 0 \quad \forall j$, and
3) $\sigma(A\backslash b) = \bigcup_j \partial D_j \cup \sigma_1(A)$, where $\sigma_1(A) \subset \sigma(A)$.

Remark 7. This theorem indicates that a subset of the spectrum of A could be included in $\sigma(A\backslash b)$ as shown in the example below.

Example 8.

$$A = \begin{pmatrix} 0 & 1 \\ 0 & 0 \end{pmatrix}, \quad b = \begin{pmatrix} 1 \\ 0 \end{pmatrix}.$$

$\sigma(A) = 0$ and $\sigma(A\backslash b) = 0 \cup |z| = 1$.

Theorem 9. *Let $A \in R^{n \times n}$, $b \in R^n$ be given.*
    We have

$$\sigma(A\backslash b) = \{\kappa + \rho(\theta)e^{i\theta} | \theta \in (0, 2\pi]\} \tag{2.9}$$

and $\sigma(A) \subset int\ \sigma(A\backslash b)$,
where κ is a constant satisfying

$$2\|A - \kappa I\|_2 \leq \|b\|$$
$$\rho(\cdot) \in C^\infty(R) \quad with \quad \rho(\theta + 2\pi) = \rho(\theta).$$

Remark 10. This result asserts that $\sigma(A\backslash b)$ is a closed infinitely differentiable curve and embraces $\sigma(A)$, if $\|b\|$ is large enough in a sense.

Theorem 11. *(The evolution of $\sigma(A\backslash b)$).* *Let A be a $n \times n$ matrix and $b \in R^n$. There exist $\mu_1, \mu_2, \ldots, \mu_s \in \sigma(A)$ and $\delta > 0$. For any $t \in (0, \delta]$, we have*

$$\sigma(A\backslash tb) = \bigcup_{j=1}^{s} \Gamma_{j,t} \cup \sigma_t(A), \tag{2.10}$$

where $\sigma_t(A) \subset \sigma(A)$, $\Gamma_{j,t} = \{\mu_j + \rho_{j,t}(\theta)e^{i\theta} | \theta \in (0, 2\pi]\}$.

Remark 12. This conclusion depicts the evolution of $\sigma(A\backslash b)$ pattern while the norm of b goes to zero. It can be shown in figure 2.

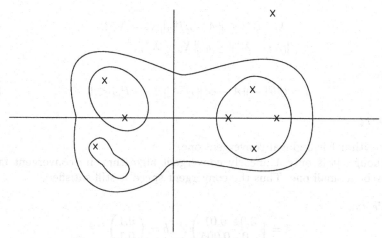

Figure 2: evolution of $\sigma(A\backslash tb)$ while $t \to 0$.
$$t_1 > t_2 > t^* = \frac{2\|A - \kappa I\|}{\|b\|} > t_3$$

3   Algorithm aspect of $\sigma(A\backslash b)$

To compute the inhomogeneous eigenvalue set is of interest for the estimation of the pseudospectra of a matrix A and b.

Algorithm 1.
For a given $A \in R^{n \times n}$, $b \in R^n$ and $\kappa \in c$ with

$$\|b\| > 2\|A - \kappa I\|_2 \tag{3.1}$$

The algorithm for the inhomogeneous eigenvalue is

1) Take an arbitrary initial guess

$$X_0 \in \mathcal{B} \equiv \left\{ X \in C^n \,\Big|\, \|X\|_2 = 1 \right\}$$

and a fixed parameter $\theta \in (0, 2\pi]$.
2) For $\ell = 1, 2, \ldots$ until converge   *do*

$$X_\ell = e^{-i\theta} U \left((A - \kappa I)X_\ell - b \right)$$
$$\lambda_\ell = \kappa + e^{i\theta} \|(A - \kappa I)X_\ell - b\|_2$$

where $U(X) = X/\|X\|$.

For the convergence of Algorithm 1, we give

Theorem 13. *Let $A \in R^{n \times n}$, $b \in R^n$ be a given matrix and vector respectively. λ^* and X^* are expected inhomogeneous eigenvalue and eigenvector. The algorithm 1 gives*

$$|\lambda_\ell - \lambda^*| \le \|A - \kappa I\|_2 \|X_\ell - X^*\|, \tag{3.2}$$
$$\|X_\ell - X^*\| \le q^\ell \|X_0 - X^*\|, \tag{3.3}$$

with

$$q = (\|b\| - \|A - \kappa I\|_2)^{-1} \|A - \kappa I\|_2 < 1. \tag{3.4}$$

Remark 14. :

1) Algorithm 1 is a global convergent one,
2) Although it is only a linearly convergent algorithm, its convergent factor may be a small one. Thus the convergent speed is still satisfied.

Example 15.

$$A = \begin{pmatrix} 0.01 & 0.01 \\ 0 & 0.005 \end{pmatrix}, \quad b = \begin{pmatrix} 0.1 \\ 0.1 \end{pmatrix}.$$

In this case, we have

$$\kappa = \frac{1}{2}\text{trace}(A) \doteq 0.75 \times 10^{-2},$$

$$\|A - \kappa I\|_2 \leq \|A - \kappa I\|_F \approx \frac{3}{4}\sqrt{2},$$

$$\|b\| = \sqrt{2} \times 10^{-1}.$$

Thus

$$\|b\| > 2\|A - \kappa I\|_2 \quad \text{is true.}$$

Finally we have

$$q \leq \frac{3}{37} \approx 0.085\,!$$

Remark 16. :

1) As required by (3.1) the convergence of algorithm 1 is ensured for the simplest geometry of the inhomogeneous eigenproblem, i.e. it contains only one star-like curve.
2) Once (3.1) does not hold, the formulation of an effective algorithm and convergence analysis become extremely difficult. Specially the bifurcation analysis and computation of the inhomogeneous eigenvalue set are still unknown and of practical interest for practical purposes.

In many applications, say a least square problem with a quadratic constraint, only the real inhomogeneous eigenvalues are expected. For this case, we give a special algorithm.

Algorithm 2:

1) To take a $X_1 \in R^n$ as an initial guess.
2) For $k = 1, 2, \ldots$ until convergent   *do*

$$y_{k+1} = AX_k - b$$

$$\mu_{k+1} = (x_k, y_{k+1})$$

$$X_{k+1} = \text{Sign}(\mu_{k+1})\frac{y_{k+1}}{\|y_{k+1}\|_2}$$

For the convergence analysis, we obtain

Theorem 17. *Let μ and X be a real inhomogeneous eigenvalue and eigenvector respectively of (2.1) (if any). $P \equiv I - xx'$ is an orthogonal projector.*

If 1) *PAP/μ is a contractor,*
     2) *X_1 is chosen sufficiently closed to X,*

then the algorithm 2 converges linearly.

4 Numerical examples

In this section two examples are computed to show the relationship between the inhomogeneous eigenvalue set and pseudospectra.

For a given $n \times n$ matrix A, the pseudospectra given by Trefethen is used for comparison. And then, we choose several right hand vector b with reasonable norm. Using Algorithm 1 we always get a inhomogeneous eigenvalue curve which matches the pseudospectra boundary very well.

Actually we don't know how the right hand vector b could be settled down in advance.

In the figures below, 'x' denote the pseudospectra, '—' curves are the computed results. In (1.4), the parameter $\varepsilon = 10^{-3}$ is considered.

Example 18.

$$A_1 = \begin{bmatrix} 0 & 1 & & & 0 \\ & \ddots & \ddots & & \\ & & \ddots & \ddots & \\ & & & \ddots & 1 \\ 0 & & & & 0 \end{bmatrix}_{32\times32}, \quad A_2 = \begin{bmatrix} 1 & 1 & 11 & 0 & 0 & 0 \\ -1 & 1 & 11 & 1 & 0 \\ & -1 & 11 & 1 & 1 \\ & & \ddots & \ddots & \\ 0 & & & -1 & 1 \end{bmatrix}_{32\times32}.$$

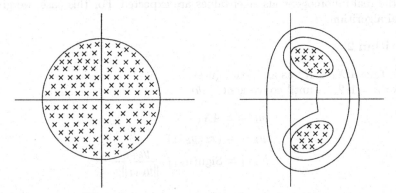

Figure 3: The inhomogeneous eigenvalue and pseudospectra.

5 Conclusion

Inhomogeneous eigenvalue problems are resulted from both practical applications and scientific research. Our computation examples show that it could be a competitive approach for computing the pseudospectra of matrices. Many questions about its geometry and bifurcation behaviors are unknown. Numerical algorithms are expected too.

References

1. Söderlind, G., Marttheij, R.M.M: Stability asymptotic estimation in non-autonomous linear differential systems. SIAM. J. Math. Anal. **16**, (1985) 69-92
2. Gander, W.: Least squares with a quadratic constraint. Numer. Math. **36**, (1981) 291-307
3. Trefethen, L.N.: Pseudospectra of matrices. Numerical Analysis (D. F. Griffith and G. A. Watson ed) Harlow Essex U.K. (1992)
4. Batterson, S.: Dynamics analysis of numerical systems. Numerical linear algebra with applications vol2. No3. (1995) pp297-309
5. Lu, X.G.: Matrix inhomogeneous eigenvalue analysis. (in Chinese) Computational Mathematics vol3. (1994) pp319-332

Multigrid for Differential-Convolution Problems Arising from Image Processing

Raymond H. Chan[1], Tony F. Chan[2] and W. L. Wan[2]

[1] Department of Mathematics, Chinese University of Hong Kong, Shatin, Hong Kong. Email: rchan@math.cuhk.edu.hk.
[2] Department of Mathematics, University of Califonia at Los Angeles, Los Angeles, CA 90095-1555. Email: chan@math.ucla.edu, wlwan@math.ucla.edu.

Abstract. We consider the use of multigrid methods for solving certain differential-convolution equations which arise in regularized image deconvolution problems. We first point out that the usual smoothing procedures (e.g. relaxation smoothers) do not work well for these types of problems because the high frequency error components are not smoothed out. To overcome this problem, we propose to use optimal fast-transform preconditioned conjugate gradient smoothers. The motivation is to combine the advantages of multigrid (mesh independence) and fast transform based methods (clustering of eigenvalues for the convolution operator). Numerical results for Tikhonov regularization with the identity and the Laplacian operators show that the resulting method is effective. However, preliminary results for total variation regularization show that this case is much more difficult and further analysis is required.

1 Introduction

In PDE based image processing, we often need to solve differential-convolution equations of the form:

$$\alpha R(u)(x) + \int_\Omega k(x - y)u(y)dy = f(x), \quad x \text{ in } \Omega, \tag{1}$$

where $u(x)$ is the recovered image, $k(x)$ is the kernel convolution function, $R(u)$ is a regularization functional and α is a positive parameter. Typical forms of $R(u)$ are:

$$R(u) = \begin{cases} u & \text{Tikhonov} \\ -\Delta u & \text{Isotropic Diffusion (ID)} \\ -\nabla \cdot (\nabla u/|\nabla u|) & \text{Total Variation (TV)}. \end{cases}$$

The discretization of (1) gives rise to a linear system of the form:

$$(\alpha A + K)u = f, \tag{2}$$

with the following properties. The matrix A, corresponding to the regularization part, is typically sparse, symmetric and positive-definite (positive semi-definite for ID and TV because the boundary condition is Neumann). The matrix K, corresponding to the convolution part, is typically ill-conditioned, symmetric and dense but with a Toeplitz structure. In this paper, we are interested in using iterative methods to solve a large system of the form (2).

The effectiveness of iterative methods depends on the choice of preconditioners. For matrix A, the commonly used preconditioners include [9]: multigrid (MG), domain decomposition (DD), incomplete LU factorization (ILU), successive over-relaxation (SOR) etc. MG or DD type preconditioners have a characteristics of optimal convergence in the sense that its convergence rate is independent of the mesh size.

For matrix K, various preconditioners have been proposed, for example, circulant preconditioners [10,6,5], sine transform preconditioners [4], cosine transform preconditioners [2] etc. For these types of preconditioners, the eigenvalues of the preconditioned system typically clustered around one which is a very desirable condition for the conjugate gradient method. Recently, a MG preconditioner [3] has also been proposed and optimal convergence is proved for a class of Toeplitz systems.

The construction of preconditioners for the sum of operators $L = \alpha A + K$, however, is difficult. Suppose M_A and M_K are two efficient preconditioners for A and K respectively. Then $M_L = \alpha M_A + M_K$ would be a good approximation to L. Unfortunately, M_L is not easily invertible in general even if M_A and M_K are.

A simple strategy is to use either M_A or M_K alone to precondition L. In [8,12], a MG preconditioner is constructed for $\tilde{L} = \alpha A + \gamma I$ which in turn is used to precondition L, hoping that the matrix K is well approximated by γI. A potential drawback is that γI may be a poor approximation to K.

In such situations, the operator splitting method of Vogel and Oman [13] may be more effective. This preconditioner approximates the inverse of L by a product of factors each involving only either A or K:

$$M = (K + \gamma I)^{1/2}(\alpha A + \gamma I)(K + \gamma I)^{1/2},$$

where γ is an appropriately chosen constant. This preconditioner is very effective for both very large and very small values of α but the performance can deteriorate for intermediate values of α.

To alleviate this problem, Chan-Chan-Wong [2] proposed a class of optimal fast-transform based preconditioners to precondition L. The main idea is to select as preconditioner the best approximation to L from a fast-transform invertible class of matrices by solving the following optimization problem:

$$\min_{M \in C} \|M - L\|_F,$$

where C is the class of matrices diagonalizable by the cosine-transform. Such optimal fast-transform based preconditioners have proven to be very effective for convolution type problems [5] and they have also been extended to elliptic problems [1]. It turns out that the optimal M for L can be computed very efficiently by exploiting the Toeplitz structure of K and the banded structure of A. Since L is not "split" in arriving at a preconditioner, the performance is not sensitive to the value of α. However, even though the performance is very satisfactory for Tikhonov and ID regularization, the convergence behavior for the TV regularization case may still depend on the mesh size. This is caused by the highly varying coefficient in the TV operator.

In view of the effectiveness of MG for A and the fast transform preconditioners for K, our idea is to combine the benefits of both. Specifically, we use fast-transform based preconditioned conjugate gradient as a smoother for MG. Our analysis and numerical results show that this is an effective smoother, whereas the standard relaxation type preconditioners are totally ineffective for convolution type problems. In this paper, we shall focus on two 1D cases: (1) $A = I$ (identity) (2) $A = -\Delta$ (Laplacian operator). In sections 2 and 3, we discuss the difficulties of using MG for $L = \alpha I + K$ and $L = -\alpha \Delta + K$ and how we tackle it through the use of fast transform based smoothers. In section 4, we discuss the total variation case. It turns out that this case is much more difficult and although we have some encouraging results, we still have not arrived at an effective method. In 5, we shall estimate the complexity of some of the methods discussed. Finally, some conclusions are made in section 6.

We remark that this paper is only a preliminary report of on-going work and much further investigation remains to be done.

2   The Case $A = I$

In this section, we shall consider operators of the form $L = \alpha I + K$, where K arises from the discretization of an integral operator of the first kind. It is well-known that K is very ill-conditioned and MG with traditional smoothers does not work well for K. The regularization term αI improves the conditioning by shifting the spectrum a distance α away from zero. It turns out that this is not enough to make MG work well. The reason is that the set of eigenvectors remains the same independent of α. We shall explain this phenomenon next.

Our observation is that common relaxation methods, for instance, Richardson, Jacobi or Gauss-Seidel method, fail to smooth the error in the geometric sense. The reason is that, unlike in the elliptic case, eigenvectors of $\alpha I + K$ corresponding to small eigenvalues are highly oscillatory while those corresponding to large eigenvalues are smooth. It is known that relaxation methods reduce the error components corresponding to large eigenvalues only and therefore they in fact remove the smooth error components. We illustrate this using Richardson iteration as an example. Let A be a symmetric positive definite matrix and let $0 < \lambda_1 \leq \cdots \leq \lambda_n$ be its eigenvalues and $\{v_k\}$ the corresponding eigenvectors. The error e^{m+1} in the $m + 1$st iteration step of the Richardson method is given by

$$e^{m+1} = (I - \frac{1}{\lambda_n}A)e^m.$$

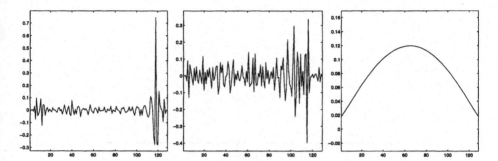

Fig. 1. Eigenvectors corresponding to (a) the smallest (b) the middle (c) the largest eigenvalue of $L = 10^{-4}I + K$. The oscillatory eigenvectors corresponding to the small eigenvalues.

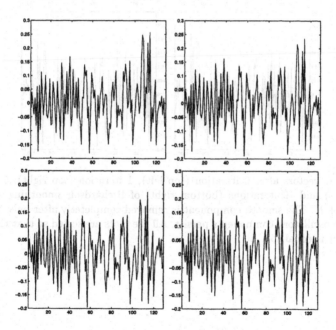

Fig. 2. Error vectors after 0 iteration (top left), 1 iteration (top right), 5 iterations (bottom left) and 10 iterations (bottom right) of Richardson smoothing applied to $L = 10^{-4}I + K$. Note that there is no smoothing effect.

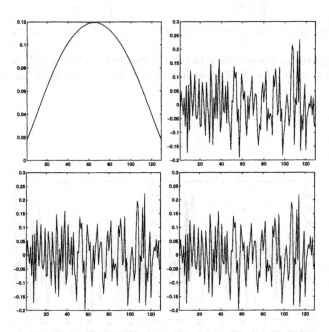

Fig. 3. Error vectors after 0 iteration (top left), 1 iteration (top right), 5 iterations (bottom left) and 10 iterations (bottom right) of Richardson smoothing applied to $L = 10^{-4}I + K$. The smooth component is removed completetly after only 1 iteration whereas the oscillatory components persist. All the plots are scaled so that the l_2-norm of the vector is equal to 1.

Let the eigendecomposition of e^m be $e^m = \sum_{k=1}^{n} \xi_k v_k$. Since $\{v_k\}$ are orthogonal by the symmetry of A, we have

$$\|e^{m+1}\|_2^2 = \sum_{k=1}^{n}(1 - \frac{\lambda_k}{\lambda_n})^2 \xi^2.$$

Note that $(1 - \lambda_k/\lambda_n) \approx 0$ when k is close to n and $(1 - \lambda_k/\lambda_n) \approx 1$ when k is close to 1. Hence, the components corresponding to large eigenvalues are reduced while those corresponding to small eigenvalues remain essentially unchanged.

We illustrate the smoothing phenomenon of the Richardson iteration applied to $L = \alpha I + K$ by a simple example. Choose $\alpha = 10^{-4}$ and $k(x) = \frac{1}{C}\exp(-x^2/0.01)$ which is known as the Gaussian blurring operator in image processing. Here $C = \int_0^1 \exp(-x^2/0.01)dx$ is the normalization constant. Let $0 < \lambda_1 \leq \cdots \leq \lambda_n$ be the eigenvalues of L and $v_1, \ldots, v_n$ be the corresponding eigenvectors. Figure 1 shows the plots of v_1, $v_{n/2}$ and v_n for $n = 128$. Relaxation methods, for example, the Richardson method, essentially reduces the error components corresponding to large eigenvalues, not necessary the high frequencies. Because of the special spectrum of L, these methods do not reduce the high frequency errors. Figure 2 shows the plots of the initial (oscillatory) error and the errors after 1, 5, 10 number of Richardson iterations. No smoothing effect can be seen. In fact, as shown in Figure 3, if the initial error consists of low frequency and a small perturbation of high frequency vectors, after one Richardson iteration, the low frequency components will be removed and the error is left with high frequency only.

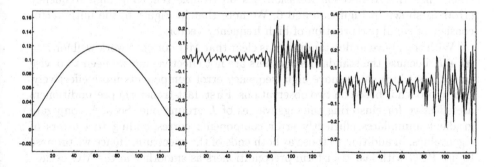

Fig. 4. Eigenvectors corresponding to (a) the smallest (b) the middle (c) the largest eigenvalue of $L = I - K$. The oscillatory eigenvectors correspond to the largest eigenvalues.

In contrast, MG converges rapidly for integral operators of the second kind of the form $L = I - K$ and this can also be explained by the smoothing argument. Figure 4 shows the eigenvectors of $L = I - K$ with K as before. Because of the minus sign, we see that eigenvectors corresponding to small eigenvalues are smooth while those of large eigenvalues are oscillatory as in the standard elliptic

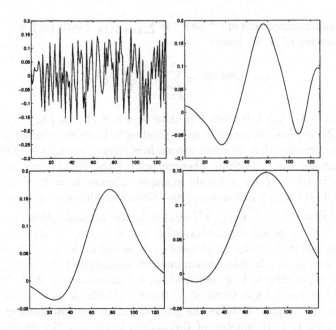

Fig. 5. Error vectors after 0 iteration (top left), 1 iteration (top right), 5 iterations (bottom left) and 10 iterations (bottom right) of Richardson smoothing applied to $L = I - K$. The oscillatory components are quickly smoothed out.

case. Thus the Richardson iteration has no trouble removing high frequency errors as shown in Figures 5 and 6. We note that in Figure 6, the initial error consists of small perturbation of high frequency vectors.

With the above understanding, it is clear that MG does not work well for $L = \alpha I + K$ because the standard smoothers are not effective and we need to devise smoothers which can remove high frequency error components more effectively. Our approach is based on two observations. First, fast-transform preconditioners are effective for clustering the eigenvalues of L around one. Second, conjugate gradient annihilates efficiently error components corresponding to clusters of eigenvalues, in addition to those at both ends of the spectrum. Hence we propose to use PCG with fast transform preconditioners as smoother in the MG cycle.

Figure 7 shows the eigenvectors of the preconditioned system using the cosine transform preconditioner. It is interesting to note that low frequency vectors are located at both ends of the spectrum while high frequency vectors concentrate at the cluster. Figure 8 shows the smoothing effect of PCG using the cosine-transform preconditioner (PCG(Cos)). We remark that MG with the optimal circulant preconditioner also produces similar plots and hence we do not show it. Table 1 shows the MG convergence (MG(*)) of different smoothers specified in the brackets. The Richardson smoother is denoted by R and the PCG smoother with the cosine-transform preconditioner is denoted by PCG(Cos). The convergence of PCG(Cos) alone is also given for comparison. Here we use

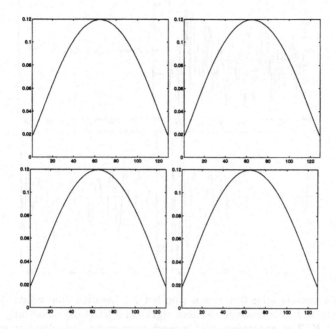

Fig. 6. Error vectors after 0 iteration (top left), 1 iteration (top right), 5 iterations (bottom left) and 10 iterations (bottom right) of Richardson smoothing applied to $L = I - K$. The smooth components remain after many iterations.

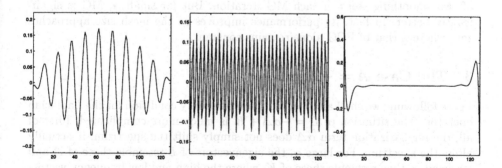

Fig. 7. Eigenvectors corresponding to (a) the smallest (b) the middle (c) the largest eigenvalue of the cosine transform preconditioned system of $L = 10^{-4}I + K$. The oscillatory eigenvectors are clustered in the middle of the spectrum.

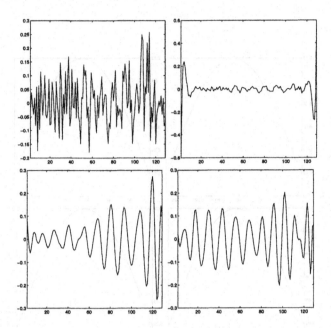

Fig. 8. Error vectors after 0 iteration (top left), 1 iteration (top right), 5 iterations (bottom left) and 10 iterations (bottom right) of PCG(Cos) smoothing applied to $L = 10^{-4}I + K$. The smoothing effect is much improved over Richardson in Figure 2.

two pre-smoothing and no post-smoothing step. The iteration is stopped when the relative residual is less than 10^{-10}. The matrix K is the Gaussian blurring operator as before. From the table, we see that PCG as smoother is much more efficient than standard relaxation methods in all cases. For large α, MG with PCG as smoother is about as efficient as PCG alone, taking into the account of two smoothing steps in each MG iteration. But for small α, MG is significantly better. In fact, its performance improves as the mesh size approaches zero whereas that of PCG alone remains constant.

3   The Case $A = -\Delta$

In the following, we shall assume Neumann boundary condition for the Laplacian operator. The situation of $L = -\alpha\Delta + K$ is much more complicated. First of all, the regularization term $\alpha\Delta$ does not simply shift the spectrum; it actually alters the spectrum. For large α, the eigenvectors of L resemble those of Δ and for small α, they resemble those of K, where the high and low frequency vectors are flipped over each other. For α in between, it is a mixture but the precise nature of the mixing is not known. We pick three different size of α to illustrate the changing spectrum of L in Figures 9, 10 and 11.

The numerical results for this case is given in Table 2. As expected, MG with standard relaxation methods as smoother deteriorates when α decreases because

α	h	1/64	1/128	1/256	1/512
	MG(R)	*	*	*	*
10^{-2}	MG(PCG(Cos))	5	4	4	4
	PCG(Cos)	8	8	8	8
	MG(R)	*	*	*	*
10^{-3}	MG(PCG(Cos))	6	5	4	4
	PCG(Cos)	11	11	11	11
	MG(R)	*	*	*	*
10^{-4}	MG(PCG(Cos))	11	7	6	6
	PCG(Cos)	18	18	18	18
	MG(R)	*	*	*	*
10^{-5}	MG(PCG(Cos))	40	18	14	11
	PCG(Cos)	33	37	36	38

Table 1. Convergence of different MG and PCG with varying α and mesh size h. $L = \alpha I + K$. * indicates more than 100 iterations. The results show that PCG(Cos) is an effective smoother.

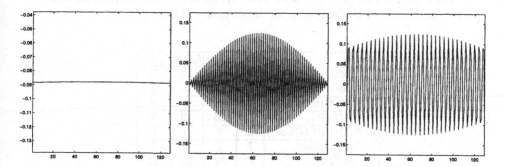

Fig. 9. Eigenvectors corresponding to (a) the smallest (b) the middle (c) the largest eigenvalue of $L = -\Delta + K$. When α is large, the eigenvectors of L resemble those of $-\Delta$.

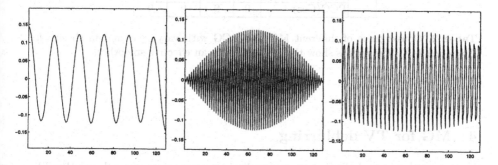

Fig. 10. Eigenvectors corresponding to (a) the smallest (b) the middle (c) the largest eigenvalue of $L = -10^{-4}\Delta + K$. For intermediate value of α, the eigenvectors corresponding to large eigenvalues resemble those of $-\Delta$ and the eigenvectors corresponding to small eigenvalues are oscillatory and resemble those of K.

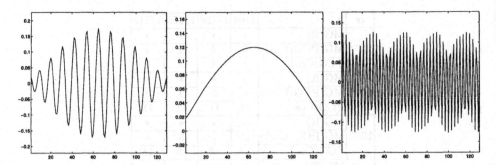

Fig. 11. Eigenvectors corresponding to (a) the smallest (b) the middle (c) the largest eigenvalue of $L = -10^{-8}\Delta + K$. When α is small, the eigenvectors of L resemble those of K.

L approaches the convolution operator K which we have shown in section 2 that standard smoothers do not work well. Again, MG(PCG(Cos)) shows better performance over PCG(Cos) alone for small values of α and h.

α	h	1/64	1/128	1/256	1/512
	MG(R)	18	19	20	21
10^{-2}	MG(PCG(Cos))	3	3	3	3
	PCG(Cos)	6	6	6	6
	MG(R)	17	18	19	20
10^{-3}	MG(PCG(Cos))	3	3	3	3
	PCG(Cos)	7	7	7	7
	MG(R)	32	32	32	32
10^{-4}	MG(PCG(Cos))	4	4	3	3
	PCG(Cos)	8	8	8	8
	MG(R)	*	*	*	*
10^{-5}	MG(PCG(Cos))	4	4	3	3
	PCG(Cos)	9	9	9	9

Table 2. Convergence of different MG and PCG with varying α and mesh size h. $L = -\alpha\Delta + K$. The results show that PCG(Cos) is an effective smoother.

4   MG for TV deblurring

In this section, we shall discuss our preliminary experience in solving the TV deblurring problem [7] by MG. The governing differential-convolution equation is slightly different from (1) and is given here:

$$\alpha R(u)(x) + \mathcal{K}^*\mathcal{K}(u) = \mathcal{K}^*z,$$

where $R(u) = -\nabla \cdot (1/|\nabla u|)\nabla u$, $\mathcal{K}u = \int_\Omega k(x-y)u(y)dx$, $\mathcal{K}^*$ is the adjoint operator of $\mathcal{K}$ and z is the observed blurred and noisy image. Basically, the convolution operator is replaced by a product of itself with its adjoint. The corresponding linear system is:

$$(\alpha A + K^T K)u = f, \tag{3}$$

which is similar to (2) with K replaced by $K^T K$. The additional challenges of solving (3) are two fold. First, the matrix A now comes from an elliptic operator with highly varying coefficient (which is $1/|\nabla u|$). It is not known if MG can handle this case efficiently. Second, the product $K^T K$ is no longer Toeplitz which complicates the implementation issue. For instance, while it is trivial to construct the Jacobi preconditioner for K, it is not so for $K^T K$ at first glance, although it turns out that it can also be done in $O(n)$ operations. Moreover, the conditioning of $K^T K$ is worse than K alone.

It turns out MG even with PCG as smoother does not work well in this case. A natural way to improve its performance is to use it as a preconditioner for conjugate gradient. However, this is not feasible as MG with PCG as smoother gives rise to a nonstationary preconditioner. One solution to this problem is based on the following observation. The success of PCG as smoother is that CG takes advantage of the clustered eigenvalues of the cosine-transform preconditioned system. We notice that it is probably advantageous but not necessary to apply CG (which gives rise to a nonstationary preconditioner) to the preconditioned system. An alternative is to use standard relaxation methods on the cosine-transform preconditioned system.

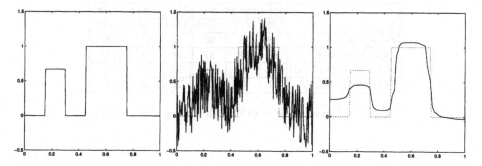

Fig. 12. (a) Original image (b) Blurred and noisy image (c) Recovered image. Gaussian blur is used and SNR=13.

We have tried out several possibilities and the results are shown in Tables 3, 4 and 5 for a TV deblurring example. The original and the blurred noisy 1D image together with the recovered image are shown in Figure 12. The signal-to-noise ratio SNR=13. Here we have used the Gaussian blur again. For each grid size h, we use the *optimal* α_{opt} for L which is chosen so that the recovered image has the specified SNR. We test three cases: $\alpha = 10 * \alpha_{opt}$, $\alpha = \alpha_{opt}$ and $\alpha = 0.1 * \alpha_{opt}$, corresponding to Tables 3-5 respectively. In each table, the second to fourth column show the convergence in the first fixed point iteration and the fifth to seventh ones show the convergence at the 11th fixed point iteration. (For our examples, the fixed point iteration has already converged at the 11th iteration.) We show these two sets of results because the coefficient $1/|\nabla u|$ is quite different for the two cases; see Figure 12. In the first fixed point iteration, the coefficient is very oscillatory whereas at the eleventh iteration, it is almost piecewise constant. With the same notation as before, the bracket right after PCG specifies the preconditioner used for CG and the bracket right after MG specifies the smoother. Here GS+Cos denotes the Gauss-Seidel (GS) method applied to the cosine-transform preconditioned system. Similarly for J+Cos where J denotes the Jacobi method.

$10 * \alpha_{opt}$	1st fixed pt. iter.			11th fixed pt. iter.		
	1/64	1/128	1/256	1/64	1/128	1/256
PCG(Cos)	42	85	108	13	49	75
PCG(MG(GS))	20	29	35	12	17	21
PCG(MG(GS+Cos))	14	28	37	4	11	12
PCG(MG(J+Cos))	17	51	79	13	17	22

Table 3. Convergence of PCG with varying h. $\alpha = 10 * \alpha_{opt}$.

α_{opt}	1st fixed pt. iter.			11th fixed pt. iter.		
	1/64	1/128	1/256	1/64	1/128	1/256
PCG(Cos)	38	81	98	42	92	106
PCG(MG(GS))	17	28	37	16	21	24
PCG(MG(GS+Cos))	14	26	34	13	15	14
PCG(MG(J+Cos))	17	45	73	20	28	28

Table 4. Convergence of PCG with varying h. $\alpha = \alpha_{opt}$.

We see that PCG(MG(GS)) and PCG(MG(GS+Cos)) are the best. They are not sensitive to α and deterioration with smaller h is slow. Besides, PCG(MG(GS+Cos)) is better than PCG(MG(GS)) for smaller α which shows that cosine transform is effective in dealing with K. However, we have not come up with an efficient implementation for these two methods. For PCG(MG(GS)), Vogel [11] has also made this observation independently. PCG(MG(J+Cos)) shows a degradation over PCG(MG(GS+Cos)), similar to that of the ordinary GS over Jacobi. We should also remark that PCG(Cos) is quite effective among the methods we have tried.

0.1 ∗ α_{opt}	1st fixed pt. iter.			11th fixed pt. iter.		
	1/64	1/128	1/256	1/64	1/128	1/256
PCG(Cos)	35	75	93	55	90	128
PCG(MG(GS))	19	35	54	15	21	25
PCG(MG(GS+Cos))	13	24	33	15	19	17
PCG(MG(J+Cos))	16	43	67	22	35	36

Table 5. Convergence of PCG with varying h. $\alpha = 0.1 \ast \alpha_{opt}$.

5   Computation complexity

Here we estimate the complexity of one iteration of some of the methods that we have described in Sections 2 and 3.

PCG(Cos): This method has been estimated in [2]. The construction of the preconditioner is $O(n)$ and the cost of the preconditioning is $O(n \log_2 n)$.

MG(R): On each level, the cost of a Richardson smoothing is essentially the cost of matrix-vector multiply. For the sparse matrix A, it can be done in $O(n_l)$ operations and for the Toeplitz matrix, it can be done in $O(n_l \log_2 n_l)$, where n_l is the size of the matrix at level l. Here we assume that K is Toeplitz at all levels. In fact, this is proved to be true in [3] if linear interpolation is used. The construction of the coarse grid matrices can also be done in $O(n)$. Thus the overall complexity of an iteration of MG(R) is $O(n \log_2 n)$.

MG(PCG(Cos)): The method is almost the same as MG(R) but with different smoother. The cost of applying the PCG(Cos) is $O(n \log_2 n)$ and hence the overall complexity is $O(n \log_2 n)$.

We remark that we have not come up with an efficient implementation of the methods in the TV case and so we do not discuss the complexity issue of those methods here.

6   Conclusions

We have shown in section 2 that standard smoothers do not work for matrices of the form $\alpha I + K$ arising from convolution operators. We have proposed to use PCG as smoother and demonstrated numerically that it is effective to reduce oscillatory errors. We have also tested the matrices of the form $-\alpha \Delta + K$ and the PCG smoother works as well.

For the TV image deblurring, the situation is complicated by the highly varying coefficient and the product of convolution operators. We have proposed several multigrid preconditioners and the numerical results are satisfactory. However the implementation issue is still left open. Further investigation is needed to devise a practical and efficient multigrid preconditioner in this case.

7   Acknowledgment

Research of R. Chan has been partially supported by HKRGC Research Grant CUHK 178/93E. Research of Tony F. Chan has been partially supported by the ONR under Contract N00014-96-1-0277 and the NSF under contract DMS-9626755. Research of W. L. Wan has been partially supported by the grants listed under the second author and the Alfred P. Sloan Foundation as a Doctoral Dissertation Fellow.

References

1. R. Chan and T. Chan. Circulant preconditioners for elliptic problems. *Numer. Linear Algebra Appl.*, 1:77–101, 1992.
2. R. Chan, T. Chan, and C. Wong. Cosine transform based preconditioners for total variation minimization problems in image processing. Technical Report 95-23, Dept. of Mathematics, UCLA, 1995.
3. R. Chan, Q. Chang, and H. Sun. Multigrid method for ill-condit-ioned symmetric Toeplitz systems. *SIAM J. Sci. Comp.*, to appear.
4. R. Chan, K. Ng, and C. Wong. Sine transform based preconditioners for symmetric Toeplitz systems. *Linear Algebra Appls.*, 232:237–260, 1996.
5. R. Chan and M. Ng. Conjugate gradient methods for Toeplitz systems. *SIAM Review*, 38:427–482, 1996.
6. T. Chan. An optimal circulant preconditioner for Toeplitz systems. *SIAM J. Sci. Stat. Comput.*, 9:766–771, 1988.
7. T. Chan and P. Mulet. Iterative methods for total variation image restoration. Technical Report 96-38, Dept. of Mathematics, UCLA, 1996.
8. M. Omen. Fast multigrid techniques in total variation-based image reconstruction. In *Proceedings of the 1995 Copper Mountain Conference on Multigrid Methods*, 1995.
9. Y. Saad. *Iterative Methods for Sparse Linear Systems*. PWS Kent Publishing Co., 1995.
10. G. Strang. A proposal for Toeplitz matrix calculations. *Stud. Appl. Math.*, 74:171–176, 1986.
11. C. R. Vogel. Private communication. March 97.
12. C. R. Vogel. A multigrid method for total variation-based image denoising. In K. Bowers and J. Lund, editors, *Computation and Control IV*. Birkhauser, 1995.
13. C. R. Vogel and M. Oman. Fast, robust total variation-based reconstruction of noisy, blurred images. 1996. Submitted to IEEE Transactions on Image Processing.

Cyclic Reduction – History and Applications

Walter Gander[1] and Gene H. Golub[2]

[1] Institut für Wissenschaftliches Rechnen, ETH Zentrum,
CH-8092 Zürich, Switzerland.
[2] Scientific Computing and Computational Mathematics,
Stanford University, USA

Abstract. We discuss the method of Cyclic Reduction for solving special systems of linear equations that arise when discretizing partial differential equations. In connection with parallel computations the method has become very important.

1 Introduction

Cyclic Reduction has proved to be an algorithm which is very powerful for solving structured matrix problems. In particular for matrices which are (block) Toeplitz and (block) tri-diagonal, the method is especially useful. The basic idea is to eliminate half the unknowns, regroup the equations and again eliminate half the unknowns. The process is continued *ad nauseum*. This simple idea is useful in solving the finite difference approximation to Poisson's equation in a rectangle and for solving certain recurrences. The algorithm easily parallelizes and can be used on a large variety of architectures [7]. New uses of Cyclic Reduction continue to be developed – see, for instance, the recent publication by Amodio and Paprzycki [1].

In this paper, we give a historical treatment of the algorithm and show some of the important applications. Our aim has not been to be exhaustive, rather we wish to give a broad view of the method.

In Section 2, we describe the algorithm for a general tri-diagonal matrix and a block tri-diagonal matrix. We show how the process need not be completed. Section 3 describes the method for solving Poisson's equation in two dimensions. We give details of the stabilized procedure developed by O. Buneman. We also describe a technique developed by Sweet [11] for making the method more useful in a parallel environment. The final Section describes the influence of the method on developing more general procedures.

2   Classical Algorithm

2.1   Scalar Cyclic Reduction

Consider the tridiagonal linear system $Ax = v$ where

$$A = \begin{pmatrix} d_1 & f_1 & & & \\ e_2 & d_2 & f_2 & & \\ & \ddots & \ddots & \ddots & \\ & & e_{n-1} & d_{n-1} & f_{n-1} \\ & & & e_n & d_n \end{pmatrix}. \tag{1}$$

The fundamental operation in Cyclic Reduction is the *simultaneous elimination of odd-indexed unknowns*. This operation may be described as a block LU-factorization of a permutation of the rows and columns of the matrix A. Let S be the permutation matrix such that

$$S(1, 2, \ldots, n)^T = (1, 3, \ldots, |2, 4, \ldots)^T$$

The permuted matrix SAS^T becomes:

$$SAS^T = \left[\begin{array}{cccc|cccc} d_1 & & & & f_1 & & & \\ & d_3 & & & e_3 & f_3 & & \\ & & \ddots & & & \ddots & \ddots & \\ \hline e_2 & f_2 & & & d_2 & & & \\ & e_4 & \ddots & & & d_4 & & \\ & & \ddots & & & & \ddots & \end{array} \right].$$

Eliminating the odd-indexed unknowns is equivalent to computing a *partial LU-factorization* [2]; viz

$$SAS^T = \left[\begin{array}{cccc|cccc} 1 & & & & 0 & & & \\ & \ddots & & & & \ddots & & \\ & & 1 & & & & 0 & \\ \hline l_1 & m_1 & & & 1 & & & \\ & l_2 & \ddots & & & \ddots & & \\ & & \ddots & & & & & 1 \end{array} \right] \left[\begin{array}{cccc|cccc} d_1 & & & & f_1 & & & \\ & d_3 & & & e_3 & f_3 & & \\ & & \ddots & & & \ddots & \ddots & \\ \hline 0 & & & & d_1^{(1)} & f_1^{(1)} & & \\ & & & & e_2^{(1)} & d_2^{(1)} & \ddots & \\ & & & & 0 & & \ddots & \ddots \end{array} \right] \tag{2}$$

For n even, this decomposition is computed for $i = 1, \ldots, \frac{n}{2} - 1$ as

$$m_i = f_{2i}/d_{2i+1}, \quad l_i = e_{2i}/d_{2i-1}, \quad l_{n/2} = e_n/d_{n-1}$$

$$f_i^{(1)} = -m_i f_{2i+1}, \quad e_{i+1}^{(1)} = -l_{i+1} e_{2i+1}$$

$$d_i^{(1)} = d_{2i} - l_i f_{2i-1} - m_i e_{2i+i}, \quad d_{n/2}^{(1)} = d_n - l_{n/2} f_{n-1}$$

Since the Schur complement is tridiagonal we may iterate and again eliminate the odd-indexed unknowns of the reduced system. G. Golub recognized (cf. [10]) that Cyclic Reduction is therefore equivalent to Gaussian elimination without pivoting on a permuted system $(PAP^T)(Px) = Pv$. The permutation matrix P reorders the vector $(1, 2, \ldots, n)$ such that the odd multiples of 2^0 come first, followed by the odd multiples of 2^1, the odd multiples of 2^2, etc.; e.g. for $n = 7$ we get the permuted system

$$\begin{pmatrix} d_1 & & & & f_1 & & \\ & d_3 & & & e_3 & f_3 & \\ & & d_5 & & & f_5 & e_5 \\ & & & d_7 & & & e_7 \\ e_2 & f_2 & & & d_2 & & \\ & & e_6 & f_6 & & d_6 & \\ & & e_4 & f_4 & & & d_4 \end{pmatrix} \begin{pmatrix} x_1 \\ x_3 \\ x_5 \\ x_7 \\ x_2 \\ x_6 \\ x_4 \end{pmatrix} = \begin{pmatrix} v_1 \\ v_3 \\ v_5 \\ v_7 \\ v_2 \\ v_6 \\ v_4 \end{pmatrix}. \tag{3}$$

The connection to Gaussian elimination allows us to conclude that if elimination with diagonal pivots is stable then Cyclic Reduction will also be stable. If A is strictly diagonal dominant or symmetric positive definite then this holds for PAP^T. No pivoting is necessary in these cases and therefore Cyclic Reduction is well defined and stable. However, fill-in is generated in the process and this increases the operation count. Cyclic Reduction requires about 2.7 times more operations than Gaussian elimination and thus has a *redundancy* of 2.7 [2].

2.2   Block Cyclic Reduction

We use the notation of Heller [8] and consider a block tridiagonal linear system $Ax = v$, i.e.

$$E_j x_{j-1} + D_j x_j + F_j x_{j+1} = v_j, \quad j = 1, \ldots, n, \quad E_1 = F_n = 0. \tag{4}$$

where x_j and $v_j \in \mathbf{R}^m$ and the blocks are $m \times m$ matrices E_j, D_j and F_j so that the dimension of the matrix A is nm.

Consider three consecutive equations which we arrange in a tableau:

Eq#	x_{2j-2}	x_{2j-1}	x_{2j}	x_{2j+1}	x_{2j+2}	rhs
$2j-1$	E_{2j-1}	D_{2j-1}	F_{2j-1}			v_{2j-1}
$2j$		E_{2j}	D_{2j}	F_{2j}		v_{2j}
$2j+1$			E_{2j+1}	D_{2j+1}	F_{2j+1}	v_{2j+1}

In order to eliminate x_{2j-1} and x_{2j+1} we multiply equation $\#(2j-1)$ with $-E_{2j}(D_{2j-1})^{-1}$, equation $\#(2j+1)$ with $-F_{2j}(D_{2j+1})^{-1}$ and add these to equation $\#(2j)$. The result is a new equation

Eq#	x_{2j-2}	x_{2j-1}	x_{2j}	x_{2j+1}	x_{2j+2}	rhs
j	E_j'	0	D_j'	0	F_j'	v_j'

where

$$E'_j = -E_{2j}(D_{2j-1})^{-1}E_{2j-1} \tag{5}$$

$$D'_j = D_{2j} - E_{2j}(D_{2j-1})^{-1}F_{2j-1} - F_{2j}(D_{2j+1})^{-1}E_{2j+1} \tag{6}$$

$$F'_j = -F_{2j}(D_{2j+1})^{-1}F_{2j+1} \tag{7}$$

$$v'_j = v_{2j} - E_{2j}(D_{2j-1})^{-1}v_{2j-1} - F_{2j}(D_{2j+1})^{-1}v_{2j+1}. \tag{8}$$

Eliminating in this way the odd indexed unknowns, we obtain a block tridiagonal reduced system. The elimination can of course only be performed if the diagonal block matrices are non-singular. Instead of inverting these diagonal blocks (or better solving the corresponding linear systems) it is sometimes possible to eliminate the odd unknowns by matrix multiplications: Multiply equation #$(2j-1)$ with $E_{2j}D_{2j+1}$, equation #$(2j)$ with $D_{2j-1}D_{2j+1}$, equation #$(2j+1)$ with $F_{2j}D_{2j-1}$ and add. If

$$D_{2j-1}D_{2j+1}E_{2j} = E_{2j}D_{2j+1}D_{2j-1} \quad \text{and}$$
$$D_{2j-1}D_{2j+1}F_{2j} = F_{2j}D_{2j-1}D_{2j+1},$$

then the matrices of the unknowns with odd indices vanish. However, if the matrices do not commute then we cannot eliminate by matrix multiplications. In case of the discretized Poisson equation (20) the matrices do commute.

Eliminating the unknowns with odd indices thus results in a new tridiagonal system for the even unknowns. The new system is half the size of the original system. We can iterate the procedure and eliminate again the odd unknowns in the new system.

2.3   Divide and Conquer Algorithm

If we reorder Equation (4) for the first step of Cyclic Reduction by separating the unknowns with odd and even indices we obtain:

$$\begin{pmatrix} D_1 & F \\ G & D_2 \end{pmatrix} \begin{pmatrix} x_o \\ x_e \end{pmatrix} = \begin{pmatrix} v_o \\ v_e \end{pmatrix} \tag{9}$$

where D_1 and D_2 are diagonal block-matrices and F and G are block-bidiagonal. Eliminating the unknowns with odd indices $x_o = (x_1, x_3, \ldots)^T$ means computing the reduced system of equations for the unknowns with even indices $x_e = (x_2, x_4, \ldots)^T$ by forming the Schur complement:

$$(D_2 - GD_1^{-1}F)x_e = v_e - GD_1^{-1}v_o. \tag{10}$$

It is also possible to eliminate the unknowns with even indices, thus forming the equivalent system:

$$\begin{pmatrix} D_1 - FD_2^{-1}G & 0 \\ 0 & D_2 - GD_1^{-1}F \end{pmatrix} \begin{pmatrix} x_o \\ x_e \end{pmatrix} = \begin{pmatrix} v_o - FD_2^{-1}v_e \\ v_e - GD_1^{-1}v_o \end{pmatrix}. \tag{11}$$

Notice that by this operation the system (11) splits in two independent block-tridiagonal systems for the two sets of unknowns x_o and x_e. They can be solved independently and this process can be seen as the first step of a divide and conquer algorithm. The two smaller systems can be split again in the same way in two subsystems which can be solved independently on different processors.

2.4 Incomplete Cyclic Reduction

Cyclic Reduction reduces $Ax = v$ in k steps to $A^{(k)}x^{(k)} = v^{(k)}$. At any level k it is possible to solve the reduced system and compute by back-substitution the solution $x = x^{(0)}$.

Hockney [9] already noticed that often the off-diagonal block-elements of the reduced system decrease in size relatively to the diagonal elements at a quadratic rate. Therefore it is possible to terminate the reduction phase earlier, neglect the off diagonal elements and solve approximately the reduced system: $y^{(k)} \approx x^{(k)}$. Back-substitution with $y^{(k)}$ yields $y \approx x$. This procedure is called *incomplete Cyclic Reduction*.

Considering the block-tridiagonal matrices

$$B^{(k)} = \left(-(D_j^{(k)})^{-1}E_j^{(k)}, 0, -(D_j^{(k)})^{-1}F_j^{(k)} \right),$$

Heller [8] proves that for diagonally dominant systems i.e. if $\|B^{(0)}\|_\infty < 1$, then

1. Cyclic Reduction is well defined and $\|x - y\|_\infty = \|x^{(k)} - y^{(k)}\|_\infty$
2. $\|B^{(i+1)}\|_\infty \leq \|B^{(i)}\|_\infty^2$
3. $\|x^{(k)} - y^{(k)}\|_\infty / \|x^{(k)}\|_\infty \leq \|B^{(k)}\|_\infty$.

S. Bondeli and W. Gander [3] considered scalar incomplete Cyclic Reduction for the the special $n \times n$ tridiagonal system $Ax = v$

$$\begin{pmatrix} a & 1 & & & \\ 1 & a & 1 & & \\ & 1 & \ddots & \ddots & \\ & & \ddots & a & 1 \\ & & & 1 & a \end{pmatrix} \begin{pmatrix} x_1 \\ x_2 \\ \vdots \\ x_{n-1} \\ x_n \end{pmatrix} = \begin{pmatrix} v_1 \\ v_2 \\ \vdots \\ v_{n-1} \\ v_n \end{pmatrix}. \tag{12}$$

Such systems have to be solved in the Fourier algorithm discussed in Section 3.4. Cyclic Reduction reduces this system in step k to $A^{(k)}x^{(k)} = v^{(k)}$ where

$$A^{(k)} = \begin{pmatrix} a_k & b_k & & & \\ b_k & a_k & b_k & & \\ & b_k & \ddots & \ddots & \\ & & \ddots & a_k & b_k \\ & & & b_k & a_k \end{pmatrix}. \tag{13}$$

As shown in [3], it is also possible to obtain an explicit expression for a_k:

1. If $|a| = 2$ then the diagonal and off-diagonal elements are

$$b_k = - \operatorname{sign}(a)/2^k \qquad (14)$$
$$a_k = \operatorname{sign}(a)/2^{k-1} . \qquad (15)$$

2. If $|a| > 2$ the diagonal elements produced by the reduction phase are given by

$$a_k = \operatorname{sign}(a)\sqrt{a^2 - 4} \coth\left(|y_0|2^k\right) , \qquad (16)$$

where

$$y_0 = \begin{cases} \ln\left(\sqrt{a^2 - 4} + a\right) - \ln 2 & \text{if } a > 2 \\ -\ln\left(\sqrt{a^2 - 4} - a\right) + \ln 2 & \text{if } a < -2 . \end{cases}$$

While $b_k \to 0$ the sequence a_k converges quadratically to

$$\operatorname{sign}(a)\sqrt{a^2 - 4}, \qquad (17)$$

and it is possible to take advantage of that fact. One can predict, depending on a and the machine precision, the number of Cyclic Reduction-steps until a_k may be assumed to be constant. E.g. for $a = 2.5$ the 14 leading decimal digits of a_k are constant for for $k \geq 5$ [3]. Thus we can apply incomplete Cyclic Reduction to solve the special system. Furthermore by making use of the explicit expressions for a_k the computation can be vectorized.

3   Poisson's Equation

We will now consider a special block tridiagonal system of equations that arises when discretizing Poisson's equation

$$\Delta u(x, y) = f(x, y) \qquad (18)$$

on some rectangular domain defined by $a < x < b$, $c < y < d$. Let

$$\Delta x = \frac{b - a}{m + 1}, \quad \Delta y = \frac{d - c}{n + 1} \qquad (19)$$

be the step sizes of the grid and denote by u_{ij} the approximation for the function value $u(x_i, y_j)$ at $x_i = a + i\Delta x$ and $y_j = c + j\Delta y$. For simplicity, we shall assume hereafter that $\Delta x = \Delta y$. If we use central difference approximations for the Laplace operator Δ and if we assume Dirichlet zero boundary conditions then we obtain the linear system

$$Tu = g \qquad (20)$$

with a block tridiagonal matrix

$$T = \begin{pmatrix} A & I & & \\ I & A & \ddots & \\ & \ddots & \ddots & I \\ & & I & A \end{pmatrix} \in \mathbf{R}^{nm \times nm}, \qquad (21)$$

$I \in \mathbf{R}^{m \times m}$ is the identity matrix and

$$A = \begin{pmatrix} -4 & 1 & & \\ 1 & -4 & \ddots & \\ & \ddots & \ddots & 1 \\ & & 1 & -4 \end{pmatrix} \in \mathbf{R}^{m \times m}. \tag{22}$$

The unknown vector u contains the rows of the matrix u_{ij} i.e.

$$u = \begin{pmatrix} u_1 \\ u_2 \\ \vdots \\ u_n \end{pmatrix} \in \mathbf{R}^{nm} \quad \text{where} \quad u_j = \begin{pmatrix} u_{1j} \\ u_{2j} \\ \vdots \\ u_{mj} \end{pmatrix} \in \mathbf{R}^m. \tag{23}$$

The right hand side is partitioned similarly:

$$g_{ij} = (\Delta y)^2 f(x_i, y_j), \quad g_j = \begin{pmatrix} g_{1j} \\ g_{2j} \\ \vdots \\ g_{mj} \end{pmatrix} \in \mathbf{R}^m \quad \text{and} \quad g = \begin{pmatrix} g_1 \\ g_2 \\ \vdots \\ g_m \end{pmatrix} \in \mathbf{R}^{nm}.$$

3.1 Cyclic Reduction with Matrix Multiplications

Elimination of the unknowns with odd indices can be done by matrix multiplications since for the system (20) the matrices $D_j = A$ and $F_j = E_j = I$ do commute. As pointed out in [5], the block matrices of the system of equations one obtains when discretizing Poisson's equation in a rectangle using the nine-point formula do also commute. Thus in this case elimination can also be done with matrix multiplication.

Assume $n = 2^{k+1} - 1$ and consider again three consecutive equations

Eq#	u_{2j-2}	u_{2j-1}	u_{2j}	u_{2j+1}	u_{2j+2}	rhs
$2j-1$	I	A	I			g_{2j-1}
$2j$		I	A	I		g_{2j}
$2j+1$			I	A	I	g_{2j+1}.

In order to eliminate u_{2j-1} and u_{2j+1} we multiply equation #($2j$) with $-A$ and add all three equations. The result is the new equation

Eq#	u_{2j-2}	u_{2j-1}	u_{2j}	u_{2j+1}	u_{2j+2}	rhs
j	I	0	$2I - A^2$	0	I	$g_{2j-1} - Ag_{2j} + g_{2j+1}$.

The recurrence relations corresponding to (5) are simpler. With $A^{(0)} = A$ and $g_j^{(0)} = g_j$ the r-th reduction step is described by

$$A^{(r+1)} = 2I - (A^{(r)})^2 \tag{24}$$

$$g_j^{(r+1)} = g_{2j-1}^{(r)} - A^{(r)} g_{2j}^{(r)} + g_{2j+1}^{(r)}, \quad j = 1, \ldots, 2^{k+1-r} - 1. \tag{25}$$

Thus after r reduction steps the remaining system of equation has the size $(2^{k+1-r} - 1) \times (2^{k+1-r-1})$:

$$u_{(j-1)2^r} + A^{(r)}u_{j2^r} + u_{(j+1)j2^r} = g_j^{(r)}, \quad j = 1, \ldots, 2^{k+1-r} - 1. \tag{26}$$

(Notice that in (26) we have assumed that $u_0 = u_{n+1} = 0$).

After k reduction steps we are left with one block $m \times m$ system

$$A^{(k)}u_{2^k} = g_1^{(k)}. \tag{27}$$

After determining u_{2^k} back-substitution is performed in which Equation (26) is solved for u_{j2^r} while $u_{(j-1)2^r}$ and $u_{(j+1)j2^r}$ are known from the previous level.

For Equation (27) and in the back-substitution phase, linear equations with the matrices $A^{(r)}$ must be solved. Furthermore transforming the right hand side (25) needs matrix-vector multiplications with $A^{(r)}$. We show in the next section how these operations can be executed efficiently.

As shown in [5] block-Cyclic Reduction can also for be applied for matrices generated by periodic boundary conditions, i.e. when the $(1, n)$ and the $(n, 1)$ block of T (20) is non-zero.

Unfortunately as mentioned in [5] the process described above is numerically unstable.

3.2   Computational Simplifications

From (24) we notice that $A^{(r)} = P_{2^r}(A)$ is a polynomial in A of degreee 2^r which as we will see can be factorized explicitly. Thus $A^{(r)}$ does not need to be computed numerically – the operator $A^{(r)}$ will be stored in factorized form. From (24) it follows that the polynomials satisfy the recurrence relation:

$$P_1(x) = x, \quad P_{2^{r+1}}(x) = 2 - P_{2^r}^2(x).$$

It is interesting that the polynomials P_{2^r} have a strong connection to the Chebychev polynomials $T_n(t)$ which are defined by expanding $\cos n\varphi$ in powers of $\cos \varphi$:

$$T_n(\cos \varphi) = \cos n\varphi.$$

Using the well known trigonometric identity

$$\cos(k + l)\varphi + \cos(k - l)\varphi = 2 \cos l\varphi \cos k\varphi \tag{28}$$

we obtain for $l = 1$ the famous three term recurrence relation ($t = \cos \varphi$)

$$T_{k+1}(t) = 2tT_k(t) - T_{k-1}(t). \tag{29}$$

More generally, relation (28) translates to the "short cut" recurrence

$$T_{k+l}(t) = 2T_l(t)T_k(t) - T_{k-l}(t).$$

Specializing by choosing $k = l$ we obtain

$$T_{2k}(t) = 2T_k^2(t) - T_0(t) = 2T_k^2(t) - 1 \tag{30}$$

almost the same recurrence as for P_{2^r}. Multiplying (30) by -2 and substituting $k = 2^r$, we obtain

$$-2T_{2^{r+1}}(t) = 2 - (2T_{2^r}(t))^2 = 2 - (-2T_{2^r}(t))^2.$$

Thus the polynomials $-2T_{2^r}(t)$ obey the same recurrence relation as P_{2^r}. However, since $P_1(x) = x$ and $-2T_1(t) = -2t$ we conclude that $x = -2t$ and therefore

$$P_{2^r}(x) = -2T_{2^r}\left(-\frac{x}{2}\right), \quad r \geq 0. \tag{31}$$

The properties of the Chebychev polynomials are well known and by (31), we can translate them to $P_{2^r}(x)$. The zeros of P_{2^r} are

$$\lambda_i^{(k)} = -2\cos\left(\frac{2i-1}{2^{r+1}}\pi\right), \quad i = 1, 2, \ldots, 2^r$$

and since for $r \geq 1$ the leading coefficient of $P_{2^r}(x)$ is -1, the factorization is

$$P_{2^r}(x) = -\prod_{i=1}^{2^r}(x - \lambda_i^{(k)}), \quad \text{thus} \quad A^{(r)} = -\prod_{i=1}^{2^r}(A - \lambda_i^{(k)}I). \tag{32}$$

Multiplication. To compute $u = A^{(r)}v = P_{2^r}(A)v$ we have now several possibilities:

1. Using (24) we could generate $A^{(r)}$ explicitly. Notice that A is tridiagonal but $A^{(r)}$ will soon be a full matrix with very large elements.
2. Using the recurrence relation for Chebychev polynomials (29) we obtain for $t = -x/2$ using (31) a three term recurrence relation for $P_n(x)$

$$P_{n+1}(x) = -xP_n(x) - P_{n-1}(x).$$

Now we can generate a sequence of vectors $z_i = P_i(A)v$ starting with $z_0 = -2v$ and $z_1 = Av$ by

$$z_i = -Az_{i-1} - z_{i-2}, \quad i = 2, 3, \ldots, 2^r.$$

Then $u = z_{2^r} = P_{2^r}(A)v = A^{(r)}v$. In [5] it is proved that the resulting algorithm *cyclic odd-even reduction and factorization* (CORF) based on this procedure is numerically unstable.
3. Using the product of linear factors (32) we compute $z_0 = -v$ and

$$z_i = (A - \lambda_i^{(k)}I)z_{i-1}, \quad i = 1, 2, \ldots, 2^r$$

and obtain $u = z_{2^r}$. This seems to be the best way to go.

Solving Block Equations. To solve $A^{(r)}v = u$ for v we have also several possibilities:

1. Generate $A^{(r)}$ explicitly and use Gaussian elimination. This simple approach is not a good idea since $A^{(r)}$ will be an ill-conditioned full matrix with large elements [5].
2. Using again the product of linear factors (32) we initialize $z_0 = -u$, solve *sequentially* the tridiagonal systems

$$(A - \lambda_i^{(k)} I)z_i = z_{i-1}, \quad i = 1, 2, \ldots, 2^r \tag{33}$$

   and obtain $v = z_{2^r}$.
3. A *parallel* algorithm was proposed using the following idea of R. Sweet [11] using a partial fraction expansion. Since the zeros of $P_{2^r}(x)$ are all simple the partial fraction expansion of the reciprocal value is

$$\frac{1}{P_{2^r}(x)} = -\frac{1}{\prod_{i=1}^{2^r}(x - \lambda_i^{(k)})} = \sum_{i=1}^{2^r} \frac{c_i^{(r)}}{x - \lambda_i^{(k)}},$$

where

$$c_i^{(r)} = \frac{1}{P_{2^r}'(\lambda_i^{(k)})} = -\prod_{\substack{j=1 \\ j \neq i}}^{2^r} \left(\lambda_j^{(k)} - \lambda_i^{(k)}\right)^{-1}.$$

Therefore we can express

$$v = \left(A^{(r)}\right)^{-1} u = (P_{2^r}(A))^{-1} u = \sum_{i=1}^{2^r} c_i^{(r)} (A - \lambda_i^{(k)} I)^{-1} u$$

where now the 2^r linear systems $(A - \lambda_i^{(k)} I)^{-1} u$ in the sum may be computed independently and in parallel. As discussed by Calvetti *et al.* in [6], partial fraction expansion may be more sensitive to roundoff errors in the presence of close poles.

3.3   Buneman's Algorithm

It has been observed that block Cyclic Reduction using the recurrence relation (25)

$$g_j^{(r+1)} = g_{2j-1}^{(r)} - A^{(r)} g_{2j}^{(r)} + g_{2j+1}^{(r)}, \quad j = 1, \ldots, 2^{k+1-r} - 1$$

to update the right hand side is numerically unstable. O. Buneman [4,5] managed to stabilize the algorithm by rearranging the computation of the right hand side in a clever way.

In his approach the right hand side is represented as

$$g_j^{(r)} = A^{(r)} p_j^{(r)} + q_j^{(r)}$$

and the vectors $p_j^{(r)}$ and $q_j^{(r)}$ are computed recursively in the following way. Initialize $p_j^{(0)} = 0$ and $q_j^{(0)} = g_j$. Then for $j = 1, \ldots, 2^{k+1-r} - 1$

1. Solve $A^{(r-1)}v = p_{2j-1}^{(r-1)} + p_{2j+1}^{(r-1)} - q_j^{(r-1)}$ for v

2. $p_j^{(r)} = p_j^{(r-1)} - v$

3. $q_j^{(r)} = q_{2j-1}^{(r-1)} + q_{2j+1}^{(r-1)} - 2p_j^{(r)}$.

There is also a simplification in the back-substituting phase. In order to compute u_{j2^r} from

$$u_{(j-1)2^r} + A^{(r)}u_{j2^r} + u_{(j+1)j2^r} = A^{(r)}p_j^{(r)} + q_j^{(r)}$$

we rearrange the equation to

$$A^{(r)}\underbrace{(u_{j2^r} - p_j^{(r)})}_{v} = q_j^{(r)} - u_{(j-1)2^r} - u_{(j+1)j2^r}$$

solve for v and obtain $u_{j2^r} = p_j^{(r)} + v$.

3.4   Fourier Algorithm

Using the eigen-decomposition of the matrix A it is possible to transform the block tridiagonal system (20) to m independent tridiagonal linear $n \times n$ systems of the form (12). These systems can then be solved e.g. in parallel by incomplete Cyclic Reduction.

The key for this procedure is the fact that the eigenvalues and -vectors of the matrix A (22) can be computed explicitly. It is well known that

$$B = \begin{pmatrix} 2 & -1 & & \\ -1 & 2 & \ddots & \\ & \ddots & \ddots & -1 \\ & & -1 & 2 \end{pmatrix} \in \mathbf{R}^{m \times m}$$

has the eigenvalues λ_i eigenvectors Q with

$$q_{ij} = \sqrt{\frac{2}{m+1}} \sin\frac{ij\pi}{m+1}, \quad \lambda_i = 2 - 2\cos\frac{i\pi}{m+1}. \tag{34}$$

This result may be generalized to matrices of the form

$$C = \begin{pmatrix} a & b & & \\ b & a & \ddots & \\ & \ddots & \ddots & b \\ & & b & a \end{pmatrix}.$$

Since $C = -bB + (a + 2b)I$ we conclude that the eigenvectors are the same and that the eigenvalues of C are $\tilde{\lambda}_i = -b\lambda_i + a + 2b$. The eigenvalues of A are obtained for $b = 1$ and $a = -4$.

Thus we can decompose the matrix A (22) as $A = Q\Lambda Q^T$. Introducing this decomposition in the system (20) and multiplying each block equation from the left by Q^T we obtain

$$Q^T u_{j-1} + \Lambda Q^T u_j + Q^T u_{j+1} = Q^T g_j, \quad j = 1, \ldots, n.$$

If we introduce new variables $\hat{u}_j = Q^T u_j$ the system (20) becomes

$$\hat{u}_{j-1} + \Lambda \hat{u}_j + \hat{u}_{j+1} = \hat{g}_j, \quad j = 1, \ldots, n.$$

Finally by permuting the unknowns by grouping together the components with the same index of $\hat{u}_j$

$$\tilde{u}_j = (\hat{u}_{i,1}, \hat{u}_{i2}, \ldots, \hat{u}_{in})^T$$

and by permuting the equations in the same way, the system (20) is transformed into m decoupled tridiagonal systems of equations

$$
\begin{pmatrix} T_1 & & & \\ & T_2 & & \\ & & \ddots & \\ & & & T_m \end{pmatrix}
\begin{pmatrix} \tilde{u}_1 \\ \tilde{u}_2 \\ \vdots \\ \tilde{u}_m \end{pmatrix}
=
\begin{pmatrix} \tilde{g}_1 \\ \tilde{g}_2 \\ \vdots \\ \tilde{g}_m \end{pmatrix}
\tag{35}
$$

$$
T_i = \begin{pmatrix} \lambda_i & 1 & & \\ 1 & \lambda_i & \ddots & \\ & \ddots & \ddots & 1 \\ & & 1 & \lambda_i \end{pmatrix} \in \mathbf{R}^n.
$$

Summarizing we obtain the following algorithm to solve Equation (20):

1. Compute the eigen-decomposition $A = Q\Lambda Q^T$ using the explicit expression (34).
2. Transform the right hand side $\hat{g}_j = Q^T g_j$, $j = 1, \ldots, n$.
3. Permute equations and unknowns and solve the m decoupled equations (35) in parallel.
4. Compute the solution by the back-transformation $u_j = Q\hat{u}_j$.

Notice that for the transformations in the second and forth step the fast Fourier transform may be used [9].

For solving the special decoupled linear systems we have the choice of several methods. First they may be solved efficiently using Gaussian elimination in parallel on different processors. Because of the special structure $[1, \lambda_k, 1]$ of the tridiagonal matrix the elements of the LU decomposition converge and may be assigned their limit [3].

A second possibility is to compute the eigen-decomposition of $[1, \lambda_k, 1] = \tilde{Q} D_k \tilde{Q}^T$. The eigenvectors are the same for all the decoupled systems, so only one decomposition has to be computed. However, this method is more expensive than Gaussian elimination. To solve one system by applying $\tilde{Q} D_k^{-1} \tilde{Q}^T$ to the right hand side we need $O(n^2)$ operations while Gaussian elimination only needs $O(n)$.

Finally we may solve these special systems in parallel using the incomplete Cyclic Reduction described in Section 2.4. The operation count is higher than for Gaussian elimination but still of $O(n)$. The advantage of this method is that one can make use of a vector arithmetic unit if it is available on a single processor [3].

4   Conclusions

Cyclic Reduction was originally proposed by Gene Golub as a very simple recursive algorithm to solve tridiagonal linear systems. This algorithm turned out to be extremely useful and efficient for modern computer architectures (vector- and parallel processing).

Applied and specialized to the block tridiagonal linear system that is obtained by discretizing Poisson's equation, it became the key for the development of *Fast Poisson Solvers*. Fast Poisson solvers have, in addition, stimulated very much the ideas and development of *Domain Decomposition* and embedding techniques.

References

1. Amodio, P., Paprzycki, M.: A Cyclic Reduction Approach to the Numerical Solution of Boundary Value ODEs. SIAM J. Sci. Comput. Vol. **18**, No. 1, (1997) 56–68
2. Arbenz, P., Hegland M.: The Stable Parallel Solution of General Banded Linear Systems. Technical Report #252, Computer Science Dept. ETH Zürich, 1996.
3. Bondeli, S., Gander, W.: Cyclic Reduction for Special Tridiagonal Matrices. SIAM J. for Matrix Analysis Vol. **15** (1994)
4. Buneman, O.: A Compact Non-iterative Poisson Solver. Rep. 294, Stanford University, Institute for Plasma Research, Stanford, Calif. (1969)
5. Buzbee, B. L., Golub, G. H., Nielson, C. W.: On Direct Methods for Solving Poisson's Equations. SIAM J. Numerical Analysis, **7** (4) (1970) 627–656
6. Calvetti, D., Gallopoulos, E., Reichel, L.: Incomplete partial fractions for parallel evaluation of rational matrix functions. Journal of Computational and Applied Mathematics **59** (1995) 349–380
7. Golub, G. H., Ortega, J. M.: Scientific Computing. An Introduction with Parallel Computing. Academic Press, Inc., (1993)
8. Heller, D.: Some Aspects of the Cyclic Reduction Algorithm for Block Tridiagonal Linear Systems. SIAM J. Numer. Anal., **13** (4) (1976) 484–496
9. Hockney, R.W.: A Fast Direct Solution of Poisson's Equation Using Fourier Analysis. J. Assoc. Comput. Mach. **12** (1965) 95–113
10. Lambiotte, J., Voigt, R.: The Solution of Tridiagonal Linear Systems on the CDC Star100 Computer. ACM Trans. Math. Softw. **1** (1975) 308–329
11. Sweet, R. A.: , A Parallel and Vector Variant of the Cyclic Reduction Algorithm. SIAM J. Sci. and Statist. Computing **9** (4) (1988) 761–765

The Role of Coordinates in Accurate Computation of Discontinuous Flow

W. H. Hui

Department of Mathematics,
Hong Kong University of Science and Technology,
Clear Water Bay, HONG KONG
email: whhui@uxmail.ust.hk

Abstract. Coordinate systems play a crucial role in the accurate computation of discontinuous flow. For two-dimensional steady supersonic flow the system based on streamlines and their orthogonals is found to be the optimal in that it is most robust and it produces infinite shock and slipline resolution. This system is also extended to two-dimensional unsteady (subsonic or supersonic) flow. However, such an optimal coordinate system is shown not to exist for general three-dimensional steady supersonic flow unless the flow velocity is perpendicular to the vorticity. Nevertheless, a system using stream surfaces as coordinate surfaces has the property of resolving slip surfaces crisply.

1 Introduction

Scientific computation is mainly about numerical solution to partial differential equations. Appropriate numerical methods for solving PDEs depend on the properties of their solutions which are, in turn, type-dependent.

Elliptic equations describe equilibrium phenomena whose solution is smooth, and hence numerical methods based on Taylor series expansion are appropriate and can be accurate. The primary difficulty in solving elliptic equations arises from the fact that signals propagate at infinite speed, making interaction of waves global.

By contrast, signals propagate at finite speed for hyperbolic equations. This implies that the interaction of waves is a local phenomena, making it easy to do computation by marching methods and making it also natural to do parallel computation. The primary difficulty in solving hyperbolic equations, however, arises from the fact that the solution will usually become discontinuous in finite time. This renders classical Taylor series-based numerical methods inappropriate; instead, shock capturing methods have been developed to cope with discontinuities during the past three decades.

This paper is devoted to accurately computing inviscid compressible flow as governed by the Euler equations of gasdynamics. As the Euler equations are hyperbolic, discontinuous flows are an important feature at high speed, and we shall show that coordinates play a crucial role in their accurate computation.

Flow discontinuities - shocks and slip surfaces - occur in nature, e.g. the spectacular bore of Tsien-Tang River in East China and the numerous rapid shear flows immediately upstream of Niagara Falls in North America. There are also abundance of discontinuous flow in man-made technology, especially at supersonic speeds (see, for instance, Ref. [1])

Accurate numerical simulation of discontinuous flow has been a research topic for the past several decades, and numerous techniques have been devised to tackle the difficulty of representing the shock and slipline (or contact line) discontinuities (for an excellent review, see [2]) which are the dominant features of the flow. After summarizing decades of research, Woodward and Collela [3] concluded that "the overall accuracy of such simulation is very closely related to the accuracy with which the flow discontinuities are represented".

It is well-known that there exist two self-contained general formulations of studying fluid flow: the Eulerian formulation observes flow at fixed locations whereas the Lagrangian formulation does so following fluid particles.

Most of the existing work (see [2] and the extensive references therein) use the Eulerian formulation in which the computational cells are fixed in space, while fluid particles move across cell interfaces. It is this convective flux that causes severe numerical diffusion in the numerical solution using Eulerian formulation. Indeed, sliplines are smeared badly and shocks are also smeared, abeit somewhat better than sliplines. Moreover, the smearing of sliplines ever increases with time and distance unless special treatments, such as artificial compression or sub-cell resolution, are introduced [4-6] which are, however, not always reliable. The primary efforts of the CFD algorithm researchers since the sixties have concentrated on developing better (more robust, accurate and efficient) ways to deal with this convective flux. Although great progress have been made and "perhaps to the point of near perfection and little return could be gained" [7], numerical diffusion still exists, causing inaccuracy and is even more difficult to handle in multi- dimensional flow problems.

Computational cells in the Lagrangian formulation, on the other hand, are literally fluid particles. Consequently, there is no convective flux across cell boundaries and the numerical diffusion is thus minimized. However, the very essence of computational cells following exactly the fluid particles can result in cell distortion so severe that the finite- difference approximations no longer resemble the original equations. To prevent this from happening, the most famous Lagrangian method in use at present time – the Arbitrary Lagrangian - Eulerian Technique (ALE) [8,9] – uses continuous re-zoning and re-mapping to the Eulerian grid. Unfortunately, this process requires interpolations of geometry and flow variables which result in loss of accuracy, manifested as numerical diffusion which ALE wants to avoid in the first place.

Recently, the author and his collaborators [10-15] have introduced the Generalized Lagrangian Formulation (GLF), which uses streamlines and their orthogonals as coordinate lines. This is shown to be superior to both the Eulerian and the Arbitrary Lagrangian- Eulerian formulation. In GLF, the fluid particles are followed in their direction of motion but not with their speeds. In addition, geometry conservation is enforced and there is thus no need to re-map to the Eulerian grid. Consequently, the associated loss of accuracy seen in ALE does not occur, nor does the numerical diffusion due to convective flux in the Eulerian formulation. This formulation then ensures crisp resolution of slipline discontinuities [11]. Furthermore, crisp shock resolution is achieved by using a shock-adaptive Godunov scheme based on GLF [14].

In section 2 we shall study the case of 2-D steady supersonic flow, whereas extension to 3-D steady supersonic flow will be given in section 3 and extension to 2-D unsteady flow given in section 4.

2   2-D Steady Supersonic Flow

The Euler equations for flow of a perfect gas obeying a γ - law may be written in conservation form as follows:

$$
\begin{cases}
\dfrac{\partial \rho}{\partial t} + \nabla \bullet (\rho \mathbf{q}) = 0 \\[2mm]
\dfrac{\partial (\rho \mathbf{q})}{\partial t} + \nabla \bullet (\rho \mathbf{q}\mathbf{q} + p\mathbf{I}) = 0 \\[2mm]
\dfrac{\partial (\rho e)}{\partial t} + \nabla \bullet (\rho H \mathbf{q}) = 0
\end{cases}
\tag{1}
$$

where $\mathbf{I}$ is the identity tensor, $\mathbf{q}$ denotes the velocity vector, p the pressure, ρ the density, $e = \dfrac{1}{2}\mathbf{q} \bullet \mathbf{q} + \dfrac{1}{\gamma - 1}\dfrac{p}{\rho}, H = e + \dfrac{p}{\rho}$, and γ is the ratio of specific heats of the gas. ($\gamma = 1.4$ is used in this paper)

Most existing work for solving the Euler equations (1) are based on the Eulerian formulation of fluid motion in Cartesian coordinates. Accordingly, for two-dimensional steady flow, (1) become

$$
\frac{\partial E}{\partial x} + \frac{\partial F}{\partial y} = 0
\tag{2}
$$

where $\qquad E = (\rho u, \rho u^2 + p, \rho u v, \rho u H)^T, \quad F = (\rho v, \rho u v, \rho v^2 + p, \rho v H)^T$ and x and y are the Cartesian coordinates, u and v are the x- and y- component of velocity, respectively, and $q = \sqrt{u^2 + v^2}$. For supersonic flow, (2) may be solved by marching in x[16].

All shock-capturing methods for solving equations (2) using the Eulerian formulation with Cartesian grid suffer from three draw-backs, namely (see Fig 1).

(a) Sliplines are smeared badly; with a second order TVD scheme it typically takes six or more computational cells to resolve a slipline, and the resolution deteriorates with marching.

(b) Shocks are also smeared, albeit somewhat better than sliplines. And

(c) The Cauchy marching problem (in the x-direction, say) becomes ill-posed and thus fails where $u/a < 1$ (a being the speed of sound), although the Mach number $M = q/a > 1$, i.e., the flow is still supersonic there.

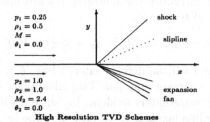

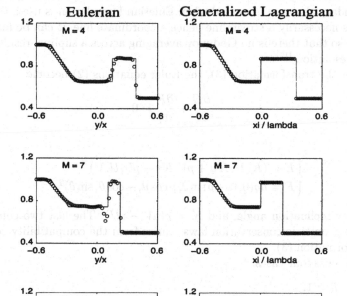

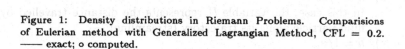

Figure 1: Density distributions in Riemann Problems. Comparisions of Eulerian method with Generalized Lagrangian Method, CFL = 0.2. —— exact; o computed.

These drawbacks can be remedied by a sequence of three ideas and techniques: (a) using streamlines as coordinate lines so as to crisply resolve sliplines, (b) using a shock- adaptive Godunov scheme so as to crisply resolve shocks, and (c) finding the orthogonals (to streamlines) so that the marching direction is aligned with the flow direction, hence the marching will always succeed wherever $M > 1$.

Mathematically, we introduce the following transformation of the independent variables from (x, y) to (λ, ξ)

$$dx = hud\lambda + U d\xi, \quad dy = hvd\lambda + V d\xi \tag{3}$$

where h is arbitrary and will be defined later. Clear, $\xi = $ constant represents a streamline, thus ξ is a stream function. Two advantages of using streamlines as coordinate lines are immediately evident: (a) since a solid boundary always coincides with a streamline hence a coordinate line, the boundary condition on the solid boundary can be satisfied exactly without first generating a grid to fit the boundary shape as is needed when Eulerian formulation is used; (b) since a slipline is necessarily a streamline hence a coordinate line, it can be made a cell interface so that there is no Godunov averaging across a slipline, resulting in its infinite resolution [11].

Under the transformation (3), the Euler equations (2) become

$$\frac{\partial E}{\partial \xi} + \frac{\partial F}{\partial \eta} = 0 \tag{4}$$

where

$$\begin{cases} E = (K, H, Ku + pV, Kv - pU, U, V)^T, \\ F = hq(0, 0, -p\sin\theta, p\cos\theta, -\cos\theta, \sin\theta)^T \end{cases} \tag{5}$$

is the flow inclination angle, and $K = \rho(uV - vU)$. The last two equations in (4) – the geometric conservation laws – arise from the compatibility conditions of transformation (3).

We discuss four cases.

Case (a): $h = 1$.

In this case, the variable λ introduced in (3) is the Lagrangian time [10−12], and the fluid particles are followed literally in their direction of motion as well as speeds. In other words, the formulation is genuinely Lagrangian, though differing from the classical Lagrangian one in that the geometric conservation laws governing the particle deformation are enforced.

Case (b): $h = 1/q$.

In this case, the variable U represents the distance travelled by the fluid particle along its trajectory (streamline).

Case (c): $h = 1/u$.

In this case, without loss of generality we can take U equal to zero, reducing (3) to the well-known von Mises transformation.

Case (d): $h = \dfrac{1}{q} \exp \left(\displaystyle\int_{\xi_0}^{\xi} \dfrac{1}{\rho q^2} \dfrac{\partial p}{\partial \xi} d\xi \right)$ (6)

where $\xi_0 = \xi_0(\lambda)$ is an arbitrary function. In this case, the coordinate lines $\lambda = $ constant and $\xi = $ constant are orthogonal [15] to each other. It can be shown that the well-posedness of the Cauchy problem, and hence numerical stability of the marching scheme, requires the velocity component in the marching direction, i.e. $\nabla\lambda$, to be supersonic. This orthogonal (λ, ξ) coordinate system, for which $\nabla\lambda \| \mathbf{q}$, is therefore the most robust system, and is to be used throughout this paper.

The above formulation using (λ, ξ) as coordinates is called the generalized Lagrangian formulation (GLF) in which the fluid particles are followed in their direction of motion but not with their speeds. It has the advantages of both the Lagrangian formulation of following the direction of particle motion and the Eulerian formulation of grid orthogonality.

With sharp slipline resolution guaranteed by the generalized Lagrangian formulation, shock resolution can then be made crispy using a shock- adaptive Godunov scheme [12,14]. This is done by splitting a shock-containing cell into subcells along the shock trajectory(ies) and treating the resulting subcells as regular cells in the Godunov scheme. In this way the Godunov averaging across a shock discontinuity is avoided, and the shock is resolved crisply. We note that shock-adaptive technique is essential to preserving coordinate orthogonality on crossing a shock [15].

In the first example, we compute the uniform supersonic flow in a channel formed by two plane boundaries, one of which is aligned with the free stream and the other making an angle of 15.322° to it. A shock is formed at once and reflected from the boundaries alternately. Fifteen cells are used in the computation and a first order Godunov scheme with the shock-adaptive technique is used. Figure 2(a) plots the computed grid and shocks. It is seen that the grid is orthogonal everywhere except, of course, on crossing the shocks where the grid lines change directions abruptly. The shocks are seen resolved sharply. The computation terminates when the flow Mach number in the flow region DEF equals 1.00719. In this problem, the exact analytic solution is reproduced by the numerical computation as is evidenced in Fig. 2(b), which plots Mach number distribution across the channel near its end.

For comparison, the same flow was computed using a non- orthogonal grid, case (b), and the computed shock waves and flow-generated grid are plotted in Fig. 2(c). Although the computed flow is also very accurate and the shocks resolved equally sharply, the computation cannot proceed beyond the second shock reflection. This is because the velocity component in the direction normal to the coordinate line = const is subsonic, although the total velocity is still supersonic. This renders the local Cauchy problem ill-posed, resulting in violent numerical instabilities. It should also be pointed out that if the von Mises grid (case (c)) is used or the Cartesian grid, Eq. (2), is used the computation cannot proceed beyond the second shock reflection either, similar to Fig. 2(c). This is because the x-component of velocity is not supersonic in the region DEF, i.e. $u/a = 0.9649$, although the total velocity is supersonic there.

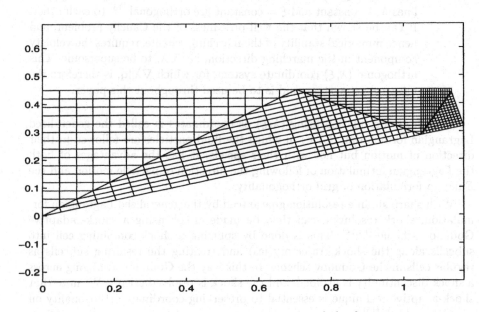

Fig. 2(a): Flow-generated grid and shocks

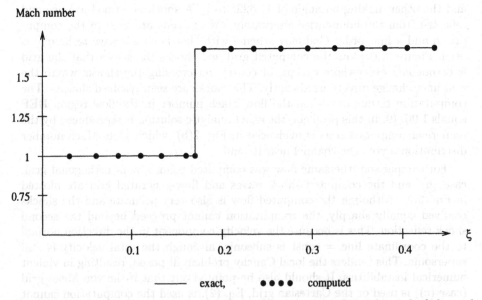

Fig. 2(b): Mach number along a coordinate line $\lambda = $ const near the end of channel

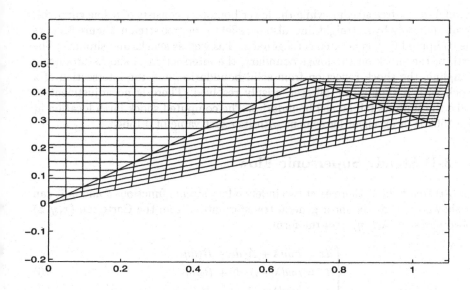

Fig. 2(c): Flow-generated grid and shocks using a non-orthogonal grid: case (b)

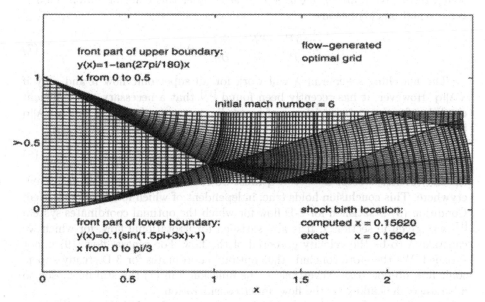

Fig. 3: Flow-generated grid, shocks and sliplines

In the second example, we use the 2nd order TVD scheme [11] and 40 cells to compute a supersonic flow with $M_\infty = 6$ through a converging channel of which the upper boundary consists of a straight line followed by another straight line parallel to the free stream, while the lower boundary consists of a sine curve followed smoothly by a straight line, also parallel to the free stream. Figure 3 shows the computed flow-generated orthogonal grid as well as shocks and sliplines. The birth of the shock on the lower boundary, the intersection of shocks producing a slipline, the shock reflection from solid boundaries, and the intersection of a shock with a slipline are all seen crisply resolved. There is no adjustment to make the sliplines fall on cell boundaries. The computed shock birth location at $= 0.1562$ is in excellent agreement with its exact value of 0.15643.

3   3-D Steady Supersonic Flow

For 3-D steady flow, there exist two independent stream functions ξ and η. It can be shown [17] that the most general transformation from the Cartesian (x, y, z) coordinates to (λ, ξ, η) is of the form

$$\begin{cases} dx = hud\lambda + A_1 d\xi + B_1 d\eta \\ dy = hvd\lambda + A_2 d\xi + B_2 d\eta \\ dz = hwd\lambda + A_3 d\xi + B_3 d\eta \end{cases} \tag{6}$$

where h is an arbitrary function, u, v, w are the x-, y- and z-component of velocity, respectively, and $q = \sqrt{u^2 + v^2 + w^2}$. Under this transformation the 3-D Euler equations (1) become

$$\frac{\partial E}{\partial \lambda} + \frac{\partial F}{\partial \xi} + \frac{\partial G}{\partial \eta} = 0. \tag{7}$$

The marching scheme in λ will work for all supersonic flow if and only if $\nabla\lambda \| \mathbf{q}$. However, it has recently been found [17] that a necessary and sufficient condition for the global existence of a coordinate system λ, ξ, η for which $\nabla\lambda \| \mathbf{q}$ is

$$q \bullet \nabla \times \mathbf{q} = 0 \tag{8}$$

i.e. the vorticity vector $\omega = \nabla \times \mathbf{q}$ must be normal to the velocity vector everywhere. This conclusion holds true, independent of which formulation is used. Condition (9) is satisfied for 2-D flow for which the optimal coordinates system [15] was given in section 2. It is also satisfied for irrotational flow, for which we can take λ to be the velocity potential of the flow. For general flow, (9) is not satisfied. We therefore conclude that optimal coordinates for 3-D steady supersonic flow do not exist. Nevertheless, the function h in (7) can still be chosen to advantage, depending on the flow under consideration.

In an important recent paper on 3-D supersonic flow Loh & Liou [18] have considered the special case of $h = 1/q$ for which λ represents the distance travelled by a fluid particle along its streamline. The marching direction $\nabla\lambda$ in this case is, however, not the flow direction. Such a coordinate system is therefore not an optimal one and the marching fails [18] in regions where the velocity component in the λ direction is subsonic, even though the flow is supersonic there.

Despite the facts that the (λ, ξ, η) system used is not an optimal one and no use was made of the shock-adaptive technique, they have successfully computed complicated 3-D supersonic flows with high degree of accuracy, especially crispy resolution of slip surfaces and automatic generation of grid by flow. In other words, the essential advantages of the GLF is retained in 3-D steady supersonic flow.

4   2-D Unsteady Flow

There are two ways to extend the GLF to 2-D unsteady flow: (A) using instantaneous streamlines as coordinate lines and (B) using pathlines as coordinate lines. It is well-known that streamlines and pathlines coincide in steady flow, but do not do so in unsteady flow. However, they are tangential to each other at the point of contact, so in a marching scheme over a short time interval, they have the same direction of motion of fluid and there is no convective flux across cell boundaries in either case. Consequently, crispy resolution of sliplines is expected in both cases.

For 2-D flow, Eq (1) in the Eulerian formulation become

$$\frac{\partial Ee}{\partial t} + \frac{\partial Fe}{\partial x} + \frac{\partial Ge}{\partial y} = 0 \tag{9}$$

where

$$\begin{cases} Ee &= (\rho, \rho u, \rho v, \rho e)^T, \\ Fe &= (\rho u, \rho u^2 + p, \rho u v, \rho u H)^T, \\ Ge &= (\rho v, \rho u v, \rho v^2 + p, \rho v H)^T \end{cases} \tag{10}$$

The most general transformation that uses streamlines as coordinate lines is

$$\begin{cases} dt &= d\tau \\ dx &= A d\tau + hu d\lambda + U d\xi \\ dy &= B d\tau + hv d\lambda + V d\xi \end{cases} \tag{11}$$

Clearly, at $t = t_0$, a curve $\xi = $ const corresponds to $dx/dy = u/v$ hence it is a (instantaneous) streamline at $t = t_0$.

The most general transformation that uses pathlines as coordinate lines is, on the other hand,

$$\begin{cases} dt &= h d\lambda + A_0 d\xi + B_0 d\eta \\ dx &= hu d\lambda + A_1 d\xi + B_1 d\eta \\ dy &= hv d\lambda + A_2 d\xi + B_2 d\eta \end{cases} \tag{12}$$

It can be easily shown that the material derivatives $\dfrac{D\xi}{Dt} = \dfrac{D\eta}{Dt} = 0$, hence $\xi = $ const and $\eta = $ const jointly represent a pathline.

As in section 2, the function h in (11) or (12) is arbitrary.

Liou [7] has used instantaneous streamline coordinates, whereas Loh and Hui [19] has used pathline coordinates, (i.e. (12) with $h = 1$ and $A_0 = B_0 = 0$), both leading to good success.

In Fig. 4a,b are plotted Mach number contours and pressure contours of a flow resulting from a $M = 1.3$ plane shock past a 25° wedge [19]. The Mach reflection is accurately computed, especially the triple-point position and the slipline are sharply resolved. The computed Mach contours are in very good agreement with the shadow-graph of the flow [1, Fig. 236] as reproduced in Fig. 4c.

Finally, in Fig. 5 are shown the contours of pressure, density and Mach number in a 2-D flow resulting from the implosion/explosion of a flow confined in a unit square walls where $p = \rho = 1, u = v = 0$ initially everywhere, except in a inside square of 0.36×0.36 oriented symmetrically at 45° where $p = 0.14, \rho = 0.125$ and $u = v = 0$. The twelve contour plots correspond (from left to right) to $t = 0.25$ to $t = 0.40$ with a uniform increment of $\Delta t = 0.05$. A crude grid of 60×60 was used in the computation [19], and it is seen to capture the flow feature very well. In particular, shocks are resolved in 2-4 points while sliplines are resolved crisply. A more extended time evolution of the density contours is given in Fig. 6. The contours are comparable to those in [20] and [21] using Eulerian method, where grids of 240×240 to 359×359 were employed.

Acknowledgments

This research is supported by a Competitive Earmark Research Grant from the University Research Grants Council of Hong Kong.

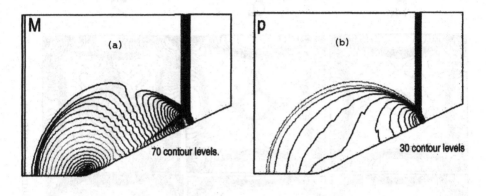

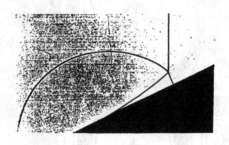

Fig. 4: Mach reflection: a $M = 1.3$ plane shock past a 25 degree wedge, 105×60 grid. (a) Mach number contours at $t = 1.2$, (b) pressure contours at $t = 1.2$, (c) shadowgraph [1, Fig. 236]

Fig. 5: Comparisons of pressure, density, and Mach number contours

Fig. 6: The 12 plots correspond to $t = 0.05$ to $t = 0.6$ with a uniform increment of $\Delta t = .05$

References

1. M. Van Dyke, (1982) "An Album of Fluid Motion, the Parabolic Press.
2. H.C. Yee (1989) "A Class of High Resolution Explicit and Implicit Shock Capturing Methods; VKI Lecture Series 1989-04, vol. 1 pp. 1-216.
3. P. Woodward and P. Collea (1984) "The Numerical Simulation of Two-Dimensional Fluid Flow with Strong Shocks; J. Comput. Phys., vol. 54, pp. 115-173.
4. A. Harten (1978) "The Artificial Compression Method for Computation of Shocks and Contact Discontinuities: III Self-Adjusting Hybrid Schemes; Math. Comp., vol. 32, pp. 363-389.
5. A. Harten (1989) "ENO Schemes with Subcell Resolution; J. Comput. Phys, vol. 83, pp. 148-184.
6. C.W. Shu and S. Osher (1989) "Efficient Implementation of Essentially Non- oscillatory Shock-Capturing Schemes II; J. Comput. Phys, vol. 83, pp. 32-78.
7. M.S. Liou (1995) "An Extended Lagrangian Method; J. Comput. Phys, vol. 118, pp. 294-309.
8. C.W. Hirt, A.A. Amsden and J.L. Cook (1974) "An Arbitrary Lagrangian-Eulerian Computing Method for All Flow Speeds; J. Comput. Phys., vol. 14, pp. 227-253.
9. W. Pracht (1975) "Calculating Three-Dimensional Fluid Flows at All Speeds with an Eulerian-Lagrangian Computing Mesh; J. Comput. Phys., vol. 17, pp. 132-159.
10. C.Y. Loh and W.H. Hui (1990) "A New Lagrangian Method for Steady Supersonic Flow Computation, Part I: Godunov Scheme; J. Comput. Phys, vol. 89, pp. 207-240.
11. W.H. Hui and C.Y. Loh (1992) "A New Lagrangian Method for Steady Supersonic Flow Computation, Part II: Slipline Resolution; J. Comput. Phys, vol. 103, pp. 450-464.
12. W.H. Hui and C.Y. Loh (1992) "A New Lagrangian Method for Steady Supersonic Flow Computation, Part III: Strong Shocks; J. Comput. Phys, vol. 103, pp. 465-471.
13. W.H. Hui and Y.C. Zhao (1992) "A Generalized Lagrangian Method for solving the Euler Equations: in Nonlinear Hyperbolic Problems: Theoretical, Applied and Computational Aspects (Edited by A. Donato and F. Oliveri), pp. 336-345.
14. C.Y. Lepage and W.H. Hui (1995) "A Shock-Adaptive Godunov Scheme Based on the Generalized Lagrangian Formulation; J. Comput. Phys, vol. 122, pp. 291-299.
15. W.H. Hui and D.L. Chu (1996) "Optimal Grid for the Steady Euler Equations; Comput. Fluid Dynamics Journal, vol. 4, pp. 403-426.
16. H.M. Glaz and A.B. Wardlaw (1985) "A High-Order Godunov Scheme for Steady Supersonic Gas Dynamics; J. Comput. Phys, vol. 58, pp. 157- 187.
17. W.H. Hui and Y. He (1997) "Hyperbolicity and Optimal Coordinates of the 3-D Steady Supersonic Euler Equations, SIAM J. Appl Math, vol. 57, no. 4, Aug 1997.
18. C.Y. Loh and M.S. Liou (1994) "A New Lagrangian Method for Three-Dimensional Steady Supersonic Flows; J. Comput. Phys, vol. 113, pp. 224-248.
19. C.Y. Loh and W.H. Hui (1996) "A new Lagrangian Method for Time-dependent Inviscid Flow Computation; (preprint).
20. T. Aki and F. Higashino (1990) "A Numerical Study on Implosion of Polygonally Interacting Shocks and Consecutive Explosion in a Box in AIP (American Institute of Physics) Conference Proceedings, vol. 208, pp. 167-172.
21. X.Y. Wang (1995), private communication.

Quadratic Programming for Positive Solutions of Linear Inverse Problems

B McNally and E R Pike

Physics Department, King's College London

Abstract. The study of ill-posed linear inverse problems, characterised by Fredholm equations of the first kind, has important applications in many areas of science and technology. Problems of this type introduce some loss of information between what we may call a generalised "object" and its "image" under the linear mapping. This loss of information often makes the attempted inversion of image to object very difficult. In practice the image is usually a discrete set of observations but the mapping may be retained essentially as a finite-rank integral operator using a fine discretisation throughout in the object space. This gives a very useful "automatic" interpolation of the recovered object at a sampling rate much higher than the conditioning of the problem would otherwise allow. Work is presented here which attempts to add in some of the lost information by making use of such *a-priori* constraints as positivity or known moments. This is achieved by the method of quadratic programming using a selection of different optimisation criteria. This paper is mostly a review of work previously published by the authors and their colleagues.

1 Introduction

The ill-conditioned equations with which we are concerned in this paper are of a type which describe, for example, restoration of diffraction-limited optical images [1], analytic continuation in high temperature superconductivity and other problems in modern theoretical physics [2] and photon correlation spectroscopy (PCS), the experimental sizing of macromolecules by the scattering of laser light [3]. These are Fredholm equation of the first kind and describe how the collected data (the image) is formed from the unknown solution (the object) and the imaging function (the kernel). A general theory of linear instrumental systems with noise is given by Bertero and Pike in [4]. If we first consider for simplicity the one-dimensional continuous case, we can denote the object by $f(x)$, the image by $g(y)$, both L^2 functions, and the kernel by $K(y,x) : L^2 \to L^2$. Then we have:

$$g(y) = \int_{-\infty}^{\infty} K(y,x) f(x) \, dx \qquad (1)$$

For example, in the case of macromolecular sizing, f is a particle-size probability distribution, g is the first-order correlation function and K describes a Laplace transform.

When the data is collected experimentally it can only be sampled at a finite number of points, which leads to the problem being discretised. The image is represented on N_i sample points with values $g(y_j)$ $1 \leq j \leq N_i$, the object on a large number N_o points with values $f(x_i)$ $1 \leq i \leq N_o$, and the integral operator represented as a matrix operator $K(y_j, x_i)$ $1 \leq j \leq N_i$ $1 \leq i \leq N_o$. This discretised Fredholm equation of the first kind, with a suitably discretised noise vector, is represented as:

$$g(y_j) = K(y_j, x_i)f(x_i) + \eta(y_j) \tag{2}$$

or more simply, setting the image, object and noise as vectors, and the imaging kernel as a matrix operator:

$$g = Kf + \eta \tag{3}$$

Computationally, since f represents a continuous function, the number of points used to represent it, N_o, has an upper limit constrained only by the available processing power. For most of the problems which we have investigated we have found a value of $N_o \sim 200$ adequate to approximate the continuous case. The number of sampled data points, N_i, is dependent on the effort required to sample the data experimentally, and the amount of "blurring" present in the problem. Ideally the number and positioning of data points will be chosen, in a given noise environment, to avoid ill-conditioning altogether, but, unfortunately, experimentors are not always reliable in these matters! Typically, N_i will be much less than N_o.

1.1   The Singular System as a General Description

The most general description of equation (1), which also leads to a method of solving for f, is the singular value decomposition (SVD). If $\mathbf{K}$ is a compact linear operator then the map $\mathbf{K} : X \mapsto Y$ gives rise to two sets of orthonormal basis functions - $\{\mathbf{u}_k\}$ and $\{\mathbf{v}_k\}$ spanning the ranges of $\mathbf{K}$ and its adjoint $\mathbf{K}^*$ respectively. There also exists a set of singular values $\{\sigma_k\}$ associated with the mapping from X to Y such that:

$$\mathbf{K}\mathbf{u}_k = \sigma_k \mathbf{v}_k$$
$$\mathbf{K}^*\mathbf{v}_k = \sigma_k \mathbf{u}_k \tag{4}$$

$$\langle \mathbf{u}_j, \mathbf{u}_k \rangle = \delta_{j,k}$$
$$\langle \mathbf{v}_j, \mathbf{v}_k \rangle = \delta_{j,k} \tag{5}$$

Thus, the "singular system" of the kernel - $\{\sigma_k, \mathbf{u}_k, \mathbf{v}_k\}$ - depends on the support of the object, the support of the image and the form of the kernel.

By analogy with Fourier analysis, the increasing index k corresponds to an increasing measure of spatial or temporal frequency. Because $\mathbf{u}_k$ is a basis set for X, and $\mathbf{v}_k$ is a basis set for Y, it is simple to decompose both the object and the image into a weighted sum of these basis functions:

$$f = \sum_{k=1}^{N_o} \langle f, u_k \rangle u_k$$

$$g = \sum_{k=1}^{N_i} \langle g, v_k \rangle v_k \tag{6}$$

Here $\langle a, b \rangle = a_1 b_1 + a_2 b_2 + \cdots + a_N b_N$ where a and b are N-dimensional vectors. The standard method of recovering f from g, first performed for the kernel describing Fourier band-limited transmission by Slepian & Pollak [5] in 1961 may be formulated as follows. From equation (6), the sampled data may also be written as:

$$g = \sum_{k=1}^{N_i} \frac{\langle g, v_k \rangle v_k \sigma_k}{\sigma_k}$$

which, from equation (4), is equivalent to:

$$g = \sum_{k=1}^{N_i} \frac{\langle g, v_k \rangle K u_k}{\sigma_k}$$

$$= K \sum_{k=1}^{N_i} \frac{\langle g, v_k \rangle u_k}{\sigma_k}$$

$$= Kf$$

equating the last two lines, this yields the inversion:

$$f = \sum_{k=1}^{N_i} \frac{\langle g, v_k \rangle u_k}{\sigma_k} \tag{7}$$

However, this solution may still be ill-conditioned due to $\sigma_k \to 0$ as the index k increases. It is clear that, if the data is oversampled, equation (7) must have the summation stopped (or rolled off) at some stage if there is to be any chance of a successful inversion. What happens in practice is that an index is chosen $(R < N_i)$ where it is decided that experimental noise has effectively hidden the basis function weighting coefficients. This acts as a regularising parameter and gives the Truncated Singular Value Decomposition solution.

$$\tilde{f} = \sum_{k=1}^{R} \frac{\langle g, v_k \rangle u_k}{\sigma_k} \tag{8}$$

There is, in fact, a veritable industry concerned with choosing the regularising parameter which we shall not discuss here, save to say that in the best experiments the noise level is known and the parameter can then be chosen without difficulty, but see Golub's contribution to this volume. In essence, SVD elucidates the underlying mathematical structure in any situation described by equation (1) in a way which is most effectively related to the physical problem involved. It gives a method of partitioning "noise space" from "signal space", using the singular value spectrum, which minimises the effects of physical noise, and is widely used in many problems in science and engineering.

2 Aiming for a Positive Solution

In many situations (e. g. Incoherent Imaging and PCS.) it is known that $\mathbf{f}$ is non-negative. However, when a TSVD solution is formed from the data, it is discovered that $\tilde{\mathbf{f}}$ contains negative regions. In the noiseless case this is solely due to the summation cutoff at index R. In the presence of noise the case is worse still, due to the weights of components with indices $k < R$ being perturbed from their true values.

Another way of comparing the object, and the TSVD approximation to it, is the following:

$$\mathbf{f} = \sum_{k=1}^{R} \frac{\langle \mathbf{g}, \mathbf{v}_k \rangle}{\sigma_k} \mathbf{u}_k + \sum_{k=R+1}^{N_i} \frac{\langle \mathbf{g}, \mathbf{v}_k \rangle}{\sigma_k} \mathbf{u}_k$$

$$\tilde{\mathbf{f}} = \sum_{k=1}^{R} \frac{\langle \mathbf{g}, \mathbf{v}_k \rangle}{\sigma_k} \mathbf{u}_k + \sum_{k=R+1}^{N_i} c_k . \mathbf{u}_k \qquad (9)$$

$$c_k \equiv 0$$

It is the choice of $c_k \equiv 0$ that produces the unwanted negative regions. However, it is only because there is no reliable information about $\{c_k\}$ which leads to the arbitrary choice of setting them to zero. On the other hand, if this causes negative regions in the reconstruction, it is immediately obvious that this choice for the values of $\{c_k\}$ is incorrect. The task, then, is to try to find a choice for $\{c_k\}$ that leads to a non-negative TSVD reconstruction.

2.1 Mathematical Programming

Mathematical programming is the general term used to describe the branch of mathematics concerned with choosing values for a set of variables, subject to various constraints placed upon them. Probably the best known subset of Mathematical Programming is called "linear programming". Problems of this type can be described by the following set of relations:

Minimise	$f = d_1\, c_1 + d_2\, c_2 + \cdots + d_n\, c_n$
	$a_{11}c_1 + a_{12}c_2 + \cdots + a_{1n}c_n = b_1$
	$a_{21}c_1 + a_{22}c_2 + \cdots + a_{2n}c_n = b_2$
Subject to the constraints	....
	$a_{m1}c_1 + a_{m2}c_2 + \cdots + a_{mn}c_n = b_m$
	$c_i \geq 0 \qquad (i = 1, \ldots, n)$

The *Simplex Method* is a procedure for solving such a set of equations. In brief, the method finds a basic feasible solution, calculates the direction that will decrease f and moves in that direction until one of the constraints is about to be violated. At that point the routine selects a new direction for decreasing f. In this manner the optimal set of $\{c_n\}$ is found that minimises f while still satisfying the linear constraints. It is a feature of linear programming that the solution is always to be found on a vertex of the n-dimensional volume defined by the constraints.

Quadratic programming has a very similar definition to that of linear programming, but is not so straightforward to solve. Using matrix notation for the definition we have:

Minimise	$F(\mathbf{c})$	$\mathbf{c} \in \mathcal{R}^n$
Subject to the constraints	$\mathbf{l} \leq \left\{ \begin{array}{c} \mathbf{c} \\ \mathbf{Dc} \end{array} \right\} \leq \mathbf{u}$	
Quadratic Programming	$F(\mathbf{c}) = \mathbf{e}^T \mathbf{c} + \frac{1}{2}\mathbf{c}^T \mathbf{Ac}$	
Least Squares	$F(\mathbf{c}) = \mathbf{e}^T \mathbf{c} + \frac{1}{2}\|\mathbf{b} - \mathbf{Ac}\|^2$	

This notation is based upon that used in [6]. In all of the work so far undertaken the routine has always been used in the least squares mode, with the choice of $\{e_n\} \equiv 0$. As will be shown in the following section, the least squares mode was ideal for minimising certain aspects of the reconstruction. In the above notation, $\mathbf{c}$ represents the set of unknown basis weights, $\{c_k\}$, and $\mathbf{D}$ is an array, formed from the $\mathbf{u}$ basis functions, which along with $\mathbf{l}$ and $\mathbf{u}$ was used to ensure positivity. The matrix $\mathbf{A}$ was formed so as to implement the various optimisation choices.

2.2   What choice for $\{c_k\}$?

Micchelli in 1986 [7] remarkably showed that there is a *unique* choice for $\{c_k\}$ that leads to a non-negative TSVD reconstruction with minimum L^2 norm. Unless this reconstruction contains inordinately detailed structure this solution can be found to a close approximation by quadratic programming and we have implemented and verified this computationally in several cases. Borwein and Lewis in 1992 [8] showed how to find these coefficients using the Fenchel dual and this has been implemented recently by de Villiers [9]. It would be tempting to hope that since there is a unique choice of $\{c_k\}$ that produces the minimum L^2 norm, it must be the original object. Unfortunately this is not the case; there is no a priori reason that the original object has to be the positive, minimum L^2 norm realisation for the first R values of $\langle \mathbf{f}, \mathbf{u}_k \rangle$. This leads to an arbitrary choice having to be made about the reconstruction. It has already been decided that $\{c_k\}$ should be chosen to ensure a non-negative TSVD solution - but that constraint is not enough to ensure a unique solution. Thus, one of many possible choices must be made for $\{c_k\}$. Let $\hat{\mathbf{f}}$ represent the TSVD solution with possibly non-zero values of $\{c_k\}$ added to ensure positivity. The various optimisation choices so far studied include:

- Minimise $\sum |\hat{\mathbf{f}}|$
- Minimise $\sum |\hat{\mathbf{f}}|^2$
- Minimise $\alpha \sum |\hat{\mathbf{f}}|^2 + (1 - \alpha) \sum |\hat{\mathbf{f}}|$
- Minimise $|\frac{\partial \hat{f}}{\partial x}|^2$
- Minimise $|\frac{\partial^2 \hat{f}}{\partial x^2}|^2$

It should be obvious that there are an endless number of these minimisation choices; each one capable of producing a different positive $\hat{\mathbf{f}}$ that exactly fits the data to within the noise level.

The first of these minimisation choices is the minimum $L^{(1)}$ norm. Unfortunately there is no guarantee of uniqueness for this solution - as there was for the minimum $L^{(2)}$ solution. That the minimum $L^{(1)}$ norm may be degenerate is a consequence of it not being a strictly convex problem. In fact, the solution can be found by using linear programming techniques (such as the Simplex Method). A degenerate solution would then lie on a hyper-plane in the $N_o - R$ dimensional space of the problem.

2.3   Including Constraints and Optimisations

As indicated in section 2.1, the constraints for the system are specified by the system of equations:

$$1 \leq \left\{ \begin{matrix} \mathbf{c} \\ \mathbf{Dc} \end{matrix} \right\} \leq \mathbf{u}$$

The first of these constraints, $1 \leq \mathbf{c} \leq \mathbf{u}$, imposes some constraint on the values that the missing higher-order basis weights are allowed to take. In all cases so far studied there was no *a-priori* knowledge to suggest there should be any limit on these values. Hence, 1 and $\mathbf{u}$ were set to $-\infty$ and ∞ respectively. $\mathbf{D}$ is known as the *"Constraint Matrix"*, and is used to specify constraints on the form of the reconstruction. Examining the case of imposing positivity on point i gives (from equation (9)):

$$\hat{f}(i) = \sum_{k=1}^{R} \frac{\langle \mathbf{g}, \mathbf{v}_k \rangle}{\sigma_k} u_k(i) + \sum_{k=R+1}^{N_i} c_k . u_k(i)$$

$$= \tilde{f}(i) + \sum_{k=R+1}^{N_i} c_k . u_k(i)$$

Introducing the constraint of $0 \leq \hat{f}(i) \leq \infty$ then gives:

$$0 \leq \qquad \hat{f}(i) \qquad \leq \infty$$

$$\Rightarrow \quad 0 \leq \tilde{f}(i) + \sum_{k=R+1}^{N_i} c_k.u_k(i) \leq \infty$$

$$\Rightarrow -\tilde{f}(i) \leq \sum_{k=R+1}^{N_i} c_k.u_k(i) \quad \leq \infty$$

Recognising that every point in the reconstruction $(1 \leq i \leq N_o)$ must be positive gives $\mathbf{1} = -\tilde{\mathbf{f}}$ and $\mathbf{u} = \infty$. Thus, the constraint matrix, $\mathbf{D}$, is such that $D_{ij} = u_j(i)$. The various minimisation schemes were included in a very similar manner to this, with some aspect of the reconstruction equalling the corresponding aspect of $\tilde{\mathbf{f}}$ plus some combination of the higher-order $\mathbf{u}$ basis functions.

2.4   Weakness of the Positivity Constraint

Some recent reports have claimed methods of superresolution (a restored image could be said to be superresolved when it contains accurate high spatial frequency components that are not detectable in the collected data) based on constraining the restoration to be non-negative. It will be shown here that the positivity constraint is *not* sufficient to recover accurately the missing spatial frequencies.

Imagine two non-negative objects, $\hat{\mathbf{f}}_1$ and $\hat{\mathbf{f}}_2$, such that for the first R singular functions $\langle \hat{\mathbf{f}}_1, \mathbf{u}_k \rangle = \langle \hat{\mathbf{f}}_2, \mathbf{u}_k \rangle$. Both $\hat{\mathbf{f}}_1$ and $\hat{\mathbf{f}}_2$ will have identical images and TSVD solutions to within the noise level. Now suppose that $\hat{\mathbf{f}}_1$ is double peaked, and $\hat{\mathbf{f}}_2$ is single peaked. Finally, take it that $\hat{\mathbf{f}}_1$ just happens to be the non-degenerate L^1 minimisation, and $\hat{\mathbf{f}}_2$ the L^2 minimisation, for the first R fixed weights.

Suppose $\hat{\mathbf{f}}_1$ is imaged, the data collected, and the TSVD solution formed. This will most probably contain negative regions, so it may be desired to find a choice for $\{c_k\}$ that ensures positivity. If the L^1 minimisation criterion is chosen then the original object is recovered exactly, since $\hat{\mathbf{f}}_1$ is defined to have this property. However, if the L^2 minimisation criterion is instead chosen, the restored object is not $\hat{\mathbf{f}}_1$, but $\hat{\mathbf{f}}_2$. But it is not enough to select the L^1 choice again in the future just because it has worked so well this time. If the experiment is repeated with $\hat{\mathbf{f}}_2$ being imaged instead of $\hat{\mathbf{f}}_1$ it turns out the L^2 is the "magic" choice that recovers the object exactly.

3   Further Constraints in Different Situations

A proper choice of minimisation criterion is dependent on the form of the original object. However, since it is the form of the object that the process is trying to recover, the choice of minimisation criterion is not much better than a pure guess. Fortunately, some inverse problems outside the realm of imaging lend themselves to further *a-priori* constraints beyond that of positivity.

For example, in PCS [3] accurate values of the first two moments of the reconstruction of $\hat{f}$ can be found independently from the data. In work on high temperature superconductivity [2], which we discuss in the next section, a reconstruction of the "Spectral Weight Function" is required from the "Matsubara Green's function". This is one of many ill-posed inverse problems of analytic continuation which arise in computational theoretical physics. In this particular case it is possible to calculate exactly the first three moments of the reconstruction. Using this extra *a-priori* information narrows down the range of possible functions that still fit the data and the *a-priori* information within the noise level.

3.1   An example from high-T_c superconductivity theory

The particular inverse problem in this situation is

$$g(y) = \int_{-\infty}^{\infty} \left[\frac{\exp(-xy)}{1 + \exp(-\beta x)} \right] f(x)\, dx \qquad y \in [0, \beta] \tag{10}$$

where $g(y)$ is the numerically calculated Matsubara Green's Function, and $f(x)$ is the Spectral Weight Function - the desired function. The "n^{th}-moment" of a function is defined as

$$\mu_n = \int_{-\infty}^{\infty} x^n f(x)\, dx \tag{11}$$

In the case studied, not only was it known that the reconstruction had to be non-negative, but also values for μ_0, μ_1 and μ_2 could be accurately precalculated. These moment constraints were built into the quadratic programming routine in much the same way as the positivity constraint. That this was able to be done relied on the linearity of the problem. Making use of equation (6) leads to:

$$\begin{aligned}
\int_{-\infty}^{\infty} x^n f(x)\, dx &= \int_{-\infty}^{\infty} x^n \sum_k \langle \mathbf{f}, \mathbf{u}_k \rangle \mathbf{u}_k\, dx \\
&= \int_{-\infty}^{\infty} x^n \left(\sum_{k=1}^{R} \langle \mathbf{f}, \mathbf{u}_k \rangle \mathbf{u}_k + \sum_{k=R+1}^{No} c_k \mathbf{u}_k \right) dx \\
&= \int_{-\infty}^{\infty} x^n\, \tilde{\mathbf{f}}\, dx + \sum_{k=R+1}^{No} c_k \int_{-\infty}^{\infty} x^n\, \mathbf{u}_k\, dx
\end{aligned} \tag{12}$$

This means that the nth moment of the final reconstruction is equal to the nth moment of the TSVD solution, plus a weighted sum of the nth moments of each of the higher-order basis functions. In practice, if m moments are to be used as constraints, and $N_o - R$ higher-order weights are to be found, the m moments of each of the $N_o - R$ functions are precalculated. These are then used as part of the general constraint matrix in the quadratic programming routine.

4 Conclusions

It has been shown that it is possible to add weighted amounts of the unmeasurable higher-order u_k basis singular functions by using the technique of quadratic programming. This can be easily adapted to fit any linear constraints - such as known coefficients for the first N singular functions in the expansion, positivity and known moment values. Once the constraints have been met it is then necessary to specify some further optimisation condition - such as the reconstruction possessing minimum L^1 or L^2 norm and, having done this, our numerical work confirms and implements the Micchelli uniqueness theorem. The choice of minimisation criterion is purely arbitrary - with no one choice working well in all cases. Once one has chosen to guess unmeasurable components of the solution to make it positive, using these methods or any of the many alternative iterative nonlinear methods, there will be an infinite family of solutions which all fit the data to the same accuracy and which may differ widely. The utmost caution is therefore required if non-linear methods are used since any apparent increase in resolution over that of the TSVD solution can always be spurious.

5 Acknowledgements

This work has been supported by the US Army Research Office under grant no DAAH04-95-1-0280.

References

1. E. G. Steward. *Fourier Optics*, Ellis Horwood Limited, 1983
2. C. E. Creffield, E. G. Klepfish, E. R. Pike and Sarben Sarkar. *Physical Review Letters*, **75**, 517-520, 1995
3. H. Z. Cummins and E. R. Pike, *NATO Advanced Study Institute Series B: Physics*, Plenum Press, New York, 1974
4. M. Bertero and E. R. Pike. Signal processing for linear instrumental systems with noise: a general theory with illustrations from optical imaging and light scattering problems in *Handbook of Statistics*, Volume 10, Eds N. K. Bose and C. R. Rao, Elsevier, Amsterdam, 1993, pp1-46
5. D. Slepian and H. O. Pollak. *Bell Systems Technical Journal*, **40**, 43-64 1961
6. *Fortran NAG Library*, **E04NCF**, pp15
7. C.A. Micchelli, P.W. Smith, J. Swetits and J.D. Ward, *Constructive Approximation*, **1**, 1985, 93-102
8. J. M. Borwein and A. S. Lewis, *Mathematical Programming*, **57**, 15-48 1992
9. G. de Villiers, in Mathematics in Signal Processing, Ed J. McWhirter *Signal Processing*, (Oxford University Press) to appear 1997.

Regularized Blind Deconvolution using Recursive Inverse Filtering

Michael K. Ng[1]*, Robert J. Plemmons[2]**, and Sanzheng Qiao[3]***

[1] ACSys, Computer Sciences Laboratory, Research School of Information Sciences and Engineering, The Australian National University, Canberra, ACT 0200, Australia. E-mail: mng@cslab.anu.edu.au

[2] Department of Mathematics and Computer Science, Wake Forest University, Winston-Salem, NC 27109. E-mail: plemmons@wfu.edu

[3] Department of Computer Science and Systems, McMaster University, Hamilton, Ontario L8S 4K1, Canada. E-mail: qiao@maccs.dcss.mcmaster.ca

Abstract. Image restoration involves the removal or minimization of degradation (blur, clutter, noise, etc.) in an image using a priori knowledge about the degradation phenomena. Blind restoration is the process of estimating both the true image and the blur from the degraded image characteristics, using only partial information about degradation sources and the imaging system. Our main interest concerns optical image enhancement, where the degradation involves a convolution process. When an otherwise collimated, coherent beam of light encounters a turbulent flow field that includes density fluctuations, its optical wavefront becomes aberrated causing the beam to be degraded. Only partial a priori knowledge about the degradation phenomena in aero-optics is generally known, so here the use of blind deconvolution methods is essential. In this paper we provide a method to incorporate truncated eigenvalue and total variation regularization into a nonlinear recursive inverse filter blind deconvolution scheme first proposed by Kundur and Hatzinakos. We call our approach the nonnegativity and support regularized recursive inverse filter (NSR-RIF) algorithm. Simulation tests are reported on optical imaging problems.

1 Introduction

A fundamental issue in image restoration is blur removal in the presence of observation noise. Noise (sometimes more appropriately called "clutter" in images) may come, for example, as the result of thermal effects, measurement errors, digitation, or may be introduced by a recording or transmission medium. In the important case where the blurring operation is spatially invariant, i.e., it operates uniformly across the object domain, then the basic restoration computation

* Research supported by the Cooperative Research Centre for Advanced Computational Systems (ACSys).

** Research sponsored by the Air Force Office of Scientific Research and by the National Science Foundation.

*** Research supported by the Natural Sciences and Engineering Council of Canada.

involved is simply a deconvolution process that faces the usual difficulties associated with ill-conditioning in the presence of noise [4,20]. Such a situation often occurs, for example, in imaging through atmospheric turbulence. In particular, the problem of imaging through a medium is encountered in many important situations, from astronomy to medicine.

The image observed from a shift invariant linear blurring process, such as an optical system, is described by how the system blurs a point source of light into a larger image. The image of a point source is called the *point spread function* PSF, which we denote by h. The observed image g is then the result of convolving the PSF h with the "true" image, say f. This blurring process is represented by the convolution equation

$$g = h \star f. \tag{1}$$

The standard deconvolution problem is to recover the image f from (1), given the observed image g and the blurring operator h. This basic problem appears in many forms in signal and image processing. There is much interest in removing blur and noise degradations from 1-D chemical spectra, as well as 2-D images from microscopes, telescopes, photographs, CT or MRI scanners, satellite sensors, and scintigrams (nuclear medicine images) [4,20].

The PSF of an imaging system can sometimes be described by a mathematical formula, e.g., in the case of an out-of-focus lens system. More often, the PSF must be estimated empirically. Empirical estimates of the PSF can sometimes be obtained by imaging a relatively bright, isolated point source. In astro-imaging the point source might be a natural guide star or a guide star artificially generated using range-gated laser backscatter, e.g, [2,6,12,16,23]. Notice here that the PSF as well as the image may be degraded by noise.

In many applications data corresponding to h is not completely known. *Blind deconvolution* is the process of estimating both the true image f and the blur h from the degraded image g. The purpose of this paper is to incorporate regularization into and refine a nonlinear recursive inverse filter blind deconvolution method first proposed by Kundur and Hatzinakos [17–19]. They call their scheme the *nonnegativity and support constrained, recursive inverse filtering method*, or NAS-RIF, for short.

Applications of blind image deconvolution in optics abound in science and engineering [4,6,8,16,20]. Our work to enhance the quality of optical images has applications in defense, and to civilian technology, including astronomical imaging, e.g., [6,16,22,24].

An algorithm for one dimensional regularized iterative blind deconvolution using truncated eigenvalue and total variation regularization in conjunction with recursive inverse filtering is developed in §2. We apply regularization to the inverse filter by using an inexpensive *eigenvalue truncation* scheme, and allow the user the option of applying *total variation regularization* to the estimated image. The method is extended to two dimensional imaging problems in §3. We call our approach the *nonnegativity and support regularized recursive inverse filter* (NSR-RIF) algorithm. Preliminary numerical tests are reported in §4 on some simulated optical imaging problems, and a comparison is made with the NAS-RIF algorithm.

2   One Dimensional Deconvolution

We begin with the one dimensional signal restoration problem, which we term as image restoration in order to be consistent with terminology used later in this paper. Consider an image vector

$$f = (f_1, ..., f_n)^T, \qquad f_i \geq 0, \quad \text{for } 1 \leq i \leq n,$$

and a blurring operator, or point spread function, represented by a $(2m+1)$-by-1 vector

$$h = (h_{-m}, ..., h_0, ..., h_m)^T.$$

Assuming $2m < n$, we can write the observed image g as the convolution of h and f, as given in (1). The standard deconvolution problem is to recover the image f from (1), given the observed image g and the blurring operator h. Blind deconvolution is the process of estimating both the true image f and the blur h from the degraded image g.

Writing (1) in discrete, i.e., matrix form, we have

$$g = Hf \tag{2}$$

where the convolution matrix

$$H = \begin{pmatrix} h_{-m} & & 0 \\ \vdots & \ddots & \\ h_m & \ddots & h_{-m} \\ & \ddots & \ddots & \ddots \\ & & h_m & \ddots & h_{-m} \\ & & & \ddots & \vdots \\ 0 & & & & h_m \end{pmatrix} \tag{3}$$

is $(n+2m)$-by-n and column circulant. Note that H is rectangular. So, we embed H in a square circulant matrix

$$C = (H_1 \ H \ H_2) \tag{4}$$

where both

$$H_1 = \begin{pmatrix} h_0 & \cdots & h_{-m+1} \\ h_1 & \cdots & h_{-m+2} \\ \vdots & \ddots & \vdots \\ h_m & \ddots & h_1 \\ 0 & \ddots & \vdots \\ \vdots & \ddots & h_m \\ 0 & \ddots & 0 \\ h_{-m} & \ddots & \vdots \\ \vdots & \ddots & 0 \\ h_{-1} & \cdots & h_{-m} \end{pmatrix} \quad \text{and} \quad H_2 = \begin{pmatrix} h_m & \cdots & h_1 \\ 0 & \ddots & \vdots \\ \vdots & \ddots & h_m \\ 0 & \ddots & 0 \\ h_{-m} & \ddots & \vdots \\ \vdots & \ddots & 0 \\ h_{-1} & \ddots & h_{-m} \\ \vdots & \ddots & \vdots \\ h_{m-2} & \cdots & h_{-1} \\ h_{m-1} & \cdots & h_0 \end{pmatrix}$$

are $(n + 2m)$-by-m and column circulant. Thus

$$H = C \begin{pmatrix} 0 \\ I_n \\ 0 \end{pmatrix}.$$

Correspondingly, we can embed f in an $(n + 2m)$-by-1 vector

$$\tilde{f} = \begin{pmatrix} \tilde{f}_a \\ \tilde{f}_b \\ \tilde{f}_c \end{pmatrix} = \begin{pmatrix} 0 \\ f \\ 0 \end{pmatrix}$$

where 0 in the last vector represents an m-by-1 null vector. Then (2) is equivalent to

$$g = C\tilde{f}. \tag{5}$$

Equation (5) is called the extended sequence convolution form of (2) and leads to the *extended sequence deconvolution method* [15]. Here, (5) is often preferable to (2), because its coefficient matrix C is a square circulant matrix, which has many desirable properties. For example, if H in (3) has full column rank, then C is invertible and its inverse is also circulant and C^{-1} can be efficiently computed using the fast Fourier transform (FFT), yielding the extended sequence deconvolution method for computing f.

In blind deconvolution, we are given the observed image g. To estimate f and H using (2), we need to estimate the length (extent, or support) n of the true image f. The bandwidth $2m+1$ of H can then be found using n and the length of g. Similarly, if we use (5) to estimate $\tilde{f}$ and C, we also need to know the size n of the nonzero vector f in $\tilde{f}$. The structure of C can then be determined by n and the length of g. Thus, the size of the true image, called the support constraint, is crucial information in blind deconvolution using either (2) or (5). In addition to the support constraint on f, nonnegativity of the pixel values, represented by the f_i, is an important requirement. This is called the *nonnegativity constraint*, and can be incorporated into deconvolution computations [4,20]. Assuming that C is invertible, we can formulate the blind deconvolution problem as:

Find C (or the first column of C) such that $\tilde{f} = C^{-1}g$ with $\tilde{f}_a = \tilde{f}_c = 0$ and $\tilde{f}_b \geq 0$.

2.1   Iterative Blind Deconvolution

Some iterative blind deconvolution methods begin with a knowledge-based (non-negativity, finite support) estimate for f, "deconvolve" $g = h \star f$ to estimate h, and then iterate in an alternating fashion to improve the estimates for both f and h [5,8,29]. These methods involve variations of direct, rather than inverse, filtering. Kundur and Hatzinakos [17–19], however, have recently presented a novel blind deconvolution method using recursive *inverse filtering*. Inverse filters are easier to implement and avoid certain inversion procedures, thus reducing the computational complexity.

The method by Kundur and Hatzinakos uses a variable finite impulse response (FIR) filter of length $2p + 1$, where p is an estimate for m in H in (3), characterized by its parameter vector:

$$u = (u_{-p}, ..., u_0, ..., u_p)^T, \qquad (6)$$

and the process iterates with respect to u. The observed image g is the input to the filter. The output is the convolution of u and g: $y = u \star g$, which is then constrained and used to approximate the true image f. Assuming the length of g is k, we can write the convolution in matrix form:

$$y = Ug,$$

where the convolution matrix

$$U = \begin{pmatrix} u_{-p} & & & & 0 \\ \vdots & \ddots & & & \\ u_p & \ddots & u_{-p} & & \\ & \ddots & \vdots & \ddots & \\ & & u_p & \ddots & u_{-p} \\ & & & \ddots & \vdots \\ 0 & & & & u_p \end{pmatrix} \qquad (7)$$

is a $(k + 2p)$-by-k banded Toeplitz matrix of bandwidth $2p + 1$. The output y is then passed through a nonlinear filter which maps y to $\tilde{y}$ such that $\tilde{y}$ has a finite support, say n, and is nonnegative. Specifically, assuming that we choose p so that $r = (k + 2p - n)/2$ is an integer and denoting $P_n = \text{diag}(p_i)$ where

$$p_i = \begin{cases} 1, & \text{if } y_{r+i} > 0; \\ 0, & \text{otherwise;} \end{cases} \quad \text{for } 1 \leq i \leq n,$$

we get, in matrix form,

$$\tilde{y} = \begin{pmatrix} 0 & 0 & 0 \\ 0 & P_n & 0 \\ 0 & 0 & 0 \end{pmatrix} y.$$

In other words, the nonlinear filter sets any component which is either outside the finite support or is negative, to zero. The resulting vector $\tilde{y}$ is an estimate for the true nonnegative image with support n. That is

$$\tilde{y} \approx \tilde{f}.$$

The error vector is given by

$$z = y - \tilde{y} = \begin{pmatrix} I_r & 0 & 0 \\ 0 & I_n - P_n & 0 \\ 0 & 0 & I_r \end{pmatrix} y \equiv Dy. \tag{8}$$

The parameter vector u in (6) for our FIR filter is determined by minimizing the error $\|z\|_2^2$. In order to avoid a trivial null vector u, an additional term

$$\gamma(\sum_{i=-p}^{p} u_i - 1)^2 = \gamma(e^T u - 1)^2$$

is incorporated into the objective function, where $\gamma \geq 0$ and $e = (1, ..., 1)^T$. Specifically, the objective function is

$$J(u) = \|z\|_2^2 + \gamma(e^T u - 1)^2. \tag{9}$$

An optimal FIR filter is found by minimizing the above objective function $J(u)$. It is proved in [19] that $J(u)$ is convex. Thus the global minimum exists and a variety of numerical optimization algorithms can be used to compute u minimizing $J(u)$. Kundur and Hatzinakos [17–19] use a nonlinear conjugate gradient method to search for the minimum. Thus they call their algorithm the *nonnegativity and support constraints recursive inverse filter* (NAS-RIF).

As described above, in matrix terminology, Kundur and Hatzinakos use a banded Toeplitz matrix U of the form (7) to approximate the deconvolution matrix. Recalling that k is the length of the input image vector g and $2p + 1$ is the filter length, their deconvolution matrix is a $(k + 2p)$-by-k banded Toeplitz matrix of lower bandwidth $2p + 1$.

Using the circulant matrix C representation, we propose to use a k-by-k circulant matrix, which we call S, to approximate the inverse of C. Of course the inverse of a circulant matrix is circulant. Specifically, we let S be the k-by-k circulant matrix with first column

$$(s_0, ..., s_p, 0, ..., 0, s_{-p}, ..., s_{-1})^T, \quad 1 \leq p < \frac{k}{2} \tag{10}$$

where p is an estimate for m in H in (3), and denote

$$S = Circ[(s_0, ..., s_p, 0, ..., 0, s_{-p}, ..., s_{-1})^T].$$

Thus, the FIR filter performs a multiplication of the circulant matrix S and the observed image g. This multiplication can be implemented efficiently using the FFT. Note that the output of the filter is a k-by-1 vector, whereas in NAS-RIF the output is $(k + 2p)$-by-1. The output $y = Sg$ is sent to the nonlinear filter and mapped to $\tilde{y}$ satisfying the nonnegativity and support constraints. Then the error $z = y - \tilde{y}$ is measured and fed back to compute an adjustment to the FIR parameter vector u to reduce the value of the objective function $J(u)$. This procedure is iterated until the objective function reaches the global minimum, or some early stopping criteria applies.

2.2   Regularizing Iterative Blind Deconvolution

Continuous deconvolution can be modeled as an integral equation of the first kind (an ill-posed inverse problem [7,9]). It is well-known that deconvolution algorithms can be extremely sensitive to noise [4,15]. For example, if the noise is additive, then the blurring process can be represented by a convolution equation of the form

$$g = h \star f + \eta,$$

which, by (5), is equivalent to

$$g = C\tilde{f} + \eta. \tag{11}$$

Assuming that C is invertible, from (11) we have

$$C^{-1}g = \tilde{f} + C^{-1}\eta.$$

In applications arising from integral equations of the first kind, the ill-conditioning of C stems from the wide range of the magnitudes of its eigenvalues [7,9]. Therefore, excess amplification of the noise at small eigenvalues can occur. Figure 1 shows a practical example of a 2-dimensional discrete point spread function from [22] in atmospheric imaging, with a general Gaussian form [15]. Figure 2 depicts the distribution of the magnitudes $|\lambda_i|$ of the eigenvalues of the convolution matrix C corresponding to the point spread function in Figure 1. We see that many of the $|\lambda_i|$ are relatively small.

Since any realistic signal processing problem involves noise, it is necessary to incorporate regularization into deconvolution to stablize the computation [7,9,13,14]. Regularization methods attempt to alleviate sensitivity to the noise by "filtering" out eigen-components of the solution belonging to the noise subspace. For some iterative methods, it has been established that early termination of the iterations accomplishes this regularization effect. That is, the eigencomponents of the signal subspace are reconstructed in the first (possibly many) iterations and, after reaching a certain approximate restoration, the components in the noise subspace begin to be reconstructed. It is at this point, where the noise begins to contaminate the reconstruction, that the iterations are halted, see, e.g., Hanke and Hansen [10].

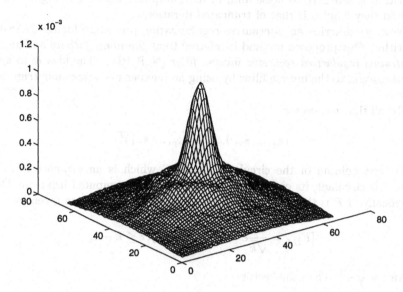

Fig. 1. Example of a discrete 2-D Gaussian blur point spread function (64 × 64)

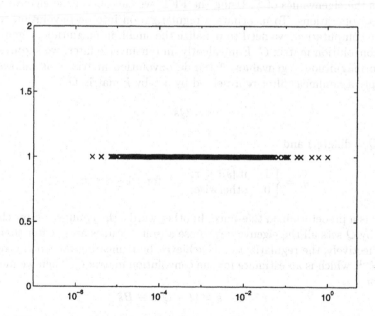

Fig. 2. Distribution of the $|\lambda_i|$ for the 129 × 129 convolution matrix C with the PSF given in Figure 1

The simulations by Kundur and Hatzinakos [18] show that their NAS-RIF algorithm is sensitive to noise and, in fact, amplifies noise. The regularization method they employ is that of truncated iterations.

Now, we describe an alternative regularization procedure for the NAS-RIF algorithm. Our proposed method is referred to as the *nonnegativity and support constraints regularized recursive inverse filter* (NSR-RIF). The idea is to apply regularization to the inverse filter by using an inexpensive *eigenvalue truncation* scheme.

Recall that the vector

$$(s_0, ..., s_p, 0, ..., 0, s_{-p}, ..., s_{-1})^T$$

is the first column of the circulant matrix S which is an estimate for C^{-1}. Since S is circulant, its eigenvalues can be efficiently computed using the DFT. Specifically, if F is the k-by-k Fourier matrix, i.e., its entries are given by

$$[F]_{j,\ell} = \frac{1}{\sqrt{k}} e^{-2\pi i j \ell / k}, \quad 0 \le j, \ell \le k - 1,$$

where $i = \sqrt{-1}$. Then the vector

$$\hat{s} = F(s_0, ..., s_p, 0, ..., 0, s_{-p}, ..., s_{-1})^T$$

contains the eigenvalues of S. Using the FFT, we can obtain the eigenvalues in $O(k \log k)$ operations. To incorporate regularization into deconvolution to stabilize the computation, we need to penalize the small, in magnitude, eigenvalues of the convolution matrix C. Equivalently, in our inverse filter, we suppress the large, in magnitude, eigenvalues of the deconvolution matrix S as follows. We first apply a nonlinear filter represented by a k-by-k matrix Q:

$$\tilde{s} = Q\hat{s}$$

where $Q = \text{diag}(q_i)$ and

$$q_i = \begin{cases} 1, & \text{if } |\hat{s}_i| \le \tau; \\ 0, & \text{otherwise}; \end{cases} \quad \text{for} \quad 1 \le i \le k,$$

where τ is a predetermined tolerance. In other words, the nonlinear filter characterized by Q sets all the eigenvalues whose absolute values are greater than τ to zero. Effectively, the regularization is achieved by truncating the small eigenvalues of S^{-1}, which is an estimate for the convolution matrix C. Then we measure the error

$$v = \hat{s} - \tilde{s} = (I - Q)\hat{s} \equiv B\hat{s}$$

and incorporate this error into the objective function $J(u)$ given by (9). Specifically, let

$$s = (s_{-p}, ..., s_{-1}, s_0, ..., s_p)^T$$

be the filter parameter vector, then our regularized objective function is reformulated now in terms of s as

$$J_{reg}(s) \equiv \|z\|_2^2 + \mu\|v\|_2^2 + \gamma(e^T s - 1)^2, \tag{12}$$

recalling that z is the error vector $y - \tilde{y}$. The parameter μ in (12) controls the degree of regularization. The filter parameter vector s is determined by minimizing our new objective function (12). Figure 3, which also involves regularization of the image calculation as described in Section 2.3, gives an overview of our overall scheme.

Now we consider the gradient of J_{reg}. Using the following equations

$$\begin{pmatrix} s_0 \\ \vdots \\ s_p \\ 0 \\ \vdots \\ 0 \\ s_{-p} \\ \vdots \\ s_{-1} \end{pmatrix} = \begin{pmatrix} 0 & I_{p+1} \\ 0 & 0 \\ I_p & 0 \end{pmatrix} \begin{pmatrix} s_{-p} \\ \vdots \\ s_{-1} \\ s_0 \\ \vdots \\ s_p \end{pmatrix}, \qquad D^T D = D, \qquad B^T B = B,$$

we can show that the gradient of $J_{reg}(s)$ is

$$\nabla J_{reg}(s) = 2Gz + 2\mu\mathrm{Re}(E^H v) + 2\gamma(e^T s - 1)e \tag{13}$$

where G is a $(2p+1)$-by-k row-circulant matrix with its first row:

$$(g_{p+1}, g_{p+2}, \ldots, g_k, g_1, \ldots, g_p),$$

and

$$E = F \begin{pmatrix} 0 & I_{p+1} \\ 0 & 0 \\ I_p & 0 \end{pmatrix}.$$

Since G is row-circulant and F is the k-by-k Fourier matrix, the gradient can be computed efficiently. Furthermore, we can show that $J_{reg}(s)$ is convex.

Theorem 1. *Let $J_{reg}(s)$ be twice differentiable in $R^{(2p+1)}$. Then J_{reg} is convex for all s in $R^{(2p+1)}$ if the Hessian $\nabla^2 J_{reg}(s)$ is positive semi-definite, and is strictly convex if the Hessian is positive definite.*

From (13), it can be shown that the Hessian of the objective function is given by

$$\nabla^2 J_{reg}(s) = 2GDG^T + 2\mu\mathrm{Re}(E^H BE) + 2\gamma ee^T, \tag{14}$$

leading to the convexity result for $J_{reg}(s)$, by Theorem 1. Thus the global minimum exists and a variety of numerical optimization algorithms can be used to compute s minimizing $J_{reg}(s)$. In our numerical examples, a nonlinear conjugate gradient method is used to search for the global minimum.

Our basic regularized iterative blind deconvolution algorithm is listed in Table 1. Figure 3 summarizes the method and also shows how additional regularization of the estimated image can be incorporated into the scheme, according to the method proposed in §2.3.

2.3   Regularizing the Estimated Image

In this section, we propose a regularization method for the image estimated by our NSR-RIF algorithm given in Table 1. This image regularization method can also be incorporated into the NAS-RIF by Kundur and Hatzinakos [17–19] to improve their restored image.

The image $\tilde{y}$ estimated by the NSR-RIF satisfies the nonnegativity and support constraints. Let $\tilde{a}$ be the nonnegative segment of size n at the center of $\tilde{y}$, i.e.,

$$\tilde{a} = (0 \quad I_n \quad 0)\tilde{y}.$$

In the one-dimensional case, we perform the regularization by solving for

$$a = (a_1, ..., a_n)^T$$

in the following penalized least squares minimization problem:

$$\min_a \left\{ \frac{1}{2}\|\tilde{a} - a\|_2^2 + \alpha \sum_{k=1}^{n-1} \sqrt{|a_{i+1} - a_i|^2 + \beta} \right\}$$

where $\alpha > 0$ controls the degree and $\beta (\geq 0)$ controls the variability of the penalty term. When $\beta = 0$ we have the usual *total variation minimization*, which has been studied extensively in recent years, e.g., [1,5,25–28]. It is known that this approach is especially effective in preserving *sharp* edges. Similarly, total variation has the advantage that it does not penalize *smooth* images. On the other hand, disadvantages include the complexity level in solving total variation minimization problems, and the fact that minimizing the variation of the pixel values a_i can sometimes cause a loss of fine detail in the image [7]. For these reasons we leave the incorporation of total variation regularization of the estimated image as a user option in our NSR-RIF scheme (see Figure 3).

Suppose that a is the solution of above minimization problem. The gradient of the associated objective function at a is

$$a + \alpha A_\beta(a)a - \tilde{a},$$

where $A_\beta(a)$ is a symmetric and tridiagonal matrix with its subdiagonal

$$\begin{pmatrix} -\dfrac{1}{\sqrt{|a_2 - a_1|^2 + \beta}} \\ \vdots \\ -\dfrac{1}{\sqrt{|a_n - a_{n-1}|^2 + \beta}} \end{pmatrix},$$

Table 1. NSR-RIF Algorithm.

Definitions:

g: the blurred and noisy signal of size k.
n: the support size.
p: the filter length is $2p + 1$.
$s(t)$: the FIR filter parameter vector of dimension $2p + 1$ at the t-th iteration.
$S(t)$: the corresponding matrix of the parameter vector $s(t)$ of the FIR filter.
$\tilde{y}(t)$: the estimate of the original signal at the t-th iteration.
$J_{reg}(t)$: the objective function at $s(t)$.
$\nabla J_{reg}(t)$: the gradient vector of $J_{reg}(t)$.
ϵ: the tolerance for the termination.

Initial Conditions:

Set $s(0)$ to all zeros with a unit spike in the middle,
or, to an estimate corresponding to the inverse of the PSF.

Iterations:

For $t = 0, 1, \ldots$
  Compute $y(t) = S(t)g$.
  Project $y(t)$ onto $\tilde{y}(t)$.
  Compute the error $z(t) = y(t) - \tilde{y}(t)$.
  Compute $\hat{s}(t) = Fs(t)$, where F is the DFT operator.
  Project $\hat{s}(t)$ onto $\tilde{s}(t)$.
  Compute the error $v(t) = \hat{s}(t) - \tilde{s}(t)$.
  If $J_{reg}(t) < \epsilon$, then stop;   **otherwise** compute $\nabla J_{reg}(t)$ using (13).
  If $t = 0$, **then** set the conjugate gradient direction vector $d(t) = - \nabla J_{reg}(t)$;
  otherwise compute $e(t) = [\nabla J_{reg}(t) - \nabla J_{reg}(t - 1)]^H \nabla J_{reg}(t) / \| \nabla J_{reg}(t - 1) \|_2^2$;
  and set the direction vector $d(t) = - \nabla J_{reg}(t) + e(t)d(t - 1)$.
  Perform a line minimization to determine δ_t such that
    $J_{reg}(s(t) + \delta_t d(t)) \leq J_{reg}(s(t) + \delta d(t)), \quad \forall \delta \in \mathcal{R}.$
  Compute $s(t + 1) = s(t) + \delta_t d(t)$.
End For

and its diagonal

$$\begin{pmatrix} \frac{1}{\sqrt{|a_2-a_1|^2+\beta}} \\ \frac{1}{\sqrt{|a_2-a_1|^2+\beta}} + \frac{1}{\sqrt{|a_3-a_2|^2+\beta}} \\ \vdots \\ \frac{1}{\sqrt{|a_{n-1}-a_{n-2}|^2+\beta}} + \frac{1}{\sqrt{|a_n-a_{n-1}|^2+\beta}} \\ \frac{1}{\sqrt{|a_n-a_{n-1}|^2+\beta}} \end{pmatrix}.$$

Thus the minimizer a is the solution of the following equation

$$(I_n + \alpha A_\beta(a))a = \tilde{a}.$$

We make two observations: first, $I_n + \alpha A_\beta(a)$ is an M-matrix (see e.g., Berman and Plemmons [3]). Thus the entries of the inverse of $(I_n + \alpha A_\beta(a))$ are nonnegative. Second, a is the result of two consecutive mappings on the filter output y, first nonnegativity and support and then total variation regularization. These two observations ensure the nonnegativity of a.

We remark that the regularized image estimate a can replace $\tilde{y}$ as an input to compute the error vector z in the recursive inverse filter algorithm. The switch in Figure 3 indicates this option.

3   2-D Problems: Atmospheric Blurring

The results of §2 extend in a natural way to 2-D image blind deconvolution. Our main interest concerns optical image enhancement. Applications of blind image deconvolution in optics can be found in many areas of science and engineering, e.g, see the recent book by Roggemann and Welsh [24]. This work to enhance the quality of optical images has applications in defense, and to civilian technology, including astronomical and medical imaging. Only partial a priori knowledge about the degradation phenomena or PSF in aero-optics is generally known, so here the use of blind convolution methods is essential. In addition, the estimated PSF is generally degraded in a manner similar to that of the observed image.

If we let $\mathcal{H}$ denote the blurring operator and η the noise process, then the image restoration problem, for example with additive noise, can be expressed as a linear operator equation

$$g = \mathcal{H}f + \eta, \tag{15}$$

where g and f denote functions containing the information of the recorded and original images, respectively. Let v and w denote two-dimensional variables. If $\mathcal{H}$ is a convolution operator, as is often the case in optical imaging, then the operator acts uniformly (i.e., in a spatially invariant manner) on f. Here, (15) can be written as

$$(\mathcal{H}f)(x) = \int h(x-y)f(y)dy. \tag{16}$$

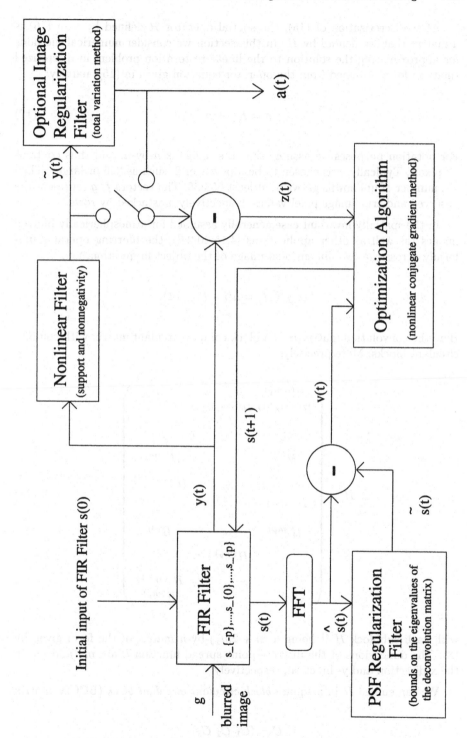

Fig. 3. NSR-RIF Algorithm

After discretization of (15), the spatial operator $\mathcal{H}$ defined by h in (16) is a matrix that we denote by H. In this section we consider numerical methods for approximating the solution to the linear restoration problem in discretized (matrix) form obtained from the operator equation given in (15), namely:

$$g = Hf + \eta. \tag{17}$$

For notation purposes we assume that the image is n-by-n, and thus contains n^2 pixels. Typically, n is chosen to be a power of 2, such as 256 or larger. Then the number of unknowns grows to at least 65,536. The vectors f, g represent the observed and true image pixel values, respectively, unstacked by rows.

In the spatially invariant case generally assumed for atmospherically blurred images (as well as other applications [4,15,20,24]), the blurring operates uniformly across the two dimensional image of the object in question, i.e.,

$$h(i, j; k, \ell) = h(i - k, j - \ell).$$

Here the convolution matrix H is a block column-circulant matrix with column-circulant blocks. More precisely,

$$H = \begin{pmatrix} H^{(-m_y)} & & & 0 \\ H^{(-m_y+1)} & H^{(-m_y)} & & \\ \vdots & \ddots & \ddots & \\ H^{(0)} & \ddots & \ddots & H^{(-m_y)} \\ \vdots & \ddots & \ddots & H^{(-m_y+1)} \\ H^{(m_y-1)} & \ddots & \ddots & \vdots \\ H^{(m_y)} & \ddots & \ddots & H^{(0)} \\ & H^{(m_y)} & \ddots & \vdots \\ & & \ddots & H^{(m_y-1)} \\ 0 & & & H^{(m_y)} \end{pmatrix} \tag{18}$$

with each subblock $H^{(j)}$ being a $(n + 2m_x)$-by-n matrix of the form given by (3). The dimensions of the discrete point spread function h are m_x and m_y in the x-direction and y-direction, respectively.

We can embed H in a square *block-circulant-circulant-block* (BCCB) matrix

$$C = (C_1 \ C_2 \ C_3)$$

where

$$C_2 = \begin{pmatrix} C^{(-m_y)} & & & & 0 \\ C^{(-m_y+1)} & C^{(-m_y)} & & & \\ \vdots & & \ddots & \ddots & \\ C^{(0)} & & \ddots & & \ddots & C^{(-m_y)} \\ \vdots & & & \ddots & & \ddots & C^{(-m_y+1)} \\ C^{(m_y-1)} & & \ddots & & \ddots & \vdots \\ C^{(m_y)} & & \ddots & & \ddots & C^{(0)} \\ & & C^{(m_y)} & \ddots & & \vdots \\ & & & & \ddots & C^{(m_y-1)} \\ 0 & & & & & C^{(m_y)} \end{pmatrix},$$

$$C_1 = \begin{pmatrix} C^{(0)} & \cdots & C^{(-m_y+1)} \\ C^{(1)} & \cdots & C^{(-m_y+2)} \\ \vdots & \ddots & \vdots \\ C^{(m_y)} & \ddots & C^{(1)} \\ 0 & \ddots & \vdots \\ \vdots & \ddots & C^{(m_y)} \\ 0 & \ddots & 0 \\ C^{(-m_y)} & \ddots & \vdots \\ \vdots & \ddots & 0 \\ C^{(-1)} & \cdots & C^{(-m_y)} \end{pmatrix} \quad \text{and} \quad C_3 = \begin{pmatrix} C^{(m_y)} & \cdots & C^{(1)} \\ 0 & \ddots & \vdots \\ \vdots & \ddots & C^{(m_y)} \\ 0 & \ddots & 0 \\ C^{(-m_y)} & \ddots & \vdots \\ \vdots & \ddots & 0 \\ C^{(-1)} & \ddots & C^{(-m_y)} \\ \vdots & \ddots & \vdots \\ C^{(m_y-2)} & \cdots & C^{(-1)} \\ C^{(m_y-1)} & \cdots & C^{(0)} \end{pmatrix}.$$

Here each subblock $C^{(j)}$ is a $(n + 2m_x)$-by-$(n + 2m_x)$ matrix of the form given by (4).

Following the BCCB matrix C representation, we use a $(n+2m_y)(n+2m_x)$-by-$(n + 2m_y)(n + 2m_x)$ BCCB matrix deconvolution matrix S to approximate the inverse of C. Note that the inverse of a BCCB matrix is also BCCB and is thus completely specified by its first column. Hence the inverse FIR filter vector is completely determined by the first column of the deconvolution matrix S.

In the extension of our NSR-RIF algorithm, we perform a multiplication of the circulant matrix S and the stacked, observed image g using the two dimensional fast Fourier transform (2-D FFT). The output y is sent to the nonlinear filter and mapped to $\tilde{y}$ to satisfy the nonnegativity and support constraints. The output, $\hat{s}$, of the 2-D FFT of the first column of S is sent to the nonlinear filter and mapped to $\tilde{s}$ for regularization. The errors $z = y - \tilde{y}$ and $v = \hat{s} - \tilde{s}$ are measured and fed back to compute an adjustment to the FIR parameter vector to reduce the value of the objective function. This procedure is iterated until the objective function reaches the global minimum, or some early stopping criteria applies. The regularization scheme for the estimated image described in §2.3 also extends to 2-D case in a natural way.

3.1   Guide Star Image

In two dimensional deconvolution problems arising in ground-based atmospheric imaging, a primary objective is to remove the blurring in an image resulting from the effects of atmospheric turbulence. The problem consists of an n-by-n image, received by a ground-based imaging system together with an n-by-n image of a guide star PSF, observed under similar circumstances. Empirical estimates of the PSF can sometimes be obtained by imaging a relatively bright, isolated point source. In astro-imaging the point source might be a natural guide star or a guide star artificially generated using range-gated laser backscatter, e.g, [2,12,23]. Notice here that the PSF as well as the image may be degraded by noise.

The imaging system detects the atmospheric distortions, represented by the PSF, using the guide star image. A wavefront sensor measures the optical distortions which can then be digitized into a blurred image of the guide star pixels. To form an initial estimate of the PSF, the rows of the blurred pixel guide star image are stacked into a column vector. The 2-D FFT of this vector provides estimates of eigenvalues, denoted by λ_i, of the convolution matrix C. We choose a truncation parameter ζ to separate these eigenvalues, in a manner similar to that used in [11,22]. The truncation parameter ζ is chosen so that those eigenvalues satisfying $|\lambda_i| \geq \zeta$ correspond to the signal subspace, while those satisfying $|\lambda_i| < \zeta$ correspond to the noise subspace. Thus, we set the initial estimate of the convolution matrix C to a BCCB matrix with eigenvalues

$$\tilde{\lambda}_i = \begin{cases} \lambda_i, & \text{if } |\lambda_i| \geq \zeta; \\ 1, & \text{if } |\lambda_i| < \zeta; \end{cases}.$$

In other words, we replace the eigenvalues corresponding to the noise subspace with ones, as discussed in [11] and [22]. Since the filter parameter vector corresponds to the inverse of the convolution matrix, it can be *initialized* by applying the 2-D inverse FFT to the vector of eigenvalues λ_i^{-1}.

To determine the truncation parameter ζ , we compute the discrete Fourier transform of the blurred and noisy image or the guide star image, and find the point at which the Fourier coefficients level off. The index where this stagnation begins indicates where the random errors (noise) start to dominate the blurred and noisy image or the guide star image. The value at this point is chosen as ζ. See [7], [11] or [22] for details of this procedure for estimating ζ.

4   Numerical Examples in Optical Imaging

In this section we present numerical tests of data samples to illustrate the effectiveness of our NSR-RIF algorithm approach to blind deconvolution. All the experiments are performed in MATLAB with machine epsilon $\mu \approx 10^{-16}$ precision on a Sun Ultra SPARC workstation.

Synthetic Data. The first data sample involves a degraded image of a simple known block object. This example consists of a simple 9×9 image (with 81 unknown pixel values) containing three blocks as shown in Figure 4. To obtain a blurred image, we used a Gaussian-type point spread function, shown in Figure 4, and convolved it with the original image. The blurred image is, in addition, polluted by Gaussian noise so that the resulting observed image has SNR = 10 DB, as shown in Figure 5 (left).

For our NSR-RIF algorithm, the best restoration is achieved at 9 iterations and we see from Figure 5 (right) that a visually appealing result is obtained.

For comparison purposes, the NAS-RIF algorithm suggested by Kundur and Hatzinakos [17–19] is applied to the degraded image. We see that the method converges to a good solution (Figure 6 (left) with 13 iterations), but then their scheme exhibits noise amplification on subsequent iterations (Figure 6 (right)).

In addition, Table 2 shows the corresponding relative errors for the above restored images. Our tests indicate that the NSR-RIF algorithm can effectively recover images even in the presence of high noise levels.

Table 2. Relative errors of the restored images for the NAS-RIF and NSR-RIF algorithms

y	$\|\|f - y\|\|_2/\|\|f\|\|_2$
Blurred and noisy image	0.7146
NSR-RIF restored image at 9 iterations (Figure 5)	0.4162
NAS-RIF restored image at 13 iterations (Figure 6)	0.4738
NAS-RIF restored image at 20 iterations (Figure 6)	0.7568

The second data sample consist a synthetically generated binary text image of the letters "HK97" as shown in Figure 7 (left). To obtain a blurred image, we used a Gaussian-type point spread function shown in Figure 7 (right) and convolved it with the original image. We show that the post-processing of the image estimate given by NSR-RIF algorithm can effectively regularize the restored image. Figure 8 shows the observed image which is blurred by the Gaussian-type point spread function shown in Figure 7 (right) and is added by Gaussian noise with SNR = 50 DB. We see from Figure 9 that the post-processing procedure is useful.

Simulated Ground-Based Telescope Data. The image boundaries in Example 1 are very regular. In the next example, we consider a 256 × 256 image with irregular boundaries. This model problem data was obtained from the US Air Force Phillips Laboratory, Lasers and Imaging Directorate, Kirtland Air Force Base, New Mexico. This model has been used for testing various image restoration algorithms, e.g., [5,11,21,22,24,27,28].

Specifically, the true object is an ocean reconnaissance satellite, which is shown in Figure 10 (left). A computer simulation algorithm at Phillips Laboratory was used to produce a degraded image of the satellite, shown in Figure 11, as would be observed from a modern ground-based telescope equipped with

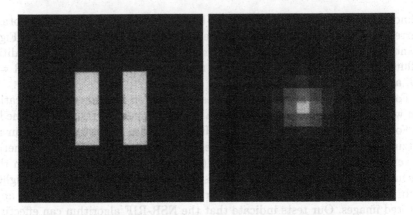

Fig. 4. Original image (left) and point spread function (right)

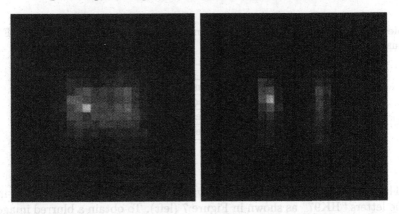

Fig. 5. Degraded image with SNR = 10 DB (left), and NSR-RIF restoration using $p = 11$, $\gamma = 0.75$, $\mu = 1.0$ and $\tau = 20$ with 9 iterations (right)

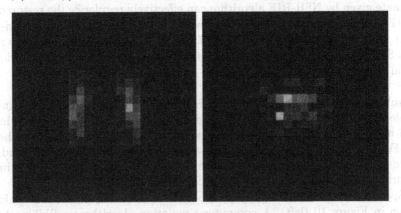

Fig. 6. For comparison purposes: NAS-RIF restoration using $p = 2$, $\gamma = 0.1$ and $\mu = 0.0$ with 13 iterations (left), and with 20 iterations (right)

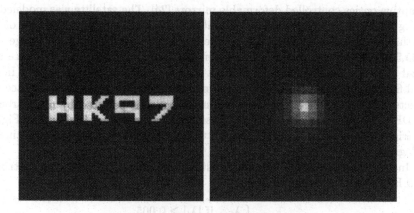

Fig. 7. Original image (left) and point spread function (right)

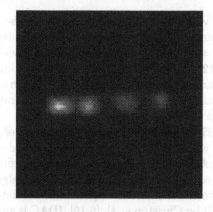

Fig. 8. Degraded image with SNR = 50 DB

Fig. 9. NSR-RIF restoration using $p = 2$, $\gamma = 0.1$, $\mu = 0.5$ and $\tau = 10$ with 68 iterations (left) and, after post-processing the image estimate (right)

adaptive-optics controlled deformable mirrors [24]. The satellite was modeled as being 12 meters in length and in an orbit 500 kilometers above the surface of the earth. The simulated charge-coupled device (CCD) for forming the image was a 65,536 pixel square array. CCD root-mean-square read-out noise variance was fixed at 15 microns per pixel to reflect a realistic state-of-the-art detector. In actual field experiments, several hundred measurement are averaged to reduce the effects of noise. In this example, the blurred image is polluted by noise around 30 DB. The guide star observed under similar circumstances is shown in Figure 10 (right). Notice the blur and noise in the image of the guide star, resulting in a degraded PSF.

In this example, we set the initial estimate of the 2-D convolution matrix C to a BCCB matrix with eigenvalues

$$\tilde{\lambda}_i = \begin{cases} \lambda_i, & \text{if } |\lambda_i| \geq 0.005; \\ 1, & \text{if } |\lambda_i| < 0.005. \end{cases}$$

The filter parameter vector is initialized by applying the 2-D inverse FFT to the vector of eigenvalues $1/\lambda_i$. Computed restorations by NSR-RIF algorithm (4 iterations, $\mu = 1.2$, $\tau = 120$ and $\gamma = 4.5$) using the knowledge of the guide star are shown in Figure 12. We also see that the guide star image and the support of the original image are useful in the blind deconvolution of the satellite image.

In summary, we have presented a new regularized iterative blind deconvolution method, based on recursive inverse filtering.

Preliminary numerical results indicate effectiveness of the method. Future work on this project will involve (1) testing the method on real astronomical data provided by the Air Force Phillips Laboratory Starfire Optical Range, (2) incorporating the use of multiple frames of data into the algorithm, and (3) comparing the performance of our NSR-RIF algorithm with the Iterative Deconvolution Algorithm (IDAC) by Christou et al, [6,16]. IDAC is an iterative direct filter non-linear deconvolution algorithm which, as with our NSR-RIF algorithm, uses an error metric minimization scheme, by means of a conjugate gradient search. Their code is available, e.g., at the European Southern Observatory in Munich, Germany.

References

1. L. Alvarez and J. Morel, *Formalization and Computational Aspects of Image Analysis*, Acta Numerica (1994), pp. 1–59.
2. T. E. Bell, *Electronics and the Stars*, IEEE Spectrum 32, no. 8 (1995), pp. 16–24 .
3. A. Berman and R. J. Plemmons, *Nonnegative Matrices in the Mathematical Sciences*, 2nd Edition, SIAM Press Classics Series no. 9, Philadelphia, 1994.
4. K. Castleman, *Digital Image Processing*, Prentice–Hall, NJ, 1996.
5. T. Chan and C. Wong, *Total Variation Blind Deconvolution* , CAM Rep. 96–45, Department of Math., UCLA.
6. J. Christou, *Blind Deconvolution Postprocessing of Images Collected by Adaptive-Optics*, SPIE Proc. 2534 (1995), pp. 226–234.

Fig. 10. Original satellite image (left) and guide star image (right)

Fig. 11. Observed image

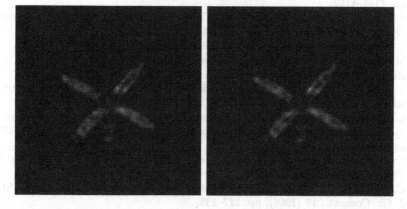

Fig. 12. Restored image with the exact support (left) and restored image with the overestimated support (right), using the guide star image to initialize NSR-RIF

7. H. Engl, M. Hanke and A. Neubauer, *Regularization of Inverse Problems*, Kluwer Academic Publishers, The Netherlands, 1996.
8. D. Fish, A. Brinicombe and R. Pike, *Blind Deconvolution by Means of the Richardson-Lucy Algorithm*, J. Optical Soc. Amer. A 12 (1995), pp. 58–65.
9. C. Groetsch, The Theory of Tikhonov Regularization for Fredholm Equations of the First Kind, Pitman Publishing, London, 1984.
10. M. Hanke and P. Hansen, *Regularization Methods for Large-scale Problems*, Surveys Math. Indust., 3 (1993), pp. 253–315.
11. M. Hanke, J. Nagy and R. Plemmons *Preconditioned Iterative Regularization for Ill-posed Problems*, Numerical Linear Algebra and Scientific Computing, Eds. L. Reichel, A. Ruttan, R. Varga, de Guyter Press, Berlin, (1993), pp. 141–163.
12. J. W. Hardy, *Adaptive Optics*, Scientific American 270, no. 6 (1994), pp. 60–65.
13. A. Houacine, *Regularized Fast Recursive Least Squares Algorithms for Adaptive Filtering*, IEEE Trans. on Signal Processing, Vol. 39, no. 4 (1991), pp. 860–871.
14. A. Houacine, *Regularized Fast Recursive Least Squares Algorithms for Finite Memory Filtering*, IEEE Trans. on Signal Processing, Vol. 40, no. 4 (1992), pp. 758–769.
15. A. Jain, Fundamentals of Digital Image Processing, Prentice-Hall, Englewood Cliffs, NJ, 1989.
16. S. Jefferies and J. Christou, *Restoration of Astronomical Images by Iterative Blind Deconvolution*, Astrophysics J., 415 (1993), pp. 862–874.
17. D. Kundur and D. Hatzinakos, *Blind Image Deconvolution*, IEEE Signal Processing Magazine, May, 1996, pp. 43–64. (See also, *Blind Image Deconvolution Revisited*, IEEE Signal Processing Magazine, November, 1996, pp. 61–63).
18. D. Kundur and D. Hatzinakos, *A Novel Blind Deconvolution Scheme for Image Restoration Using Recursive Filtering*, preprint, 1996.
19. D. Kundur and D. Hatzinakos, *On the Global Asymptotic Stability of the NAS-RIF Algorithm for Blind Image Restoration*, Proc. ICIP96, Inter. Conf. on Image Proc., Lausanne, Switzerland, September 1996.
20. R. Lagendijk and J. Biedmond, *Iterative Identification and Restoration of Images*, Kluwer Press, Boston, 1991.
21. J. Nagy, R. Plemmons and T. Torgersen, *Iterative Image Restoration using Approximate Inverse Preconditioning*, IEEE Trans. on Image Processing, 5, no. 7 (1996), pp. 1151–1162.
22. J. Nagy, P. Pauca, R. Plemmons and T. Torgersen, *Space-varying Restoration of Optical Images*, Preprint, August (1996).
23. J. Nelson, *Reinventing the Telescope*, Popular Science, 85 (1995), pp. 57–59.
24. M. Roggemann and B. Welsh, *Imaging Through Turbulence*, CRC Press, Boca Raton, FL, 1996.
25. L. Rudin, S. Osher and E. Fatemi, *Nonlinear Total Variation Based Noise Removal Algorithms*, Physica D., 60 (1992), pp. 259–268.
26. L. Rudin and S. Osher, *Total Variation Based Image Restoration with Free Local Constraints*, Proc. IEEE International Conf. on Image Processing, II (1994), pp. 31–35.
27. C. Vogel, *Solution of Linear Systems arising in Nonlinear Image Deblurring*, This Proceedings, Springer, 1997.
28. C. Vogel and M. Oman, *Iterative Methods for Total Variation Denoising*, SIAM J. Sci. Comput., 17 (1996), pp. 227–238.
29. Y. You and M. Kaveh *A Regularization Approach to Joint Blur Identification and Image Restoration*, IEEE Trans. on Image Proc., 5 (1996), pp. 416–427.

On Optimal Backward Perturbation Bounds

Ji-guang Sun *

Department of Computing Science, Umeå University, S-901 87 Umeå, Sweden

Abstract. In general, a computed solution to a problem solves many perturbed problems. The optimal backward perturbation bound for the problem with respect to the computed solution is a measure of the nearness between the perturbed problems and the original problem. An explicit expression of the optimal backward perturbation bound may provide a convenient way of testing the stability of practical algorithms. This paper surveys some new results on optimal backward perturbation bounds for linear systems, underdetermined systems, and linear least squares problems.

1 Introduction

Consider the linear system $Ax = b$. Let $\tilde{x}$ be a computed solution to the system. In general, the residual $b - A\tilde{x} \neq 0$ but it is small, and there may be many ΔA such that $(A + \Delta A)\tilde{x} = b$, where ΔA are called backward perturbations. Define $\eta(\tilde{x})$ by

$$\eta(\tilde{x}) = \min\{\|\Delta A\|/\alpha : (A + \Delta A)\tilde{x} = b\}, \tag{1.1}$$

where α is any positive scalar (for instance, $\alpha = 1$ or $\alpha = \|A\|$). The quantity $\eta(\tilde{x})$ is called the optimal backward perturbation bound (OBPB) for the system $Ax = b$ with respect to the computed solution $\tilde{x}$, or the backward error of the computed solution $\tilde{x}$.

An algorithm for computing the solution x to the system $Ax = b$ is called backward stable if, for any A, it produces a computed $\tilde{x}$ with a small $\eta(\tilde{x})$ (See, e.g., [2], [8]). Consequently, to find an explicit expression of the OBPB $\eta(\tilde{x})$ may be very useful for testing the stability of practical algorithms.

The following is a classical result on OBPBs for linear systems.

Theorem 1.1 (Rigal and Gaches [12], Kovarik [11]). *Let $A \in \mathcal{R}^{n \times n}$, $b \in \mathcal{R}^n$, and let $\tilde{x}$ be a computed solution to the linear system $Ax = b$. Let $\| \cdot \|$ denote a vector norm and its subordinate matrix norm. Then the OBPB $\eta_\infty(\tilde{x})$ defined by*

$$\eta_\infty(\tilde{x}) = \min\{\epsilon : (A + \Delta A)\tilde{x} = b + \Delta b, \|\Delta A\| \leq \epsilon\|E\|, \|\Delta b\| \leq \epsilon\|f\|\} \tag{1.2}$$

* This work was supported by the Swedish Natural Science Research Council under Contract M-AA/MA 06952-303 and the Department of Computing Science, Umeå University.

has the expression

$$\eta_\infty(\tilde{x}) = \frac{\|\hat{r}\|}{\|E\|\|\tilde{x}\| + \|f\|},\tag{1.3}$$

where $\hat{r} = b - A\tilde{x}$.

An extension of (1.2)–(1.3) can be found in [9].

Take $E = A$ and $f = b$ in (1.2)–(1.3), the corresponding $\eta_\infty(\tilde{x})$ is the optimal relative backward perturbation bound. As Pete Stewart [13] comments that if the optimal relative backward perturbation bound $\eta_\infty(\tilde{x})$ is a modest multiple of the rounding unit for the computer in question, then the computation has proceeded stably. Conversely, if the computation has proceeded stably then $\eta_\infty(\tilde{x})$ will be satisfactorily small. Thus the formula (1.3) provides a very convenient way of testing the stability of practical algorithms.

Note that an OBPB is also called a normwise backward error in the literature (see, e.g., [4], [6]–[10]). The classical backward error analyses of Wilkinson [24], [25] provide upper bounds on normwise and/or componentwise backward errors of computed solutions to many problems in matrix computations.

In recent years, the study of OBPBs for the problems in matrix computations is developed (see, e.g., [8]–[10], [15]–[23]). The purpose of this paper is to survey some new results on OBPBs for linear systems, underdetermined systems, and linear least squares problems.

This paper is organized as follows. In §2 we give a general formulation of OBPBs. In §3 we extend (1.2)–(1.3) to more general forms for general linear systems, and discuss structured linear systems. In §4 and §5 we survey new results on OBPBs for underdetermined systems and linear least squares problems, respectively. Finally, some conclusions are drawn in §6.

By using componentwise analysis one may obtain a more satisfactory backward error measure. See Higham [8, Chapters 7, 19, 20] for systematical expositions on the study of the componentwise backward errors.

Throughout this paper we use $\mathcal{R}^{m \times n}$ to denote the set of real $m \times n$ matrices, and $\mathcal{R}^m = \mathcal{R}^{m \times 1}$. $\emptyset$ stands for the empty set. $A^\dagger$ denotes the Moore-Penrose inverse, $P_A = AA^\dagger$ the orthogonal projection onto the column space of A, and $P_A^\perp = I - P_A$. $\| \cdot \|_F$ stands for Frobenius norm, and $\| \cdot \|_2$ for the Euclidean vector norm and the spectral norm. For an arbitrary vector norm $\nu(\cdot)$ on $\mathcal{R}^n$ the dual norm $\nu^D(\cdot)$ is defined by [14, p.57]

$$\nu^D(y) = \max_{\nu(x)=1} |y^T x|, \quad y \in \mathcal{R}^n.$$

Moreover, $\lambda_{\min}(A)$ denotes the smallest eigenvalue of a symmetric matrix A, and $\sigma_{\min}(A) \equiv \left(\lambda_{\min}(A^T A)\right)^{1/2}$ denotes the smallest singular value of a matrix A.

2   A General Formulation of OBPBs

Generally speaking, most problems in matrix computations can be cast in the form of solving an equation $r(a_1, \ldots, a_k; x) = 0$ with $a_1 \in \mathcal{A}_1, \ldots, a_k \in \mathcal{A}_k$, where $\mathcal{A}_1, \ldots, \mathcal{A}_k$ and $\mathcal{X}$ are finite dimensional normed linear spaces. For example, $r(A, b; x) = b - Ax$ for the linear system. Let $\tilde{x}$ be a computed solution to the equation $r(a_1, \ldots, a_k; x) = 0$. In general, the residual $r(a_1, \ldots, a_k; \tilde{x}) \neq 0$ but it is small, and there may be many $(\Delta a_1, \ldots, \Delta a_k)$ such that $r(a_1 + \Delta a_1, \ldots, a_k + \Delta a_k; \tilde{x}) = 0$, where $(\Delta a_1, \ldots, \Delta a_k)$ are called backward perturbations.

If we only consider backward perturbations in a_1, then after the manner of (1.1) we can define the OBPB $\eta(\tilde{x})$ by

$$\eta(\tilde{x}) = \min \left\{ \frac{\nu(\Delta a_1)}{\alpha_1} \ : \ r(a_1 + \Delta a_1, a_2, \ldots, a_k; \tilde{x}) = 0 \right\},$$

where $\nu(\cdot)$ is any norm on $\mathcal{A}_1$, α_1 is any positive scalar (for instance, $\alpha_1 = 1$ or $\alpha_1 = \nu(a_1)$).

In the general case, there are various ways to define OBPBs. For simplicity, we assume $k = 2$. The following definitions are advisable:

(i) After the manner of (1.2), we can define the OBPB $\eta_\infty(\tilde{x})$ by

$$\eta_\infty(\tilde{x}) = \min\{\epsilon \ : \ r(a_1 + \Delta a_1, a_2 + \Delta a_2; \tilde{x}) = 0, \ \nu_i(\Delta a_i) \leq \epsilon \alpha_i, \ i = 1, 2\}, \quad (2.1)$$

where $\nu_1(\cdot)$ and $\nu_2(\cdot)$ are any norms on $\mathcal{A}_1$ and $\mathcal{A}_2$, respectively, and α_1, α_2 are any positive scalars (for instance, $\alpha_i = 1$ or $\alpha_i = \nu_i(a_i)$ for $i = 1, 2$). The OBPB $\eta_\infty(\tilde{x})$ is usually used in the literature (see, e.g., [4], [8]–[12]).

(ii) Define the OBPB $\beta^{(\omega)}(\tilde{x})$ by [18]

$$\beta^{(\omega)}(\tilde{x}) = \min \left\{ \left\| \begin{pmatrix} \nu_1(\Delta a_1) \\ \omega \nu_2(\Delta a_2) \end{pmatrix} \right\| \ : \ r(a_1 + \Delta a_1, a_2 + \Delta a_2; \tilde{x}) = 0 \right\}, \quad (2.2)$$

where $\nu_1(\cdot)$ and $\nu_2(\cdot)$ are any norms on $\mathcal{A}_1$ and $\mathcal{A}_2$, respectively, $\| \cdot \|$ is any norm on $\mathcal{R}^2$, and ω is a positive parameter.

(iii) Define the OBPB $\eta^{(\theta)}(\tilde{x})$ by

$$\eta^{(\theta)}(\tilde{x}) = \min\{\nu(\Delta a_1, \theta \Delta a_2) \ : \ r(a_1 + \Delta a_1, a_2 + \Delta a_2; \tilde{x}) = 0\}, \quad (2.3)$$

where $\nu(\cdot)$ is any norm on $\mathcal{A}_1 \otimes \mathcal{A}_2$, and θ is a positive parameter. The OBPB $\eta^{(\theta)}(\tilde{x})$ is also used in the literature [17]–[21], [23].

It is worth pointing out that the parameters ω and θ in (2.2) and (2.3) allow us some flexibility. We now note some examples:

(E–1) Let α_1, α_2 be any positive scalars (for instance, $\alpha_i = 1$ or $\alpha_i = \nu_i(a_i)$ for $i = 1, 2$). Taking $\omega = \alpha_1/\alpha_2$ and $\|\cdot\| = \|\cdot\|_p$ (the p-norm, $p \geq 1$) in (2.2), and multiplying $\beta^{(\omega)}(\tilde{x})$ by $1/\alpha_1$, yields the OBPB

$$\eta_p(\tilde{x}) \equiv \frac{1}{\alpha_1} \beta^{(\alpha_1/\alpha_2)}(\tilde{x})$$

$$= \min \left\{ \left\| \begin{pmatrix} \nu_1(\Delta a_1)/\alpha_1 \\ \nu_2(\Delta a_2)/\alpha_2 \end{pmatrix} \right\|_p : r(a_1 + \Delta a_1, a_2 + \Delta a_2; \tilde{x}) = 0 \right\}. \tag{2.4}$$

Particularly, taking $p = 1, 2, \infty$ in (2.4), yields the OBPBs $\eta_1(\tilde{x}), \eta_2(\tilde{x}), \eta_\infty(\tilde{x})$, respectively, where $\eta_\infty(\tilde{x})$ coincides with (2.1).

(E–2) Taking $\theta = 1$ in (2.3), and multiplying $\eta^{(\theta)}(\tilde{x})$ by $1/\alpha$, yields the OBPB

$$\eta^*(\tilde{x}) \equiv \frac{1}{\alpha} \eta^{(1)}(\tilde{x})$$

$$= \min \left\{ \frac{\nu(\Delta a_1, \Delta a_2)}{\alpha} : r(a_1 + \Delta a_1, a_2 + \Delta a_2; \tilde{x}) = 0 \right\}, \tag{2.5}$$

where α is any positive scalar (for instance, $\alpha = 1$, or $\alpha = \nu(a_1, a_2)$).

(E–3) Taking $\theta \to \infty$ forces $\Delta a_2 = 0$ in (2.3), yields the OBPB where only a_1 is perturbed.

Let $\nu(\cdot)$ be a norm on $\mathcal{A} \otimes \mathcal{A}_2$. Assume that the norms $\nu_i(\cdot)$ are the restrictions of $\nu(\cdot)$ on $\mathcal{A}_i$ for $i = 1, 2$, and write $\nu_i(\cdot)$ as $\nu(\cdot)$, that is, $\nu(a_1) = \nu(a_1, 0)$ and $\nu(a_2) = \nu(0, a_2)$. Moreover, assume the norm $\nu(\cdot)$ has the property that if $\nu(a_i) \leq \nu(\hat{a}_i)$ $(i = 1, 2)$ then $\nu(a_1, a_2) \leq \nu(\hat{a}_1, \hat{a}_2)$. The following result reveals the relations between $\eta^*(\tilde{x})$ and $\eta_p(\tilde{x})$ for $p = 1, 2, \infty$.

Theorem 2.1 (Sun [18]). *Let $\eta_p(\tilde{x})$ $(p = 1, 2, \infty)$ be the OBPBs defined by (2.4), and $\eta^*(\tilde{x})$ be the OBPB defined by (2.5), where we take $\alpha_1 = \nu(a_1), \alpha_2 = \nu(a_2)$, and $\alpha = \nu(a_1, a_2)$. Then*

$$\eta_\infty(\tilde{x}) \leq \eta_1(\tilde{x}) \leq 2\eta_\infty(\tilde{x}), \quad \frac{1}{\sqrt{2}}\eta_1(\tilde{x}) \leq \eta_2(\tilde{x}) \leq \eta_1(\tilde{x}),$$

$$\frac{1}{\sqrt{2}}\eta_2(\tilde{x}) \leq \eta_\infty(\tilde{x}) \leq \eta_2(\tilde{x}), \tag{2.6}$$

and

$$\frac{\min\{\nu(a_1), \nu(a_2)\}}{\nu(a_1, a_2)} \eta_\infty(\tilde{x}) \leq \eta^*(\tilde{x}) \leq \frac{\max\{\nu(a_1), \nu(a_2)\}}{\nu(a_1, a_2)} \eta_1(\tilde{x}). \tag{2.7}$$

The second inequality of (2.7) implies $\eta^*(\tilde{x}) \leq \eta_1(\tilde{x})$, and the first inequality implies that if any one of $\nu(a_1)$ and $\nu(a_2)$ is much smaller than the other, then $\eta^*(\tilde{x})$ is bounded from below by $\tau\eta_\infty(\tilde{x})$, where $\tau > 0$ is a very small scalar. This means that in some cases the quantity $\eta^*(\tilde{x})$ may be much smaller than $\eta_p(\tilde{x})$

for $p = 1, 2, \infty$. Note that a very small OBPB $\eta^*(\tilde{x})$ may be uninformative for the following reason: In the case that there is a great disparity between $\nu(a_1)$ and $\nu(a_2)$, it may be making a large relative perturbation in the small one of a_1 and a_2 while the optimal backward perturbation $(\Delta a_{1*}, \Delta a_{2*})$ is very small compared with (a_1, a_2). Consequently, for testing the stability of an algorithm, we rather take $\eta_p(\tilde{x})$ for some p in practice.

3 Linear Systems

In §3.1 we shall be concerned with explicit expressions of the OBPBs $\beta^{(\omega)}(\tilde{x})$ and $\eta^{(\theta)}(\tilde{x})$ of a general linear system $Ax = b$, where $A \in \mathcal{R}^{n \times n}$, $\tilde{x}$ is a computed solution. In §3.2 we consider structured linear systems.

3.1 General Linear Systems

By the definition (2.2), the OBPB $\beta^{(\omega)}(\tilde{x})$ is defined by

$$\beta^{(\omega)}(\tilde{x}) = \min\left\{\left\|\begin{pmatrix} \nu_1(\Delta A) \\ \omega\nu_2(\Delta b) \end{pmatrix}\right\| : (A + \Delta A)\tilde{x} = b + \Delta b\right\}, \quad \omega > 0, \qquad (3.1)$$

where $\nu_1(\cdot)$ is any norm on $\mathcal{R}^{n \times n}$, $\nu_2(\cdot)$ is any norm on $\mathcal{R}^n$, and $\|\cdot\|$ is any norm on $\mathcal{R}^2$.

We first make some assumptions on the norms. Assume that $\nu_2(\cdot)$ is any norm on $\mathcal{R}^n$, $\nu_2^D(\cdot)$ is the dual norm of $\nu_2(\cdot)$ on $\mathcal{R}^n$, and $\nu_1(\cdot)$ is any norm on $\mathcal{R}^{n \times n}$ satisfying

$$\nu_1(xy^T) = \nu_2(x)\nu_2^D(y) \quad \forall x, y \in \mathcal{R}^n, \qquad (3.2)$$

and

$$\nu_1(Mx) \leq \nu_1(M)\nu_2(x) \quad \forall M \in \mathcal{R}^{n \times n}, x \in \mathcal{R}^n. \qquad (3.3)$$

We now note two examples of the norms $\nu_1(\cdot)$ on $\mathcal{R}^{n \times n}$ and $\nu_2(\cdot)$ on $\mathcal{R}^n$ satisfying (3.2)–(3.3). Example 1. $\nu_2(\cdot)$ is any norm on $\mathcal{R}^n$, and $\nu_1(\cdot)$ is its subordinate matrix norm on $\mathcal{R}^{n \times n}$. Example 2. $\nu_2(\cdot) = \|\cdot\|_2$ on $\mathcal{R}^n$, and $\nu_1(\cdot)$ is any orthogonally invariant norm on $\mathcal{R}^{n \times n}$.

The following result gives an explicit expression of the OBPB $\beta^{(\omega)}(\tilde{x})$.

Theorem 3.1 (Sun [18]) *Let $\tilde{x}$ be a computed solution to $Ax = b$, and let $\beta^{(\omega)}(\tilde{x})$ be the OBPB defined by (3.1), in which $\nu_1(\cdot)$ and $\nu_2(\cdot)$ are the norms satisfying (3.2)–(3.3), and $\|\cdot\|$ is any norm on $\mathcal{R}^2$. Then*

$$\beta^{(\omega)}(\tilde{x}) = \frac{\nu_2(\tilde{r})}{\left\|\begin{pmatrix} \nu_2(\tilde{x}) \\ 1/\omega \end{pmatrix}\right\|^D}, \qquad (3.4)$$

where $\hat{r} = b - A\tilde{x}$, and $\|\cdot\|^D$ is the dual norm of $\|\cdot\|$ on $\mathcal{R}^2$.

The result (3.4) implies the formula (1.3) for the OBPB $\eta_\infty(\tilde{x})$ defined by (1.2). In fact, if the norm $\nu_2(\cdot)$ of theorem 3.1 is any norm $\|\cdot\|$ on $\mathcal{R}^n$, $\nu_1(\cdot)$ is its subordinate norm $\|\cdot\|$ on $\mathcal{R}^{n\times n}$, and the norm $\|\cdot\|$ of Theorem 3.1 is the norm $\|\cdot\|_\infty$ on $\mathcal{R}^2$, then

$$\frac{1}{\|E\|}\beta^{(\|E\|/\|f\|)}(\tilde{x}) = \eta_\infty(\tilde{x}),$$

and the formula (3.4) reduces to the formula (1.3).

By (2.3), the OBPB $\eta^{(\theta)}(\tilde{x})$ of the linear system $Ax = b$ is defined by

$$\eta^{(\theta)}(\tilde{x}) = \min\{\nu(\Delta A, \theta\Delta b) \; : \; (A + \Delta A)\tilde{x} = b + \Delta b\}, \quad \theta > 0, \qquad (3.5)$$

where $\nu(\cdot)$ is any norm on $\mathcal{R}^{n\times(n+1)}$.

Assume that $\nu(\cdot)$ is any norm on $\mathcal{R}^{n\times(n+1)}$ satisfying the following properties: There is a norm $\mu(\cdot)$ on $\mathcal{R}^{n+1}$ such that the restriction of $\mu(\cdot)$ on $\mathcal{R}^n$ is also a norm (we write it as the same $\mu(\cdot)$), and

$$\mu(Ly) \leq \nu(L)\mu(y) \quad \forall L \in \mathcal{R}^{n\times(n+1)}, y \in \mathcal{R}^{n+1}, \qquad (3.6)$$

and

$$\nu(xy^T) = \mu(x)\mu^D(y) \quad \forall x \in \mathcal{R}^n, y \in \mathcal{R}^{n+1}, \qquad (3.7)$$

where $\mu^D(\cdot)$ is the dual norm of $\mu(\cdot)$ on $\mathcal{R}^{n+1}$.

The following result gives an explicit expression of the OBPB $\eta^{(\theta)}(\tilde{x})$.

Theorem 3.2 (Sun [18]). *Let $\tilde{x}$ be a computed solution to $Ax = b$, and let $\eta^{(\theta)}(\tilde{x})$ be the OBPB defined by (3.5). Moreover, let the norm $\nu(\cdot)$ in (3.5) and an associated norm $\mu(\cdot)$ satisfy (3.6)–(3.7). Then*

$$\eta^{(\theta)}(\tilde{x}) = \frac{\mu(\hat{r})}{\mu\begin{pmatrix} \tilde{x} \\ -1/\theta \end{pmatrix}}, \qquad (3.8)$$

where $\hat{r} = b - A\tilde{x}$.

If $\mu(\cdot) = \|\cdot\|_2$ on $\mathcal{R}^{n+1}$, and $\nu(\cdot) = \|\cdot\|_F$ on $\mathcal{R}^{n\times(n+1)}$ in Theorem 3.2, then the formula (3.8) can be written as

$$\eta^{(\theta)}(\tilde{x}) = \frac{\theta\|\tilde{x}\|_2}{\sqrt{1 + \theta^2\|\tilde{x}\|_2^2}} \cdot \frac{\|\hat{r}\|_2}{\|\tilde{x}\|_2}. \qquad (3.9)$$

3.2   Structured Linear Systems

It is worth pointing out that if the coefficient matrix A has some special structure, and we are interested in the requirement that the perturbed matrices $A + \Delta A$ have this structure too, then the problem of finding an explicit expression of the corresponding OBPB (3.5) (or (3.1)) maybe becomes very complicated. This issue of restricted backward perturbations for structured systems is considered by Bunch [2], Bunch, Demmel, and Van Loan [3], Higham [6], Higham and Higham [9], Varah [22], and Sun [18], [20]. The paper [9] provides a framework for dealing with structured backward perturbations. [22] gives estimates of an appropriate OBPB for the Toeplitz systems. [20] describes a technique for obtaining upper and lower bounds for some appropriate OBPBs for the Vandermonde systems.

We now consider the symmetric linear system $Ax = b$ (where $A = A^T$) which is much easier to handle than the Toeplitz systems and Vandermonde systems.

Let $\tilde{x}$ be a computed non-zero solution. Define the OBPB $\eta_S^{(\theta)}(\tilde{x})$ for the symmetric linear system by symmetric backward perturbations, that is, $\eta_S^{(\theta)}(\tilde{x})$ is defined by

$$\eta_S^{(\theta)}(\tilde{x}) = \min\{\|(\Delta A, \theta \Delta b)\|_F \; : \; (A + \Delta A)\tilde{x} = b + \Delta b, \; \Delta A^T = \Delta A\}, \quad \theta > 0. \tag{3.10}$$

The following result gives an explicit expression of $\eta_S^{(\theta)}(\tilde{x})$.

Theorem 3.3 (Sun [18]). *Let* $\eta_S^{(\theta)}(\tilde{x})$ *be the OBPB of the symmetric linear system* $Ax = b$ *defined by (3.10). Then*

$$\eta_S^{(\theta)}(\tilde{x}) = \frac{\theta\|\tilde{x}\|_2}{\sqrt{1 + \theta^2\|\tilde{x}\|_2^2}} \cdot \sqrt{\left(\frac{\|\hat{r}\|_2}{\|\tilde{x}\|_2}\right)^2 + \frac{\theta^2[(\|\tilde{x}\|_2\|\hat{r}\|_2)^2 - (\tilde{x}^T\hat{r})^2]}{\|\tilde{x}\|_2^2(2 + \theta^2\|\tilde{x}\|_2^2)}}, \tag{3.11}$$

where $\hat{r} = b - A\tilde{x}$.

If the symmetric backward perturbations ΔA in (3.10) are replaced by general backward perturbations, then the corresponding OBPB is the OBPB $\eta^{(\theta)}(\tilde{x})$ defined by (3.5) with $\nu(\cdot) = \|\cdot\|_F$, and an explicit expression of $\eta^{(\theta)}(\tilde{x})$ is given by (3.9). Comparing (3.9) with (3.11), and using the inequalities

$$0 \le \frac{\theta^2[\|\tilde{x}\|_2^2\|\hat{r}\|_2^2 - (\tilde{x}^T\hat{r})^2]}{2 + \theta^2\|\tilde{x}\|_2^2} \le \|\hat{r}\|_2^2,$$

we get

$$\eta^{(\theta)}(\tilde{x}) \le \eta_S^{(\theta)}(\tilde{x}) \le \sqrt{2}\eta^{(\theta)}(\tilde{x}) \quad \forall \theta > 0,$$

which shows that enforcing symmetry of ΔA when A is symmetric does increase $\eta^{(\theta)}(\tilde{x})$ by at most a factor $\sqrt{2}$ for any $\theta > 0$. This is an extension of a result due to Higham [6], and Bunch, Demmel and Van Loan [3].

It is worth pointing out that if $Ax = b$ is a Toeplitz system or a Vandermonde system, and $\tilde{x}$ is a computed solution to the system, then the OBPB defined by using structured backward perturbations can be much larger than the OBPB defined by using general backward perturbations [22], [20].

4   Underdetermined Systems

A linear system $Ax = b$ with $A \in \mathcal{R}^{m \times n}$ and $b \in \mathcal{R}^m$ is called underdetermined whenever $m < n$. It is known that such a system either has no solution or has an infinity of solutions [5].

Suppose that the underdetermined system $Ax = b$ has solutions. In §4.1 we consider the problem of finding any solution to the system. This problem can be called the UD problem. In §4.2 we consider the problem of finding the minimum 2-norm solution to the system. This problem can be called the MUD problem. A subject in matrix computations is to study numerical methods for computing the minimum 2-norm solution to the underdetermined system [5], [8].

4.1   The UD Problem

Let $\tilde{x} \in \mathcal{R}^n$ be a computed solution to the underdetermined system $Ax = b$. After the manner of (1.2), (3.1) and (3.5) we can define the OBPBs $\eta_{\text{UD}(\infty)}(\tilde{x})$, $\beta_{\text{UD}}^{(\omega)}(\tilde{x})$ and $\eta_{\text{UD}}^{(\theta)}(\tilde{x})$. Note that the OBPB $\eta_{\text{UD}(\infty)}(\tilde{x})$ has the same expression as (1.3), the OBPB $\beta_{\text{UD}}^{(\omega)}(\tilde{x})$ has the same expression as (3.4), and $\eta_{\text{UD}}^{(\theta)}(\tilde{x})$ has the same expression as (3.8). Particularly, if we define the OBPB $\eta_{\text{UD}}(\tilde{x})$ by

$$\eta_{\text{UD}}(\tilde{x}) = \min\{\|\Delta A\|_F : (A + \Delta A)\tilde{x} = b\}, \tag{4.1}$$

then

$$\eta_{\text{UD}}(\tilde{x}) = \frac{\|\hat{r}\|_2}{\|\tilde{x}\|_2} \quad \text{with} \quad \hat{r} = b - A\tilde{x}. \tag{4.2}$$

The formula (4.2) can be derived by taking $\nu(\cdot) = \|\cdot\|_F$ and $\theta \to \infty$ forced $\Delta b = 0$ in (3.5) and (3.8).

4.2   The MUD Problem

In this subsection we consider the backward perturbations ΔA and Δb that $\tilde{x}$ is the minimum 2-norm solution to $(A + \Delta A)x = b + \Delta b$, and give explicit expressions of some corresponding OBPBs.

We first consider a special case where only the coefficient matrix A is perturbed. Let $\mathcal{D}_{\text{MUD}}$ be the subset of $\mathcal{R}^{m \times n}$ defined by

$$\mathcal{D}_{\text{MUD}} = \{\Delta A : \tilde{x} \text{ is the minimum } 2-\text{norm solution to } (A + \Delta A)x = b\}. \tag{4.3}$$

Since the set $\mathcal{D}_{\mathrm{MUD}}$ is not necessarily closed when $m > 1$, we define the corresponding OBPB $\eta_{\mathrm{MUD}}(\tilde{x})$ by using the infimum of $\|\Delta A\|_F$ on $\mathcal{D}_{\mathrm{MUD}}$, that is,

$$\eta_{\mathrm{MUD}}(\tilde{x}) = \inf_{\Delta A \in \mathcal{D}_{\mathrm{MUD}}} \|\Delta A\|_F. \tag{4.4}$$

Moreover, observe the following facts [21]:

b	$\tilde{x}$	$\mathcal{D}_{\mathrm{MUD}}$	$\eta_{\mathrm{MUD}}(\tilde{x})$
0	0	$\mathcal{R}^{m \times n}$	0
0	non-zero	$\emptyset$	—
non-zero	0	$\emptyset$	—

Hence, for deriving an explicit expression of $\eta_{\mathrm{MUD}}(\tilde{x})$ we assume $b \neq 0$ and $\tilde{x} \neq 0$.

Let $\overline{\mathcal{D}}_{\mathrm{MUD}}$ denote the closure of the set $\mathcal{D}_{\mathrm{MUD}}$. Then by (4.4)

$$\eta_{\mathrm{MUD}}(\tilde{x}) = \min_{\Delta A \in \overline{\mathcal{D}}_{\mathrm{MUD}}} \|\Delta A\|_F.$$

In [21] we have proved that $\overline{\mathcal{D}}_{\mathrm{MUD}}$ can be expressed by

$$\overline{\mathcal{D}}_{\mathrm{MUD}} = \left\{ \Delta A = \begin{pmatrix} \beta_1/\tilde{\xi}_1 & z_1^T \\ 0 & Z_2 \end{pmatrix} - A \; : \; \begin{array}{l} z_1 \in \mathcal{R}^{n-1}, Z_2 \in \mathcal{R}^{(m-1)\times(n-1)}, \\ \mathrm{rank}(z_1, Z_2^T) \leq m - 1 \end{array} \right\}.$$

This result is used to derive the following explicit expression of $\eta_{\mathrm{MUD}}(\tilde{x})$.

Theorem 4.1 (Sun and Sun [21]). *Let $A \in \mathcal{R}^{m \times n}$ $(m < n)$, non-zero $b \in \mathcal{R}^m$, and non-zero $\tilde{x} \in \mathcal{R}^n$ be given, and let $\hat{r} = b - A\tilde{x}$. Then the OBPB $\eta_{\mathrm{MUD}}(\tilde{x})$ defined by (4.3)–(4.4) can be expressed by*

$$\eta_{\mathrm{MUD}}(\tilde{x}) = \sqrt{\frac{\|\hat{r}\|_2^2}{\|\tilde{x}\|_2^2} + \sigma_{\min}^2((I_n - \tilde{x}\tilde{x}^\dagger)A^T)}. \tag{4.5}$$

Note that Chris Paige develops an alternative proof of Theorem 4.1 based on constructive transformations and reductions.

Further, we define the OBPB $\eta_{\mathrm{MUD}}^{(\theta)}(\tilde{x})$ by

$$\eta_{\mathrm{MUD}}^{(\theta)}(\tilde{x}) = \inf \left\{ \|(\Delta A, \theta \Delta b)\|_F \; : \; \begin{array}{l} \tilde{x} \text{ is the minimum 2-norm solution} \\ \text{to } (A + \Delta A)x = b + \Delta b \end{array} \right\}, \quad \theta > 0. \tag{4.6}$$

By applying Theorem 4.1, we have the following result.

Theorem 4.2 (Sun and Sun [21]). *Let $A \in \mathcal{R}^{m \times n}$ $(m < n)$, $b \in \mathcal{R}^m$, and non-zero $\tilde{x} \in \mathcal{R}^n$ be given, and let $\hat{r} = b - A\tilde{x}$. If $b + A\tilde{x}/(\theta^2\|\tilde{x}\|_2^2) \neq 0$, then*

$$\eta_{\mathrm{MUD}}^{(\theta)}(\tilde{x}) = \sqrt{\frac{\theta^2\|\tilde{x}\|_2^2}{1 + \theta^2\|\tilde{x}\|_2^2} \cdot \frac{\|\hat{r}\|_2^2}{\|\tilde{x}\|_2^2} + \sigma_{\min}^2((I_n - \tilde{x}\tilde{x}^\dagger)A^T)}. \tag{4.7}$$

It is worth pointing out the following facts:

(1) Taking $\theta \to \infty$ forces $\Delta b = 0$ in (4.6) and (4.7), we get the expression (4.5) of the OBPB $\eta_{\text{MUD}}(\tilde{x})$ defined by (4.3)–(4.4).

(2) Theorems 4.1 and 4.2 remain valid when $A \in \mathcal{R}^{m \times n}$ with $m \geq n$ [21]. Note that in such a case $\sigma_{\min}((I_n - \tilde{x}\tilde{x}^\dagger)A^T) = 0$, and

$$\eta_{\text{MUD}}(\tilde{x}) = \frac{\|\hat{r}\|_2}{\|\tilde{x}\|_2}, \qquad \eta_{\text{MUD}}^{(\theta)}(\tilde{x}) = \frac{\theta\|\tilde{x}\|_2}{\sqrt{1 + \theta^2\|\tilde{x}\|_2^2}} \cdot \frac{\|\hat{r}\|_2}{\|\tilde{x}\|_2}.$$

(3) Let $A \in \mathcal{R}^{m \times n}$ $(m < n)$, non-zero $b \in \mathcal{R}^m$, and non-zero $\tilde{x} \in \mathcal{R}^n$ be given, and let $\eta_{\text{UD}}(\tilde{x})$ and $\eta_{\text{MUD}}(\tilde{x})$ be the OBPBs defined by (4.1) and (4.4), respectively. Then by the definitions (or from the expressions (4.2) and (4.5)), $\eta_{\text{UD}}(\tilde{x}) \leq \eta_{\text{MUD}}(\tilde{x})$. Note that sometimes $\eta_{\text{MUD}}(\tilde{x})$ is much larger than $\eta_{\text{UD}}(\tilde{x})$, and there are examples where $\eta_{\text{MUD}}(\tilde{x})/\eta_{\text{UD}}(\tilde{x})$ can be arbitrarily large [21].

Furthermore, we define the OBPB $\eta_{\text{MUD}(\infty)}(\tilde{x})$ by

$$\eta_{\text{MUD}(\infty)}(\tilde{x}) = \inf \left\{ \epsilon : \begin{array}{l} \tilde{x} \text{ is the minimum 2–norm solution} \\ \text{to } (A + \Delta A)x = b + \Delta b, \text{ and} \\ \|\Delta A\|_F \leq \epsilon\alpha, \ \|\Delta b\|_2 \leq \epsilon\beta \end{array} \right\}, \qquad (4.8)$$

where α, β are any positive scalars (for instance, $\alpha = \|A\|_F$ and $\beta = \|b\|_2$). The definition (4.8) is a slight modification of that in [8, p.423, Research Problem 20.2]. An explicit expression of the OBPB $\eta_{\text{MUD}(\infty)}(\tilde{x})$ can be found in [18]. Note that the expression of $\eta_{\text{MUD}(\infty)}(\tilde{x})$ is much more complicated than (4.7). In practice, the formula (4.7) of the OBPB $\eta_{\text{MUD}}^{(\theta)}(\tilde{x})$ is sufficient for testing practical algorithms.

5   Linear Least Squares Problems

Let $A \in \mathcal{R}^{m \times n}$ and $b \in \mathcal{R}^m$ be given. In §5.1 we consider the linear least squares (LS) problem

$$\min_{x \in \mathcal{R}^n} \|b - Ax\|_2 \quad \text{with} \quad A \in \mathcal{R}^{m \times n} \text{ and } b \in \mathcal{R}^m. \qquad (5.1)$$

Observe the following facts [1], [5]: (i) If $\text{rank}(A) = n$, then there is only one solution to the LS problem (5.1) and so it must have minimum 2-norm. (ii) If $\text{rank}(A) < n$, then (5.1) has an infinity of solutions, that is, the set

$$\mathcal{X} = \{\hat{x} \in \mathcal{R}^n : \|b - A\hat{x}\|_2 = \min_{x \in \mathcal{R}^n} \|b - Ax\|_2\}$$

has an infinity of elements. Since $\mathcal{X}$ is closed convex, the set $\mathcal{X}$ has a unique element having minimum 2-norm. A subject in matrix computations is to study numerical methods for finding the unique minimum 2-norm solution to the LS problem (1.1) [1], [5]. This problem can be called the MLS problem.

5.1   The LS Problem

Let $A \in \mathcal{R}^{m \times n}$, $b \in \mathcal{R}^m$, and let $\tilde{x} \in \mathcal{R}^n$ be a computed solution to the LS problem (5.1). In this section we consider the backward perturbations ΔA and Δb that $\tilde{x}$ is a solution to the LS problem

$$\min \|(b + \Delta b - (A + \Delta A)x\|_2,$$

and give explicit expressions of some corresponding OBPBs.

First of all, we consider a special case where only the coefficient matrix A is perturbed. In 1977, Stewart [13] first discovers two perturbations ΔA of A such that the vector $\tilde{x}$ exactly minimizes $\|b - (A + \Delta A)x\|_2$, so the two backward perturbations are candidates for being of minimal norm. The problem of finding the minimal backward perturbation is also discussed by Higham [7]. In 1991, Sun [15] finds a backward perturbation ΔA_* and shows that the scalar $\|\Delta A_*\|_F$ (expressed below by (5.4)) is an improved backward perturbation bound; Waldén, Karlson, and Sun [23] prove that $\|\Delta A_*\|_F$ is the OBPB in the meaning of the Frobenius norm.

Let $\mathcal{D}_{\mathrm{LS}}$ be the subset of $\mathcal{R}^{m \times n}$ defined by

$$\mathcal{D}_{\mathrm{LS}} = \left\{ \Delta A \in \mathcal{R}^{m \times n} : \|b - (A + \Delta A)\tilde{x}\|_2 = \min_{x \in \mathcal{R}^n} \|b - (A + \Delta A)x\|_2 \right\}, \quad (5.2)$$

and let $\eta_{\mathrm{LS}}(\tilde{x})$ be the OBPB defined by

$$\eta_{\mathrm{LS}}(\tilde{x}) = \min_{\Delta A \in \mathcal{D}_{\mathrm{LS}}} \|\Delta A\|_F. \quad (5.3)$$

Observe the following facts [18]:

b	$\tilde{x}$	$\mathcal{D}_{\mathrm{LS}}$	$\eta_{\mathrm{LS}}(\tilde{x})$
0	0	$\mathcal{R}^{m \times n}$	0
non-zero	0	$\{\Delta A = -P_b A + P_b^\perp Z \ : \ Z \in \mathcal{R}^{m \times n}\}$	$\|A^T b\|_2 / \|b\|_2$

Hence, we shall assume $\tilde{x} \neq 0$ in the rest of this subsection.

The following result gives an explicit expression of $\eta_{\mathrm{LS}}(\tilde{x})$.

Theorem 5.1 (Waldén, Karlson, and Sun [23]). *Let $A \in \mathcal{R}^{m \times n}, b \in \mathcal{R}^m, \tilde{x} \in \mathcal{R}^n$ with $\tilde{x} \neq 0$, and let $\hat{r} = b - A\tilde{x}$. Moreover, let*

$$\lambda_* = \lambda_{\min} \left(AA^T - \frac{\hat{r}\hat{r}^T}{\|\tilde{x}\|_2^2} \right).$$

Then the OBPB $\eta_{LS}(\tilde{x})$ can be expressed by

$$\eta_{LS}(\tilde{x}) = \begin{cases} \dfrac{\|\hat{r}\|_2}{\|\tilde{x}\|_2} & \text{if } \lambda_* \geq 0, \\[3mm] \left[\left(\dfrac{\|\hat{r}\|_2}{\|\tilde{x}\|_2}\right)^2 + \lambda_*\right]^{1/2} & \text{if } \lambda_* < 0. \end{cases} \tag{5.4}$$

The key step for deriving the formula (5.4) is to give the following expression of the set $\mathcal{D}_{LS}$ [23]:

$$\mathcal{D}_{LS} = \{\hat{r}\tilde{x}^\dagger - vv^T(A + \hat{r}\tilde{x}^\dagger) + (I - vv^\dagger)Z(I - \tilde{x}\tilde{x}^\dagger) : v \in \mathcal{R}^m, Z \in \mathcal{R}^{m\times n}\}.$$

As Nick Higham has pointed out that the formulae given in Theorem 5.1 are unsuitable for computation because they can suffer from catastrophic cancellation when $\lambda_* < 0$. By Nick Higham's suggestion, the following expression of $\eta_{LS}(\tilde{x})$, which permits a stable numerical computation, should be used:

$$\eta_{LS}(\tilde{x}) = \min\left\{\frac{\|\hat{r}\|_2}{\|\tilde{x}\|_2}, \sigma_{\min}([A, R])\right\} \quad \text{with} \quad R = \frac{\|\hat{r}\|_2}{\|\tilde{x}\|_2}(I - \hat{r}\hat{r}^\dagger). \tag{5.5}$$

Further, we define the OBPB $\eta_{LS}^{(\theta)}(\tilde{x})$ by

$$\eta_{LS}^{(\theta)}(\tilde{x}) = \min\left\{\|(\Delta A, \theta\Delta b)\|_F : \begin{array}{l} \|b + \Delta b - (A + \Delta A)\tilde{x}\|_2 \\ = \min_{x\in\mathcal{R}^n}\|b + \Delta b - (A + \Delta A)x\|_2 \end{array}\right\}, \quad \theta > 0. \tag{5.6}$$

By applying Theorem 5.1, we have the following result.

Theorem 5.2 (Waldén, Karlson, and Sun [23]). *Let $A, b, \tilde{x}$ and $\hat{r}$ be as in Theorem 5.1, and let $\eta_{LS}^{(\theta)}(\tilde{x})$ be the OBPB defined by (5.6). Then*

$$\eta_{LS}^{(\theta)}(\tilde{x}) = \begin{cases} \dfrac{\|\hat{r}\|_2}{\|\tilde{x}\|_2}\sqrt{\tau} & \text{if } \lambda_\tau \geq 0, \\[3mm] \left[\left(\dfrac{\|\hat{r}\|_2}{\|\tilde{x}\|_2}\right)^2\sqrt{\tau} + \lambda_\tau\right]^{1/2} & \text{if } \lambda_\tau < 0, \end{cases} \tag{5.7}$$

where

$$\lambda_\tau = \lambda_{\min}\left(AA^T - \tau\frac{\hat{r}\hat{r}^T}{\|\tilde{x}\|_2^2}\right), \quad \tau = \frac{\theta^2\|\tilde{x}\|_2^2}{1 + \theta^2\|\tilde{x}\|_2^2}.$$

By Nick Higham's suggestion, the following expression of $\eta_{LS}^{(\theta)}(\tilde{x})$, which permits a stable numerical computation, should be used:

$$\eta_{LS}^{(\theta)}(\tilde{x}) = \min\left\{\frac{\|\hat{r}\|_2}{\|\tilde{x}\|_2}, \sigma_{\min}([A, R_\tau])\right\} \quad \text{with} \quad R_\tau = \sqrt{\tau}\frac{\|\hat{r}\|_2}{\|\tilde{x}\|_2}(I - \hat{r}\hat{r}^\dagger). \tag{5.8}$$

The formulae (5.4) and (5.7) have been extended to the multiple right-hand side LS problem by [17].

5.2 The MLS Problem

Let $\mathcal{D}_{LS}$ be the set defined by (5.2). We now define the subset $\mathcal{D}_{MLS}$ of $\mathcal{D}_{LS}$ by

$$\mathcal{D}_{MLS} = \{\Delta A : \tilde{x} \text{ is the minimum } 2-\text{norm solution to } \min_{x \in \mathcal{R}^n} \|b - (A + \Delta A)x\|_2\}.$$
(5.9)

Since the set $\mathcal{D}_{MLS}$ is, in most cases, a proper and not closed subset of $\mathcal{D}_{LS}$, we define the OBPB $\eta_{MLS}(\tilde{x})$ by using the infimum of $\|\Delta A\|_F$ on $\mathcal{D}_{MLS}$, that is,

$$\eta_{MLS}(\tilde{x}) = \inf_{\Delta A \in \mathcal{D}_{MLS}} \|\Delta A\|_F.$$
(5.10)

Moreover, observe the following facts [18]:

b	$\tilde{x}$	$\mathcal{D}_{MLS}$	$\eta_{MLS}(\tilde{x})$
0	0	$\mathcal{R}^{m \times n}$	0
0	non-zero	$\emptyset$	—
non-zero	0	$\{\Delta A = -P_b A + P_b^{\perp} Z \ : \ Z \in \mathcal{R}^{m \times n}\}$	$\|A^T b\|_2 / \|b\|_2$

Hence, for studying the OBPB $\eta_{MLS}(\tilde{x})$ we assume $b \neq 0$ and $\tilde{x} \neq 0$.

The following result gives an explicit expression of the OBPB $\eta_{MLS}(\tilde{x})$.

Theorem 5.3 (Sun [19]). *Let $A \in \mathcal{R}^{m \times n}$ $(m > n)$, non-zero $b \in \mathcal{R}^m$ and non-zero $\tilde{x} \in \mathcal{R}^n$ be given, and let $\mathcal{D}_{LS}$, $\eta_{LS}(\tilde{x})$, $\mathcal{D}_{MLS}$, $\eta_{MLS}(\tilde{x})$ be defined by (5.2), (5.3), (5.9), (5.10), respectively. Then the set $\mathcal{D}_{LS}$ is the closure of $\mathcal{D}_{MLS}$, and $\eta_{MLS}(\tilde{x}) = \eta_{LS}(\tilde{x})$, where $\eta_{LS}(\tilde{x})$ has the expression (5.5).*

Further, we have the following

Theorem 5.4 (Sun [19]). *Let $A, b, \tilde{x}$ and $\tilde{r}$ be as in Theorem 5.3, and let the OBPB $\eta_{MLS}^{(\theta)}(\tilde{x})$ be defined by*

$$\eta_{MLS}^{(\theta)}(\tilde{x}) = \inf\left\{\|(\Delta A, \theta \Delta b)\|_F \ : \ \begin{array}{l} \tilde{x} \text{ is the minimum } 2-\text{norm solution} \\ \text{to } \min_x \|(b + \Delta b) - (A + \Delta A)x\|_2 \end{array}\right\}, \quad \theta > 0.$$

Then $\eta_{MLS}^{(\theta)}(\tilde{x}) = \eta_{LS}^{(\theta)}(\tilde{x})$, where $\eta_{LS}^{(\theta)}(\tilde{x})$ is the OBPB defined by (5.6), and it has the expression (5.8).

6 Conclusions

We have computable expressions of the OBPBs for general linear systems, underdetermined systems (Problems UD and MUD), and linear least squares problems (Problems LS and MLS). The expressions provide a convenient way of testing the stability of practical algorithms. Several questions merit further investigation.

For example: (1) How to find explicit expressions or theoretical bounds on the OBPBs for some structured linear system? (2) How to find explicit expressions for componentwise and/or row-wise OBPBs for underdetermined systems and linear least squares problems?

References

[1] Å. Björck, *Numerical Methods for Least Squares Problems*, Society for Industrial and Applied Mathematics, Philadelphia, PA, USA, 1996.

[2] J. R. Bunch, The weak and strong stability of algorithms in numerical linear algebra, *Linear Algebra Appl.*, 88/89(1987), 49–66.

[3] J. R. Bunch, J. W. Demmel, and C. F. Van Loan, The strong stability of algorithms for solving symmetric linear systems, *SIAM J. Matrix Anal. Appl.*, 10(1989), 494–499.

[4] F. Chaitin-Chatelin and V. Frayssé, *Lectures on Finite Precision Computations*, Society for Industrial and Applied Mathematics, Philadelphia, PA, USA, 1996.

[5] G. H. Golub and C. F. Van Loan, *Matrix Computations*, Third Edition, Johns Hopkins University Press, Baltimore and London, 1996.

[6] N. J. Higham, Matrix nearness problems and applications. In *Applications of Matrix Theory*, M. J. C. Gover and S. Barnett, editors, Oxford University Press, Oxford, UK, 1989, pages 1–27.

[7] N. J. Higham, Computing error bounds for regression problems, in P. J. Brown and W. A. Fuller, editors, *Statistical Analysis of Measurement Error Models and Applications, Contemporary Mathematics 112*, pp. 195–208, American Mathematical Society, Providence, RI, 1990.

[8] N. J. Higham, *Accuracy and Stability of Numerical Algorithms*, Society for Industrial and Applied Mathematics, Philadelphia, PA, USA, 1996.

[9] D. J. Higham and N. J. Higham, Backward error and condition of structured linear systems, *SIAM J. Matrix Anal. Appl.*, 13(1992), 162–175.

[10] D. J. Higham and N. J. Higham, Componentwise perturbation theory for linear systems with multiple right-hand sides, *Linear Algebra Appl.*, 174(1992), 111–129.

[11] Z. V. Kovarik, Compatibility of approximate solution of inaccurate linear equations, *Linear Algebra Appl.*, 15(1976), 217–225.

[12] J. L. Rigal and J. Gaches, On the compatibility of a given solution with the data of a linear system, *J. Assoc. Comput. Mach.*, 14(1967), 543–548.

[13] G. W. Stewart, Research, development, and LINPACK, in *Mathematical Software III*, J. R. Rice, ed., Academic Press, New York, 1977, 1–14.

[14] G. W. Stewart and J. -G. Sun, *Matrix Perturbation Theory*, Academic Press, New York, 1990.

[15] J. -G. Sun, An improved backward perturbation bound for the linear least squares problem (manuscript), 1991. In Report UMINF 96.15, ISSN-0348-

0542, Department of Computing Science, Umeå University, 1996, pages 35–43.

[16] J. -G. Sun, Backward perturbation analysis of certain characteristic subspaces, *Numer. Math.*, 65(1993), 357–382.

[17] J. -G. Sun, Optimal backward perturbation bounds for the linear least-squares problem with multiple right-hand sides, *IMA J. Numer. Anal.*, 16 (1996), 1–11.

[18] J. -G. Sun, Optimal backward perturbation bounds for linear systems and linear least squares problems, UMINF 96.15, ISSN-0348-0542, Department of Computing Science, Umeå University, 1996.

[19] J. -G. Sun, On optimal backward perturbation bounds for the linear least squares problem, *BIT*, 37(1997), 179–188.

[20] J. -G. Sun, Optimal backward perturbation bounds for Vandermonde systems, UMINF 97.02, ISSN-0348-0542, Department of Computing Science, Umeå University, 1997.

[21] J. -G. Sun and Z. Sun, Optimal backward perturbation bounds for underdetermined systems, *SIAM J. Matrix Anal. Appl.*, 18(1997), 393–402.

[22] J. M. Varah, Backward error estimates for Toeplitz systems, *SIAM J. Matrix Anal. Appl.*, 15(1994), 408–417.

[23] B. Waldén, R. Karlson, and J. -G. Sun, Optimal backward perturbation bounds for the linear least squares problem, *Numer. Linear Algebra Appl.*, 2(1995), 271–286.

[24] J. H. Wilkinson, *Rounding Errors in Algebraic Processes*. Notes on Applied Science No. 32, Her Majesty's Stationery Office, London, 1963.

[25] J. H. Wilkinson, *The Algebraic Eigenvalue Problem*, Oxford University Press, Oxford, 1965.

Solution of Linear Systems Arising in Nonlinear Image Deblurring

C. R. Vogel[1]

Department of Mathematical Sciences, Montana State University, Bozeman, MT
59717-0240 USA. e-mail: vogel@math.montana.edu
www: http://www.math.montana.edu/~vogel

Abstract. This paper deals with the solution of large linear systems which arise when certain optimization methods are applied in image deblurring. An unconstrained penalized least squares minimization problem with total variation penalty is considered. Also addressed are several penalized least squares problems with nonnegativity constraints imposed on the solution. In each case, quasi-Newton techniques yield large structured linear systems. We investigate various preconditioning strategies to efficiently solve these linear systems.

1 Introduction

Consider the following model for discrete, noisy, blurred two-dimensional images

$$z_{ij} = \int\int_\Omega k(x_i - x', y_j - y')\, u(x', y')\, dx'\, dy' + \epsilon_{ij} \qquad (1)$$

$$\stackrel{\text{def}}{=} (Ku)(x_i, y_j) + \epsilon_{ij}, \qquad 1 \le i \le n_x,\ 1 \le j \le n_y,$$

where Ω is the unit square in $\mathbb{R}^2$. Our goal is to estimate the true image u, given the observed data z_{ij} and the point spread function $k(x, y)$.

To achieve this goal, assume a discretization $\mathbf{u} \in \mathbb{R}^n$, $n = n_x n_y$, of the solution u. We will address computational aspects of a fairly broad class of penalized least squares minimization problems

$$\min_{\mathbf{u} \in C} \frac{1}{2}\|K\mathbf{u} - \mathbf{z}\|^2 + \alpha J(\mathbf{u}), \quad \alpha > 0. \qquad (2)$$

Included in this class are the standard quadratic unconstrained minimization problems associated with Tikhonov regularization with the identity, in which case the penalty functional is (a discretization of) the squared L^2 norm

$$J(u) = \frac{1}{2}\int\int_\Omega u^2, \qquad (3)$$

and Tikhonov regularization with the squared Sobolev H^1 penalty

$$J(u) = \frac{1}{2}\int\int_\Omega |\nabla u|^2 = \frac{1}{2}\int\int_\Omega u_x^2 + u_y^2. \qquad (4)$$

In both these cases the solution can be obtained by solving a single linear system $H\mathbf{u} = -\mathbf{g}$, where $\mathbf{g}$ denotes the gradient of the cost functional in (2), and the Hessian matrix has the form

$$H = K^*K + \alpha L. \tag{5}$$

With standard (e.g., midpoint) discretization of integral operator in (1) and lexicographical ordering of the unknowns, the matrix K^*K is block Toeplitz with Toeplitz blocks, and in case (3), the matrix L is proportional to the identity I. The Hessian matrix can then be efficiently inverted using the preconditioned conjugate gradient (PCG) algorithm with circulant preconditioners [3,4,2]. In case (4), L is the discretization of the negative Laplacian with natural (i.e., no flux, or homogeneous Neumann) boundary conditions. Standard finite element or finite difference discretizations yield a matrix L with a block Toeplitz structure, and PCG with a circulant preconditioner can again be used to efficiently invert the Hessian.

Also included in the class (2) are discretizations of nonquadratic regularization functionals like total variation (TV)

$$J(u) = \int\int_\Omega |\nabla u| = \int\int_\Omega \sqrt{u_x^2 + u_y^2}. \tag{6}$$

When certain quasi-Newton methods like the "lagged diffusivity fixed point iteration" [8] are applied, one obtains a sequence of linear systems $Hs = -g$ where the Hessian again has the form (5). However, the matrix L now arises from the discretization of a *nonconstant coefficient* differential operator and the inversion of the Hessian is a more difficult matter. A class of preconditioners for such matrices based on the concept of "operator splitting" was introduced in [9]. In the next section, we will sketch the implementation of the operator splitting preconditioner.

We will also consider nonnegativity constraints in problem (2), in which case $\mathcal{C} = \{\mathbf{u} : u_i \geq 0, 1 \leq i \leq n\}$. Of the many numerical schemes to handle problems of this form, we will apply a projected Newton method due to Bertsekas [1, p. 76]. For our test problems, we have found it to be very robust and rapidly convergent. In place of the Hessian in (5), at each Newton iteration one must invert a matrix of the form

$$\overline{H} = D_A + D_I H D_I, \tag{7}$$

where D_I and D_A are diagonal matrices. Unfortunately, this can destroy the special structure of H. For instance, if H were Toeplitz, $\overline{H}$ will no long be Toeplitz. In section 3, we discuss the choice of the diagonal matrices D_I and D_A and the efficient inversion of matrices of the form $\overline{H}$.

In the final section, we present a comparison of the reconstructed images obtained by solving several unconstrained and constrained minimization problems of the form (2). We also present measures of the effectiveness of the preconditioners that were applied to the linear systems arising from the nonquadratic minimization problems.

2   Unconstrained TV Minimization

In this section, we consider the unconstrained version of problem (2) with the modified TV penalty functional

$$J(u) = \int_\Omega \sqrt{|\nabla u|^2 + \beta^2} = \int_\Omega \sqrt{u_x^2 + u_y^2 + \beta^2}. \tag{8}$$

When $\beta = 0$, this reduces to the usual TV functional (6). The "lagged diffusivity" fixed point iteration (see [8,10]) can be expressed in quasi-Newton form $u^{\nu+1} = u^\nu + s$, where s solves the linear system $H(u^\nu)s = -g(u^\nu)$ with gradient $g(u) = K^*Ku - K^*z + \alpha L(u)$ and approximate Hessian $H(u) = K^*K + \alpha L(u)$. Here $L(u)$ is the diffusion operator

$$L(u)v = -\nabla \cdot \left(\frac{1}{\sqrt{u_x^2 + u_y^2 + \beta^2}} \nabla u \right) \tag{9}$$

$$= -\frac{\partial}{\partial x} \left(\frac{u_x}{\sqrt{u_x^2 + u_y^2 + \beta^2}} \right) - \frac{\partial}{\partial y} \left(\frac{u_y}{\sqrt{u_x^2 + u_y^2 + \beta^2}} \right) \tag{10}$$

with associated natural (no flux, or zero normal derivative) boundary conditions. Standard discretization methods yield a matrix L which is positive semidefinite and block tridiagonal. Its spectrum is much like that of the discrete Laplacian, but the Toeplitz structure is lost. There may be dramatic variation in matrix entries along a given diagonal, so circulant preconditioners may not be effective.

It may be possible to simultaneously transform both matrices K^*K and L in (5) into sparse form, e.g., using a wavelet basis. We will not pursue this approach here. Instead, we will consider PCG with a preconditioner based on the "operator splitting" scheme introduced in [9]. This preconditioner takes the form

$$C = 1/\gamma(\tilde{K}^*\tilde{K} + \gamma I)^{1/2}(\gamma I + \alpha L)(\tilde{K}^*\tilde{K} + \gamma I)^{1/2}, \tag{11}$$

where γ is a positive parameter, and $\tilde{K}$ is a circulant approximation to the block Toeplitz matrix K. Note that C is symmetric and positive definite. The middle term,

$$\gamma I + \alpha L \tag{12}$$

is sparse with respect to standard bases. Matrices of this form also arise when TV methods are applied to image denoising, see [8]. Due to the large variation in the coefficients, the inversion of these matrices is not a trivial task. However, efficient inversion techniques are available, e.g., PCG with a specially designed multigrid preconditioner. The other term

$$(\tilde{K}^*\tilde{K} + \gamma I)^{1/2} \tag{13}$$

in (11) is a circulant matrix. This can be inverted using the fast Fourier transform (FFT). The construction of $\tilde{K}$ is described in [5] as well as in [10]. Implementation is similar to that of the preconditioners described in [2,4].

3   Nonnegativity Constraints

In this section we consider minimization of the quadratic functional

$$\frac{1}{2}\|K\mathbf{u} - \mathbf{z}\|^2 + \alpha\frac{1}{2}\mathbf{u}^T L\mathbf{u} \tag{14}$$

under the constraints $\mathbf{u}_i \geq 0$, $i = 1,\ldots,n$. The matrix L may depend on $\mathbf{u}$, e.g., when it arises in a quasi-Newton scheme for a TV-penalized least squares problem. It may also have a simpler form, e.g., $L = I$ in case (3) or L equal to the discretized negative Laplacian in case (4). The cost functional again has a Hessian of the form (5).

Of the many techniques for the constrained minimization of (14), we will consider the projected Newton method of Bertsekas [1, p. 76]. In the applications to which we have applied this method, the convergence has been quite rapid. Given a tolerance τ_k, define the active set

$$A = \{i : 0 \leq \mathbf{u}_i \leq \tau_k \text{ and } \mathbf{g}_i > 0\}. \tag{15}$$

The system to be solved at each Newton iteration has the form

$$\overline{H}\mathbf{s} = -\mathbf{g}, \tag{16}$$

where $\mathbf{g}$ denotes the unconstrained gradient, and $\overline{H}$ takes the form

$$\overline{H} = D_A + D_I H D_I. \tag{17}$$

The matrix D_I is the diagonal projection matrix associated with the inactive set, having diagonal entries

$$[D_I]_{ii} = \begin{cases} 0, \text{ if } i \in A, \\ 1, \text{ if } i \notin A. \end{cases} \tag{18}$$

Similarly, D_A is diagonal with diagonal entries equal to zero for indices not in the active set. For indices $i \in A$, D_A has positive diagonal entries. Their precise values seem to have little effect on the convergence of the projected Newton iteration, but they may have a significant impact on the convergence of iterative methods for the linear system (16). In case $L = I$, H has eigenvalues which cluster at α, cf., (5). The same holds for $D_I H D_I$ when D_I is nondegenerate. So that the eigenvalues of $\overline{H}$ also cluster at α, we choose

$$[D_A]_{ii} = \alpha, \quad \text{if } i \in A. \tag{19}$$

Nash and Sofer [7] propose a family of pseudo-inverse preconditioners for consistent linear systems having coefficient matrices of the form PHP, where P is an orthogonal projection matrix. The simplest member of this family is $C^\dagger = PH^{-1}P$. Motivated by this we take as an inverse preconditioner for $\overline{H}$ the matrix

$$\overline{C}^{-1} = D_A^\dagger + D_I C^{-1} D_I, \tag{20}$$

where C is a preconditioner for H. In case (3), $H = K^*K + \alpha I$, and one can take C to be a circulant preconditioner.

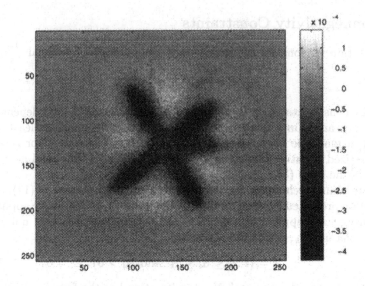

Fig. 1. Reconstructed image obtained using unconstrained penalized least squares with L^2 penalty, cf. (2)-(3) with $C = \mathbb{R}^n$.

4   Numerical Results

We begin this section with a comparison of reconstructed images obtained using various penalty terms, without and with nonnegativity constraints. The reconstructions are obtained from simulated satellite data generated from a model (1). For a detailed description of this data set, see [6]. This data set has also served to test several other image deblurring algorithms. See for example [4,5]. The true image is shown in Figure 5. Figures 1 and 2 show the reconstructions obtained using unconstrained least squares with squared L^2 penalty and modified TV penalty, respectively. The reconstruction in Figure 3 was obtained using nonnegatively constrained least squares with squared L^2 penalty. Figure 4 shows a reconstruction using nonnegatively constrained least squares with a modified TV penalty term. Note that "reverse grayscale" was used to display these images.

Next, we examine the performance of various linear equation solvers. The number of unknowns in each system is $n = n_x n_y = 256^2 \approx 6.5 \times 10^4$. Figure 6 compares the performance of CG without and with preconditioning for a linear system obtained from unconstrained TV minimization. The preconditioner is based on operator splitting, cf., (11). For details like the selection of parameters α and γ, see [10]. Note that the PCG convergence rate is about four times faster than the rate for CG without preconditioning. From the computational cost analysis in [10], which is based on an FFT count, PCG is twice as efficient as CG in this case.

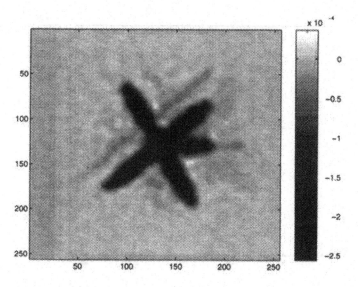

Fig. 2. Reconstructed image obtained using unconstrained penalized least squares with a modified TV penalty, cf., (2),(8) with $C = \mathbb{R}^n$.

Figure 7 compares the performance of CG vs. PCG for a linear system obtained when Bertsekas' projected Newton algorithm was used to minimize (14) with nonnegativity constraints and with $L = I$, corresponding to squared L^2 penalty. A variant of the Nash-Sofer inverse preconditioner, cf., (20), was used. In this case, preconditioning was much less effective. Slightly more than half as many PCG iterations were required to reduce the residual by three orders of magnitude. Unfortunately, the cost per iteration (again in terms of the number of FFT's) using preconditioning was twice the cost without preconditioning. Hence, the total cost with PCG was slightly *more* than the total computational cost without PCG. It should be noted that the CG iteration count of 39 required to meet the 10^{-3} relative error tolerance is much smaller than the system size $n \approx 65,000$.

References

1. D. P. Bertsekas, *Constrained Optimization and Lagrange Multiplier Methods*, Academic Press, New York, 1982.
2. R. H. Chan, T. F. Chan and C. K. Wong, *Cosine transform based preconditioners for total variation minimization problems in image processing*, UCLA Math. Dept. CAM Report 95-23 (1995).
3. R. H. Chan and M. K. Ng, *Conjugate Gradient Method for Toeplitz Systems*, SIAM Review, Vol. 38, No. 3, Sept. 1996, pp. 427-482.
4. R. H. Chan, M. K. Ng and R. J. Plemmons, *Generalization of Strang's preconditioner with applications to Toeplitz least squares problems*, Numerical Lin. Alg. and Applications , Vol. 3, 1996, pp. 45–64.

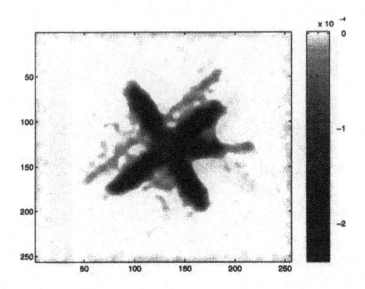

Fig. 3. Reconstructed image obtained using nonnegatively constrained penalized least squares with L^2 penalty, cf. (2)-(3) with $\mathcal{C} = \{\mathbf{u} : u_i \geq 0, 1 \leq i \leq n\}$.

5. M. Hanke and J. G. Nagy, *Restoration of Atmospherically Blurred Images by Symmetric Indefinite Conjugate Gradient Techniques*, Inverse Problems, Vol. 12 (1996), pp. 157-173.
6. J. G. Nagy, R. J. Plemmons, and T. C. Torgersen, *Iterative image restoration using approximate inverse preconditioners*, IEEE Trans. on Image Proc., Vol. 5, No. 7, July 1996, pp. 1151-1162.
7. S. G. Nash and A. Sofer, *Preconditioning reduced matrices*, SIAM J. Matrix Anal. Appl., Vol. 17 (1996), pp. 47–68.
8. C. R. Vogel and M. E. Oman, *Iterative Methods for Total Variation Denoising*, SIAM J. Sci. Comput., Vol. 17 (1996), pp. 227-238.
9. C. R. Vogel and M. E. Oman, *Fast numerical methods for total variation minimization in image reconstruction*, proceedings of SPIE 1995, San Diego, *Advanced Signal Processing Algorithms*, Vol. 2563, edited by F. T. Luk.
10. C. R. Vogel and M. E. Oman, *Fast, robust total variation–based reconstruction of noisy, blurred images*, submitted to IEEE Trans. on Image Proc., available at web site http://www.math.montana.edu/~vogel

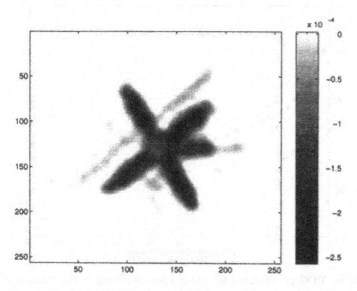

Fig. 4. Reconstructed image obtained using nonnegatively constrained penalized least squares with a modified TV penalty, cf. (2), (8) with $\mathcal{C} = \{\mathbf{u} : u_i \geq 0, 1 \leq i \leq n\}$.

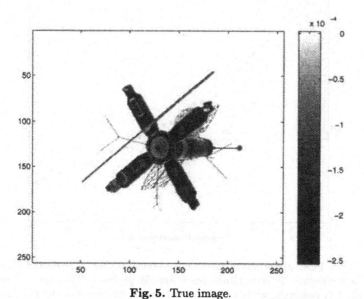

Fig. 5. True image.

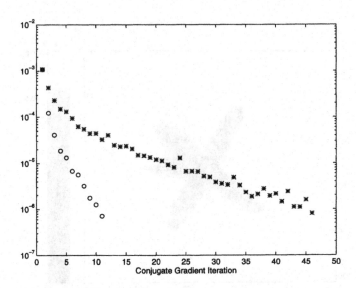

Fig. 6. CG vs. PCG performance for linear systems arising in the solution of an unconstrained least squares problem with TV penalty. The stars (∗) represent norm of the CG residuals, while the circles (o) denote the norm of the PCG residuals. The preconditioner is based on operator splitting.

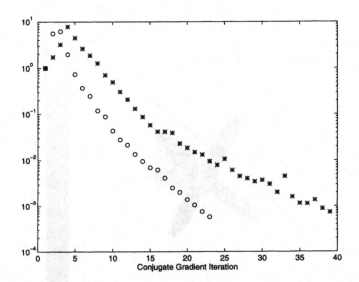

Fig. 7. CG vs. PCG performance for linear systems arising in projected Newton solution of a nonnegatively constrained least squares problem. The stars (∗) represent norm of the CG residuals, while the circles (o) denote the norm of the PCG residuals. A variant of the Nash-Sofer preconditioner was used.

Part II
Contributed Papers

Threshold Ordering for Preconditioning Nonsymmetric Problems

Michele Benzi[1], Hwajeong Choi[2], and Daniel B. Szyld[2]*

[1] CERFACS, 42 Ave. G. Coriolis, 31057 Toulouse Cedex, France (benzi@cerfacs.fr)
[2] Department of Mathematics, Temple University, Philadelphia, Pennsylvania 19122-6094, USA (choi@math.temple.edu, szyld@math.temple.edu)

Dedicated to Gene H. Golub on the occasion of his 65th birthday

Abstract. The effect of a threshold variant TPABLO of the permutation (and partitioning) algorithm PABLO on the performance of certain preconditionings is explored. The goal of these permutations is to produce matrices with dense diagonal blocks, and in the threshold variant, with large entries in the diagonal blocks. Experiments are reported using matrices arising from the discretization of elliptic partial differential equations. The iterative solvers used are GMRES, QMR, BiCGStab and CGNR. The preconditioners are different incomplete factorizations. It is shown that preprocessing the matrices with TPABLO has a positive effect on the overall performance, resulting in better convergence rates for highly nonsymmetric problems.

1 Introduction

For the solution of sparse nonsymmetric linear systems of equations of the form

$$Av = b, \tag{1}$$

it is customary nowadays to use Krylov-based iterative methods, such as GMRES [16], QMR [9], etc. Usually, one needs to precondition the system, e.g., using an incomplete factorization [10], [13], [14]. As is well known, the ordering of the variables influences the effect of incomplete factorization preconditioners, and thus the number of steps the iterative methods takes to converge [6].

In this contribution, we consider a symmetric permutation of the matrix A in (1), i.e., of the form $P^T AP$, and then solve the equivalent system

$$P^T APw = P^T b, \quad \text{with } v = Pw$$

by several iterative methods and preconditionings.

The symmetric permutation we use is performed by the algorithm TPABLO (Threshold PArameterized BLock Ordering) [3]. This algorithm produces a permuted matrix with dense diagonal blocks, while the entries outside the diagonal

---

* Supported by the National Science Foundation grant DMS-9625865.

blocks have magnitude below a prescribed threshold. It is based on the algorithm PABLO [11]. We briefly describe these algorithms in Sec. 2, but refer the reader to [11] and [3] for their full description and complexity analysis. Both algorithms are linear in the order of the matrix and the number of nonzeros, but in the threshold variant, the constant could be large.

The algorithm PABLO, with some modifications, has been used to establish diagonal blocks used as a block diagonal (or block Jacobi) preconditioner in several contexts [3], [7], [12], [18]. This is the first time where other types of preconditioners are studied.

The effect of the threshold reordering on the performance of preconditioned Krylov subspace solvers is illustrated in Sec. 3, using matrices arising from the discretization of certain elliptic PDEs in two dimensions. Additional experiments will be reported in a forthcoming paper.

2   The Permutation Algorithm

We begin by reviewing the algorithm PABLO [11] on which TPABLO is based. These algorithms work with the graph of the matrix, choosing one node at a time, adding it to a group of nodes which would form the blocks along the diagonal, if the new node satisfies certain criteria. The amount of work is controlled by not searching through all the available nodes, but only through a subset of eligible nodes, those which are adjacent to some nodes in the current set, i.e., in the group of nodes of the block along the diagonal being formed. This feature makes the algorithms linear in the order of A and in the number of its nonzeros [3], [11], adding relatively little computational time to the overall solution method.

Given an $n \times n$ matrix $A = (a_{ij})$, let $G = (V, E)$ be its associated graph, i.e., $V = \{v_1, \ldots, v_n\}$ is the set of n vertices and E is the set of edges, where $(v_i, v_j) \in E$ if and only if $a_{ij} \neq 0$; see, e.g., [5]. Given this graph, PABLO constructs q subgraphs $G_k = (V_k, E_k)$, $k = 1, \ldots, q$. The number of subgraphs, q, i.e., the number of the corresponding diagonal blocks, is not known a priori, but is determined by the algorithm and it depends on the structure of the graph G and the input parameters. In the version presented in [7], q is set by the user to match the number of available processors. In the algorithm PABLO (as well as in the threshold variant used in this paper), a first node is taken from the stack of unmarked nodes and this node starts a new current set of vertices P, then additional nodes are taken from the stack and added to the set P if they satisfy certain criteria, or sent back to the stack if not. Two criteria are used in PABLO to determine if a node v should be added to a current set $P \subset V$, corresponding to a diagonal block, by measuring how full $\{v\} \cup P$ is, and how much the vertices in $\{v\} \cup P$ are connected to each other. The first criterion is measured by the ratio α of the total number of edges corresponding to $\{v\} \cup P$ to the number of edges that subgraph would have if it were complete (corresponding to a full submatrix). If it is satisfied then the new node is added. The second test is that the new node v must be adjacent to at least a certain proportion β of nodes in the subgraph corresponding to P and more than outside it. For further details, see [11].

In the algorithm TPABLO used in this paper, a third additional criterion is used to decide if a new vertex v is added to the current subgraph being formed. Let γ be the given threshold, let P be the set of nodes of the current subgraph, and v_j be the vertex being tested for addition to P (corresponding to the jth row and column of the matrix). The node v_j is added to P if, in addition to either of the two criteria in PABLO, the following holds:

$$|a_{ij}| > \gamma \quad \text{or} \quad |a_{ji}| > \gamma \quad \text{for at least one } i \in P.$$

The use of this additional criterion produces a permuted matrix in which every entry in the off-diagonal blocks is smaller than the threshold in absolute value. Entries in the diagonal blocks may still have magnitude smaller than the threshold γ, these are not discarded. In addition, the user can specify a minimum and a maximum block size; see further [3].

We point out that the three main parameters of TPABLO, i.e., the two connectivity parameters α and β, and the threshold γ, together with the parameter for maximum size of each block, provide the user with great flexibility. Thus, one may wish to find permutations such that the resulting matrix has denser or sparser diagonal blocks, or ones whose off-diagonal blocks have smaller and smaller entries.

We conclude this brief description of TPABLO by noting that there is no guarantee that the blocks along the diagonal obtained by this algorithm be nonsingular. In fact, it would be easy to construct examples for which the resulting block diagonal matrix is singular. Nevertheless, we should note that this problem did not arise in our experiments.

3   Numerical Experiments

We consider the following partial differential equation in $\Omega = (0,1) \times (0,1)$

$$-\epsilon\Delta u + \frac{\partial e^{xy}u}{\partial x} + \frac{\partial e^{-xy}u}{\partial y} = g \tag{2}$$

with Dirichlet boundary conditions. Equation (2) has been used many times as a model problem in the literature; see, e.g., [13]. The problem is discretized using centered differences for both the second order and first order derivatives with grid size $h = 1/33$, leading to a block tridiagonal linear system of order $N = 1024$ with $NZ = 4992$ nonzero coefficients. The right-hand side is chosen so that the solution v to the discrete system is one everywhere. The parameter $\epsilon > 0$ controls the difficulty of the problem —the smaller is ϵ, the harder it is to solve the discrete problem by iterative methods. For our experiments, we generated ten linear systems of increasing difficulty, corresponding to $\epsilon^{-1} = 100, 200, \ldots, 1000$. The coefficient matrix A becomes increasingly nonsymmetric as ϵ gets smaller.

The Krylov subspace methods tested were GMRES(20) [16], BiCGStab [17], QMR [9] and CGNR (conjugate gradient method applied to the normal equations $A^T A v = A^T b$). The preconditioners used were ILU(0) [10], ILUT [14] and, for use with CGNR, a drop tolerance-based incomplete QR factorization (see, e.g., [13] and [1] for information on incomplete QR preconditioning). In all our experiments we used $v_0 = 0$ as initial guess and we stopped the iterations when the 2-norm of the 'true' residual $b - A v_k$ had been reduced to less than 10^{-9}, with a maximum of 1000 iterations allowed.

For the TPABLO ordering, we set $\alpha = \beta = 0$, a maximum block size of 100, and threshold $\gamma = 0.005$ for $\epsilon^{-1} = 100, 200, 300$ and $\gamma = 0.01$ for $\epsilon^{-1} > 300$. We show in Fig. 1 the sparsity patterns corresponding to (a) the natural ordering and (b) the TPABLO ordering for $\epsilon^{-1} = 1000$.

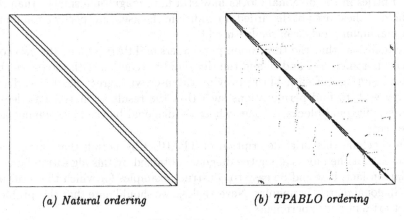

(a) Natural ordering   (b) TPABLO ordering

Fig. 1. Matrix patterns

Table 1. Fill-in for the drop tolerance-based ILUT and IQR preconditioners (n/o = natural ordering, TP = TPABLO)

ϵ^{-1}	ILUT(5,10^{-2})		ILUT(10,10^{-3})		IQR(10^{-2})	
	n/o	TP	n/o	TP	n/o	TP
100	12540	11258	21319	15520	9783	12718
200	14458	13507	22935	19465	16742	14259
300	14706	14139	23640	20850	19108	15050
400	14782	14333	23876	21541	20093	15559
500	14796	14425	24374	21970	20718	16167
600	14807	14439	24571	21986	20697	16360
700	14828	14523	24633	21953	21537	16260
800	14852	14467	24677	22082	24362	16845
900	14863	14002	24773	22023	25680	17135
1000	14856	14041	24855	22039	24727	17429

In Table 1 we show the amount of fill-in incurred by the incomplete factorizations ILUT and IQR for typical values of the parameters, using the natural ordering and TPABLO. The figures refer to the number of nonzeros in the incomplete L and U factors for ILUT, and in the incomplete R factor for QR. Note that the amount of fill-in increases almost monotonically with ϵ^{-1}, and that TPABLO reduces the amount of fill-in in all cases, except one.

In Tables 2 and 3 we show the number of iterations for various combinations of Krylov solvers and preconditioners, with the natural ordering and with TPABLO. It is clear that as ϵ decreases, the rates of convergence tend to deteriorate with both orderings, but the deterioration is much slower in the TPABLO ordering. This phenomenon is particularly striking for CGNR/IQR. As a result, TPABLO, which gives worse results than the natural ordering for the 'easy' problems (ϵ large), becomes clearly better for the 'hard' problems (ϵ small).

Table 2. Number of iterations for different preconditioners and iterative solvers (n/o = natural ordering, TP = TPABLO).   A † means that convergence was not attained in 1000 iterations.

	ILU(0)						ILUT(5,10^{-2})						ILUT(10,10^{-3})					
	GMR		BST		QMR		GMR		BST		QMR		GMR		BST		QMR	
ϵ^{-1}	n/o	TP	n/o	TP	n/o	TP	n/o	TP	n/o	TP	n/o	TP	n/o	TP	n/o	TP	n/o	TP
100	17	36	12	15	20	27	7	9	4	5	7	10	5	7	3	4	6	8
200	39	90	21	29	37	50	11	11	6	6	11	11	7	8	4	5	7	9
300	113	220	36	44	58	71	14	13	9	7	15	14	8	9	5	6	9	10
400	457	637	63	57	93	94	18	15	10	9	19	16	11	10	6	6	11	10
500	†	†	87	81	198	124	26	19	13	11	23	21	14	12	8	6	14	12
600	†	†	156	102	†	176	33	25	16	12	27	23	16	14	9	8	17	14
700	†	†	333	143	†	210	40	23	20	12	34	22	20	13	13	7	21	14
800	†	†	606	148	†	235	59	50	26	21	41	38	27	15	15	9	23	16
900	†	†	778	177	†	296	†	†	†	†	†	†	32	17	16	9	27	18
1000	†	†	†	205	†	338	†	†	†	†	†	†	38	18	18	10	32	19

Table 3. Number of iterations for CGNR preconditioned with incomplete QR factorization (n/o = natural ordering, TP = TPABLO).

order	100	200	300	400	500	600	700	800	900	1000
n/o	24	22	27	53	84	118	119	112	149	190
TP	23	22	22	24	25	25	23	27	28	29

Additional experiments were performed with CGNR/IQR where the columns of the coefficient matrix A were permuted according to the minimum degree heuristic applied to the structure of $A^T A$. The purpose of this heuristic is to preserve sparsity in the triangular factor R (which corresponds to the Cholesky factor of $A^T A$); see [1]. With this ordering, the convergence rate of CGNR/IQR was about as good as with TPABLO. However, fill-in in the incomplete factor R was consistently found to be about 20% higher with minimum degree than with TPABLO.

4 Concluding Remarks

The results of our experiments show that TPABLO can enhance the robustness and performance of Krylov subspace methods preconditioned with incomplete factorizations. The beneficial effect of TPABLO for the class of problems considered in this paper can be explained as a 'damping' of the instability of the incomplete LU factorizations of such matrices. This instability was first demonstrated in [8]; see also [4] and [15] for some illuminating examples of the same phenomenon. In this context, the 'instability' refers to the fact that the norm of $(LU)^{-1}$ can be extremely large. Hence, the ILU preconditioning matrix $M = LU$ is extremely ill conditioned and so is the preconditioned matrix $M^{-1}A$, although the original problem is actually rather well conditioned. As shown in [8] for the case of ILU(0), the ill conditioning is caused by the long recurrences in the forward and backward triangular solves when the preconditioning is applied. When A is highly indefinite or has large nonsymmetric part, even the more accurate ILUT factorization can exhibit this kind of instability, causing the preconditioned iteration to fail [4]. In this case, a common course of action is to increase the accuracy of the preconditioner by allowing more fill-in in the incomplete factors, which increases the cost of forming and applying the preconditioner. Another possibility would be to switch to sparse approximate inverse preconditioners, which do not involve triangular solves and are therefore immune from the type of instability described above; see, e.g., [2] and [4]. However, these preconditioners are more expensive to compute and use than incomplete factorizations.

Another alternative is to apply some reordering to the original matrix. This leaves the conditioning of the matrix unchanged, but may have a beneficial effect on the conditioning of the preconditioner. It appears that TPABLO can improve the stability of the incomplete factorization without increasing the amount of fill-in, in fact even reducing it. The price for this is the cost of the preprocessing itself, which may be negligible if a sequence of linear systems with the same matrix and different right-hand sides must be solved. A more detailed analysis will be presented, together with additional experiments, in a forthcoming paper.

Acknowledgment. We would like to thank Miroslav Tůma for providing some of the codes which were used for the numerical experiments.

References

1. Benzi M., Tůma M.: A comparison of preconditioning techniques for general sparse matrices. In S. Margenov and P. Vassilevski (Eds.), Iterative Methods in Linear Algebra, II, IMACS Series in Computational and Applied Mathematics Vol. 3, pages 191–203 (1996)

2. Benzi M., Tůma M.: A sparse approximate inverse preconditioner for nonsymmetric linear systems. SIAM J. Sci. Comput. (1998) to appear

3. Choi H., Szyld D. B.: Application of threshold partitioning of sparse matrices to Markov chains. Proceedings of the IEEE International Computer Performance and Dependability Symposium, IPDS'96, Urbana-Champaign, Illinois, September 4–6, 1996, pages 158–165, IEEE Computer Society Press, Los Alamitos, California (1996)

4. Chow E., Saad Y.: Approximate inverse techniques for block-partitioned matrices. SIAM J. Sci. Comput. (1997) to appear

5. Duff I., Erisman A. M., Reid J. K: Direct Methods for Sparse Matrices. Clarendon Press, Oxford, 1986

6. Duff I., Meurant G. A.: The effect of ordering on preconditioned conjugate gradients. BIT **29** (1989) 635–657

7. Dutto, L. C., Habashi W. G., Fortin M.: Parallelizable block diagonal preconditioners for the compressible Navier-Stokes equations. Computer Methods Appl. Mech. Engng. **117** (1994) 15–47

8. Elman H. C.: A stability analysis of incomplete LU factorizations. Math. Comp. **47** (1986) 191–217

9. Freund R., Nachtigal N. M.: QMR: A quasi-minimal residual method for non-Hermitian linear systems. Num. Math. **60** (1991) 315–339

10. Meijerink J. A., and van der Vorst, H.: An iterative solution method for linear systems of which the coefficient matrix is a symmetric M-matrix. Math. Comp. **31** (1977) 148–162

11. O'Neil J., Szyld D. B.: A block ordering method for sparse matrices. SIAM J. Sci. Stat. Comput. **11** (1990) 811–823

12. Paloschi J. R.: Testing a new parallel preconditioner on linear systems arising from flowsheeting simulation. Proceedings of ESCAPE7, Trondheim, May 1997, to appear.

13. Saad Y.: Preconditioning techniques for nonsymmetric and indefinite linear systems. J. Comp. Appl. Math. **24** (1988) 89–105

14. Saad Y.: ILUT: A dual threshold incomplete LU factorization. Num. Lin. Alg. Applic. **1** (1994) 387–402

15. Saad Y.: Preconditioned Krylov subspace methods for CFD applications. In W. G. Habashi (Ed.), Solution Techniques for Large-Scale CFD Problems, J. Wiley, Chichester, pages 139–158 (1995)

16. Saad Y., Schultz M. H.: GMRES: A generalized minimal residual algorithm for solving nonsymmetric linear systems. SIAM J. Sci. Stat. Comput. **7** (1986) 856–869

17. van der Vorst, H.: Bi-CGSTAB: A fast and smoothly converging variant of Bi-CG for the solution of nonsymmetric linear systems. SIAM J. Sci. Stat. Comput. **12** (1992) 631–644

18. Yang G., Dutto L. C., Fortin M.: Inexact block Jacobi-Broyden methods for solving nonlinear systems of equations. SIAM J. Sci. Comput. (1997) to appear

Efficient and Accurate Parameter Estimation for Linear, Bilinear and Nonlinear Systems using Walsh Functions

W.F. Blyth

Department of Mathematics, Royal Melbourne Institute of Technology,
GPO Box 2476V, Melbourne, Australia 3001

Abstract. In 1975, Chen and Hsiao used Walsh functions to estimate system parameters from the input/ouput records. We use a spectral method: the Walsh coefficients are used directly to obtain the algebraic equations that are solved to find the estimates of the parameters.

Usually, Walsh function methods are second order: that is, as the number of terms m of the Walsh series is doubled, the error is reduced by a factor of four. We propose using Richardson extrapolation, and verify, by numerical experiment with $m = 4, 8, 16, 32$, that our algorithm is second order and that Richardson extrapolation provides greatly improved accuracy.

Recently Jha et al. (1992) applied a recursive algorithm to a bilinear system and to a nonlinear system. We show how to apply and suitably scale the problem to use the first 4, 8, 16 input/output records to obtain the estimates of the parameters. We obtain results that are orders of magnitude better than those of Jha et al..

1 The Walsh functions

The Walsh functions were first introduced by Walsh [1]. Following Paley [2] we define the Walsh functions in terms of the Rademacher functions [3]. For details and properties of Walsh functions and series see [4] or [5].

The collection of Walsh functions form an orthonormal complete collection of functions for the space $L^2[0, 1)$ of square Lebesque integrable functions. Thus

$$f(x) = c_0 + c_1 W_1(x) + c_2 W_2(x) + \cdots$$

for every $f \in L^2[0, 1)$, where the coefficients are given by $c_i = \int_0^1 f(x) W_i(x) \, dx$.

Fine [6] showed that if x belongs to the subinterval $S = [(i - 1)/2^n, i/2^n]$, where $i, n \in \mathbb{N}$, $m = 2^n$ then the partial sum of the first m terms of the Walsh Fourier series is equal to the L^1 mean, $\bar{f}_i$, defined by $\bar{f}_i = \int_S f \, dx \,/|S|$ of f over this subinterval. This is a very attractive and important property of Walsh series and this is why, in applications and in the rest of this paper, Walsh series are always truncated to some $m = 2^n$ terms. In many applications, the L^1 means are not available and the $\bar{f}_i$ are approximated by the values of the function f at the midpoints of the subintervals. However, in this paper, we approximate the average values by using the trapezoidal rule. Either of these approximations introduces an second order error. The Walsh coefficients can be obtained from the $\bar{f}_i$ by using the well-know Fast Walsh Transform (see [4] or [5]).

Fine [6] derived the (infinite) Walsh series for the integrals of the of the Walsh functions. However these results for truncated series were re-derived and used in [7,8]. This approximate integration can be representated in matrix form as follows. If

$$f(x) \approx \sum_{i=0}^{m-1} c_i W_i(x) \text{ and } \int_0^x f(t)\, dt \approx \sum_{i=0}^{m-1} b_i W_i(x)$$

then

$$\begin{bmatrix} b_0 \\ \vdots \\ b_{m-1} \end{bmatrix} = P_m^T \begin{bmatrix} c_0 \\ \vdots \\ c_{m-1} \end{bmatrix} \text{ where } P_m^T = \begin{bmatrix} P_{m/2}^T & \frac{1}{2m} I_{m/2} \\ \hline \frac{-1}{2m} I_{m/2} & O_{m/2} \end{bmatrix}, \ P_2^T = \begin{bmatrix} \frac{1}{2} & \frac{1}{4} \\ -\frac{1}{4} & 0 \end{bmatrix}$$

and I_m, O_m denote order m unit and zero matrices respectively. In the following, the Einstein summation convention is used (and i takes the integer values between 0 and $m-1$).

2 Parameter Estimation: a Walsh Spectral Method

Our algorithm is a modification of that introduced by Chen and Hsiao [9] so we explain our algorithm by applying it to the first of the Illustrative Examples of [9] and report the performance on the other Illustrative Examples of [9]. In this section, t is always in the unit interval and the discrete values of t are the initial value $t = 0$ and all the multiples of $1/m$ to $t = 1$.

2.1 A First Order Linear Example

Consider estimating the a and b from the input/output records where the underlying differential equation is

$$\dot{y}(t) + ay(t) = bu(t) \quad , \qquad t \in [0,1) \tag{1}$$

and u is the unit step function. Integrate once, to obtain an integral equation where no derivatives appear:

$$\int_0^t \dot{y}(\tau)\, d\tau + a \int_0^t y(\tau)\, d\tau = b \int_0^t u(\tau)\, d\tau \tag{2}$$

giving

$$y(t) - y(0) + a \int_0^t y(\tau)\, d\tau = b \int_0^t u(\tau)\, d\tau \ . \tag{3}$$

Note that u is the known input function, so its Walsh series can be calculated. In this case, $u = W_0$ and integrating this using the P integration matrix gives the truncated Walsh series of $t = h_i W_i(t)$ where the h_i are known (the first column of P^T). The y is the measured output function with a Walsh series $y(t) = c_i W_i(t)$. The average value on each subinterval is obtained using the average of the endpoint values. This is just a trapezoidal approximation. Provided that the number, m, of Walsh series terms is a power of two, the Walsh coefficients are obtained from the (fast) Walsh transform of these average values of y.

We now replace the y and u in (3) by their Walsh series and perform the integrations using the integration matrix, P, to give

$$c_i W_i(t) - y(0) W_0(t) + a c_j P_{ji} W_i(t) = b h_i W_i(t) \ . \tag{4}$$

In [9], two sample values for t are chosen so that two equations in a and b are obtained. Here, we use a simpler spectral approach and equate the coefficients of the Walsh functions to obtain

$$c_i - y(0) \, \delta_{0i} + a c_j P_{ji} = b h_i \qquad i = 0, 1, \ldots, m - 1 \ . \tag{5}$$

This gives an overdetermined system of equations for a and b. We choose to reduce to two equations to solve by choosing $i = 0$ and $i = 1$: the lowest "frequency" terms. In our numerical experiments, we compared the results of all possible choices of which two equations to solve, and found that the results (if obtainable) were very similar. From experience with Walsh function methods, the low "frequency" terms are usually more accurate. Another reason for this choice is that we expect that this would be a better choice in the presence of noise - this is a current topic of investigation (with I. Atkinson).

Our experience (see [10–13]) with the development of a-priori error estimates and with numerical experiments with Walsh functions and integral equations shows that, usually, Walsh function methods are second order: that is, as m is doubled, the error is reduced by a factor of four. We propose using Richardson extrapolation. We verify, by numerical experiment with $m = 4, 8, 16, 32$, that the algorithm is second order and that Richardson extrapolation provides greatly improved accuracy. We used Mathematica and set $a = b = 4$ to generate the output from which parameters a and b were estimated. To simulate practice and to mimic the outputs used by [9], we rounded all of the outputs (i.e. the y values) to three decimal points. Our results are summarized in Table 1.

Table 1. Results and Errors for $\dot{y} + ay = bu$, $a = b = 4$

m	$Est.a$	Est. b	Error in a	Error in b	Rich. a	Rich. b
4	3.69618	3.69664	0.303815	0.303356		
8	3.91451	3.9159	0.0854886	0.084095	3.98729	3.98899
16	3.97621	3.97702	0.0237894	0.0229841	3.99678	3.99739
32	3.99378	3.99435	0.00622129	0.00564957	3.99963	4.00013

2.2   A Two Dimensional Linear Example

The second illustrative example of [9] is the zero input system

$$\dot{x}(t) = Ax(t) \quad , \qquad A = \begin{bmatrix} 0 & 1 \\ -2 & -2 \end{bmatrix} \tag{6}$$

with initial conditions $x_1(0) = 1$ and $x_2(0) = -1$. We follow the method described above. Here, we let $x_1(t) = c_{1i}W_i(t)$ and $x_2(t) = c_{2i}W_i(t)$. On substituting the Walsh series and integrating the differential equation once, we obtain

$$c_{il}W_l(t) = a_{ij}c_{jk}P_{kl}W_l(t) + x_i(0)W_0(t) . \tag{7}$$

We equate the coefficients of the Walsh functions and choose $l = 0, 1$ for $i = 0, 1$ which gives four equations to solve for the four parameters, the a_{ij}. We use Mathematica to generate the output values and round them to four decimal places. Our results are summarized in Table 2 where Rich. 8,4 indicates that Richardson extrapolation was used with the $m = 8$ and $m = 4$ results. It is clear that high accuracy is obtained from using low m and Richardson extrapolation: this is also computationally efficient.

Table 2. Results for $\dot{x} = Ax$

m	Est.a_{11}	Est. a_{12}	Est. a_{21}	Est. a_{22}
4	-0.0204646	0.989706	-1.97984	-2.00041
8	-0.00558527	0.997082	-1.99471	-1.99987
16	-0.00161043	0.999085	-1.99873	-1.99999
32	-0.000550433	0.999655	-1.9991	-1.99955
True Values	0	1	-2	-2
Rich. 8,4	-0.000625506	0.99954	-1.99967	-1.99969
Rich. 16,8	-0.000285478	0.999753	-2.00007	-2.00003
Rich. 32,16	-0.000197102	0.999845	-1.99923	-1.99941

2.3   A Second Order Linear Example

The Chen and Hsiao third problem (their "Laboratory test") is

$$y'' + a_1 y' + a_2 y = b_1 u' + b_2 u , \qquad u(t) = H(t) - H(t - 1/4) \tag{8}$$

where the output is $y(t)$, initially at rest, $u(t)$ is the input and $H(t)$ is the Heaviside step function. The method is the same as described above except that integration has to be performed twice (and for the Walsh series, this was done by multiplying by the integration matrix twice). All of the input/output values were rounded to four decimal places. We achieve resonable accuracy, although in this case (with discontinous u and delta function u'), for low m, Richardson

Table 3. Results for $y'' + a_1 y' + a_2 y = b_1 u' + b_2 u$

m	Est.a_1	Est. a_2	Est. b_1	Est. b_2
4	1.99013	1.97852	1.01038	1.96607
8	1.98046	1.98416	1.00273	1.97315
16	2.01603	2.00924	1.0005	2.01645
32	2.02436	2.01467	0.999961	2.02649
True Values	2	2	1	2

extrapolation is not effective - the solution is degraded (but still more accurate than the Chen and Hsiao results). See Table 3 for our results.

Note that the result for $m = 32$ is not very accurate - this is due to the errors in the data. This is confirmed by recalculating using high accuracy for the data which gives very high accuracy for $m = 32$:

$$(a_1, a_2, b_1, b_2) = (2, 2, 1, 2) \approx (2.00033, 2.00000, 1.00016, 2.00000) .$$

3 Parameter Estimation: as the Data is Available

In 1992, Jha et al. [14] developed a recursive Walsh function algorithm which was applied to a bilinear system and to a nonlinear system. Our algorithm is that of Section 2 above, with the further adaption that we scale the problems to use the first 4, 8,16 input/output records to obtain the estimates of the parameters. We obtain results that are orders of magnitude better than those of [14].

3.1 A Bilinear Example

Consider the bilinear (in y and u) example of [14] where we wish to estimate the values of the parameters a, b, c from the input/output records whose differential equation is

$$\dot{y}(t) = a y(t) + b u(t) + c y(t) u(t) \quad , \qquad t \in [0, 1) \tag{9}$$

with $u = \exp(-t/2)$, $a = b = 1$, $c = 2$ and $y(0) = 0$. Following Jha et al., we obtain the values of the product term (by multiplication!) and obtain the average value (on each subinterval - we use the trapezoidal approximation) of this product and hence the Walsh series from the Walsh transform as usual. The differential equation is integrated once to obtain an integral equation where no derivatives appear. Using Walsh series $y(t) = y_i W_i(t)$ and similarly for u and $p = yu$, equating Walsh coefficients and remembering that $y(0) = 0$ gives

$$y_i = a y_j P_{ji} + b u_j P_{ji} + c p_j P_{ji} \qquad i = 0, 1, \ldots, m - 1 . \tag{10}$$

Since we solve for a, b, c, we obtained three equations to solve by choosing $i = 0, 1, 2$. We reproduced the data of [14] by using Mathematica and a step size of $1/40$.

As a comparison with Jha's result across the whole interval of t, we used all of this data for $m = 4$ where the (composite) trapezoidal approximation was used for the average values on each subinterval (with 10 of the steps of size $1/40$). Similarly for $m = 8$, and a Richardson extrapolation was performed. High accuracy was obtained as summarized in Table 4.

Table 4. Results for the bilinear problem for the whole interval

m	Est.a	Est. b	Est. c
4	0.945174	0.892086	2.0873
8	0.987194	0.974003	2.01945
Rich. 8,4	1.0012	1.00131	1.99683
True Values	1	1	2
Jha's Values	1.2543	1.00898	1.89265

However Jha et al. took a recursive approach. In that spirit, we suitably scale the problem to use the initial values and the first m input/output records (as they become available) to obtain the estimates of the parameters. Since m steps of size $1/40$ are to be used, $t = m/40$ has to be scaled to 1. This leads to the scaled equations which are the same as (10) except that the parameters are replaced by the parameter times $m/40$. Thus the original equations (10) are solved and the parameter estimates are obtained by multiplying the results by $40/m$. Note that the results (see Table 5) are excellent. We do not expect to improve these results by Richardson extrapolation since the step size is the same for each calculation.

Table 5. Results for the bilinear problem, as data is available

m	Est.a	Est. b	Est. c
4	1.10933	0.999844	1.88372
8	0.994558	0.999728	2.00487
16	1.00111	0.999729	1.99795
32	1.00013	0.999706	1.99918
True Values	1	1	2

3.2   A Nonlinear Example

Consider the nonlinear (in y and u) example, modified slightly from [14], where we wish to estimate the values of the parameters e, f, h from the input/output records whose differential equation is

$$\dot{y}(t) = ey(t) + fy^{3/2}(t)u^{1/2}(t) + hu(t) \ , \qquad t \in [0,1) \qquad (11)$$

with $u = \exp(-t/2)$, $e = -1$, $f = 2$, $h = 1.5$ and $y(0) = 0$. Following Jha et al., we use steps of size $1/20$. The procedure is the same as above and our results are summarized in Table 6. Note that these results are again excellent.

Table 6. Results for the nonlinear problem, as data is available

m	Est.e	Est. f	Est. h
4	-0.935548	1.89895	1.49454
8	-0.974799	1.97083	1.49581
16	-0.990336	1.99046	1.497
True Values	-1	2	1.5

References

1. J.L. Walsh, J.L.: A closed set of normal orthogonal functions. Amer. J. Math. **45** (1923) 5–24
2. Paley, R.E.A.C.: A remarkable series of orthogonal functions (I & II). London Math. Soc. **34** (1932) 241–264 & 265–279.
3. Rademacher, H.: Einige sätze über Reihen von Allgemein Orthogonalen funktionen. Math. Annal. **87** (1922) 112–138.
4. Schipp, F., Wade, W.R., Simon, P., Pal, J.: *Walsh series: an introduction to dyadic harmonic analysis*. Adam Hilger, New York (1990)
5. Golubov, B.E., Efimov, A., Skvortsov, V.: *Walsh series and transforms*. Kluwer, Netherlands (1991)
6. Fine, N.J.: On the Walsh Functions. Trans. Amer. Math. Soc. **65** 322–414
7. Corrington, M.S.: A solution of differential and integral equations with Walsh functions. IEEE Trans. Circuit Theory CT-20 **5** (1973) 470–476
8. Chen, C.F., Hsiao, C.H., (1975), A Walsh series direct method for solving variation problems. J. Franklin Inst. **300** 4 (1975) 265–280
9. Chen, C.F., Hsiao, C.H.: Time-domain synthesis via Walsh functions. Proc. IEE **122** 5 (1975) 565–570
10. Sloss, B.G., Blyth, W.F.: A-priori error estimates for Corrington's Walsh function method. J. Franklin Inst. **331B** (1994) 273–283
11. Sloss, B.G., Blyth, W.F.: Corrington's Walsh function method applied to a nonlinear integral equation. J. Integral Eq. **6** (1994) 239–256
12. Uljanov, V., Blyth, W.F.: Numerical solution of Urysohn integral equations using Walsh functions. *The Role of Mathematics in Modern Engineering* (ed. Easton, A.K., Steiner, J.M.). Studentlitteratur, Lund, Sweden (1996) 621–628
13. Uljanov, V., Blyth, W.F.: Numerical solution of functional differential equations using Walsh functions. *Computational Techniques and Applications: CTAC-93* (ed. Stewart, D., Singleton, D., Gardiner, D.). World Scientific (1994) 454–461
14. Jha, A.N., Saxena, A.S., Rajaman, V.S.: Parameter estimation algorithms for bilinear and non-linear systems using Walsh functions—recursive approach. Int. J. Systems Sci. **23** (1992) 283–290

3-D Jacobi Rotation Method for Eigenvalue Problems of Matrices

Da Yong Cai[1] and Weimin Xue[2]

[1] Department of Applied Mathematics
Tsinghua University, Beijing 100084, China
[2] Department of Mathematics, Hong Kong Baptist University
Kowloon Tong, Hong Kong

Abstract. In this paper the formulation of the 3-D rotation method for the eigenvalue problems of matrices is presented. The numerical results for symmetric matrices with clustered spectrum and some non-symmetric matrices are summarized.

1 Introduction

The eigen-problems for matrices are important for both scientific research and engineering purposes. Up to now, the most effective algorithm is the QR method with shift strategy. The computations show that the QR method is less favorable if the matrix has clustered spectrum. Therefore it is still worth-while to explore some new approaches for the eigenvalues of matrices with clustered spectrum.

On the other hand, from the real Schur decomposition theorem[11], we know that there exists an orthogonal matrix $Q \in \mathbf{R}^{n \times n}$ such that

$$Q^T A Q = T, \tag{1}$$

where $A \in \mathbf{R}^{n \times n}$ is given and $T \in \mathbf{R}^{n \times n}$ is a block triangular matrix. Each main diagonal block of T is either $\mathbf{R}^{1 \times 1}$ or $\mathbf{R}^{2 \times 2}$. To our best knowledge, it is still a task to find the numerically stable algorithm to implement the Schur decomposition effectively.

The Jacobi method is one of the oldest numerical method for the eigenvalue problems of symmetric matrices. It radiated with a brief light in the 1950s, and, in recent years, has comeback [1] [2] [4] [7] because its adaptability for parallel computers and possibility to be generalized.

The basic idea of the Jacobi method is to find a 2×2 Jacobi rotator Q for the 2×2 symmetric matrix A such that $Q^T A Q$ is diagonal. The Jacobi rotation method is essentially not suitable for non-symmetric matrices.

To extend the Jacobi methods, Bojanczyk et. al. [1] studied the computation of the Euler angles of a symmetric 3×3 matrix. Motivated by their work, we propose a 3-D Jacobi rotation method. The basic idea of this new algorithm is to find two 3×3 Jacobi rotators Q_1 and Q_2 for the 3×3 matrix A such that the matrix A becomes a Hessenberg matrix after two rotations, i.e.

$$Q_2^T Q_1^T A \, Q_1 Q_2 = \begin{bmatrix} \tilde{a}_{11} & \tilde{a}_{12} & \tilde{a}_{13} \\ 0 & \tilde{a}_{22} & \tilde{a}_{23} \\ 0 & \tilde{a}_{32} & \tilde{a}_{33}. \end{bmatrix}$$

In this paper we discuss the formulation of the 3-D Jacobi rotation in an algebraic deduction, which is different to the geometrical motivation in [1], and report some numerical results for symmetric matrices with clustered spectrum and also some non-symmetric matrices.

2   3-D Jacobi Rotation

In order to compare the 3-D Jacobi method presented in this paper with the classical Jacobi method, we review the basic idea of the Jacobi method. Let

$$Q = \begin{bmatrix} c & -s \\ s & c \end{bmatrix}, \tag{2}$$

where $c^2 + s^2 = 1$. It is easy to see that Q is a rotator, if

$$c = \frac{1}{\sqrt{1+t^2}} \quad and \quad s = \frac{t}{\sqrt{1+t^2}},$$

for any parameter t. We denote that

$$\tilde{A} = (\tilde{a}_{ij})_{2\times 2} = Q^T A \, Q.$$

Thus one may derive that

$$\tilde{a}_{21} = [\, a_{21} + (a_{22} - a_{11})t - a_{12}t^2 \,]/(1+t^2).$$

Let $\tilde{a}_{21} = 0$. Then the parameter t is determined by the following equation:

$$a_{12}t^2 - (a_{22} - a_{11})t - a_{21} = 0,$$

or

$$a_{21}\hat{t}^2 + (a_{22} - a_{11})\hat{t} - a_{12} = 0,$$

where $\hat{t} = t^{-1}$. If a_{21} is already zero, the Jacobi rotation is not necessary to be applied. If $a_{21} \neq 0$, the parameter $\hat{t}$ for Jacobi rotator Q is determined by

$$\hat{t} = [\, (a_{11} - a_{22}) \pm \sqrt{(a_{11} - a_{22})^2 + 4a_{21}a_{12}} \,]/(2a_{21}). \tag{3}$$

For the case that A is symmetric, the parameter $\hat{t}$ is always real, then the Jacobi method works. But for the non-symmetric case, the determination of a real Jacobi rotation will be easily broken down. For instance, the roots in (3) are complex numbers, if A is skew-symmetric and the item $(a_{11} - a_{22})$ is small.

Thus the Jacobi rotation method is essentially not suitable for non-symmetric matrices.

Now let us turn our interests to the 3-D Jacobi rotation. The basic idea is to find a 3×3 Jacobi rotator Q for the 3×3 matrix A such that $Q^T A Q$ is a Hessenberg matrix. Let $A = (a_{ij})_{3 \times 3}$ with non-zero a_{21} and a_{31}. The first Jacobi rotator Q_1 is defined by

$$Q_1 = \begin{bmatrix} c_1 & -s_1 & 0 \\ s_1 & c_1 & 0 \\ 0 & 0 & 1 \end{bmatrix}, \tag{4}$$

where $c_1{}^2 + s_1{}^2 = 1$. It is easy to see that Q_1 is a rotator, if

$$c_1 = \frac{1}{\sqrt{1 + t_1{}^2}} \quad and \quad s_1 = \frac{t_1}{\sqrt{1 + t_1{}^2}},$$

for any parameter t_1, which is determined later. We denote that

$$B = (b_{ij})_{3 \times 3} = Q_1^T A Q_1. \tag{5}$$

It is easy to see that

$$\begin{aligned}
b_{11} &= [a_{11} + (a_{12} + a_{21})t_1 + a_{22}t_1{}^2]/\alpha_1, \\
b_{12} &= [a_{12} + (a_{22} - a_{11})t_1 - a_{21}t_1{}^2]/\alpha_1, \\
b_{21} &= [a_{21} + (a_{22} - a_{11})t_1 - a_{12}t_1{}^2]/\alpha_1, \\
b_{22} &= [a_{22} - (a_{12} + a_{21})t_1 + a_{11}t_1{}^2]/\alpha_1, \\
b_{13} &= [a_{13} + a_{23}t_1]/\sqrt{\alpha_1}, \\
b_{23} &= [a_{23} - a_{13}t_1]/\sqrt{\alpha_1}, \\
b_{31} &= [a_{31} + a_{32}t_1]/\sqrt{\alpha_1}, \\
b_{32} &= [a_{32} - a_{31}t_1]/\sqrt{\alpha_1}, \\
b_{33} &= a_{33},
\end{aligned} \tag{6}$$

where

$$\alpha_1 = 1 + t_1{}^2 \neq 0.$$

The second Jacobi rotator Q_2 is defined by

$$Q_2 = \begin{bmatrix} c_2 & 0 & -s_2 \\ 0 & 1 & 0 \\ s_2 & 0 & c_2 \end{bmatrix}, \tag{7}$$

where $c_2{}^2 + s_2{}^2 = 1$. It suffices to set

$$c_2 = \frac{1}{\sqrt{1 + t_2{}^2}} \quad and \quad s_2 = \frac{t_2}{\sqrt{1 + t_2{}^2}},$$

for an undetermined parameter t_2. We denote that

$$\tilde{A} = (\tilde{a}_{ij})_{3 \times 3} = Q_2^T B Q_2. \tag{8}$$

It is easy to get that

$$\tilde{a}_{21} = [b_{21} + b_{23}t_2]/\sqrt{\alpha_2},$$
$$\tilde{a}_{32} = [b_{32} + b_{12}t_2]/\sqrt{\alpha_2},$$
$$\tilde{a}_{31} = [b_{31} + (b_{33} - b_{11})t_2 - b_{13}t_2^2]/\alpha_2,$$

where

$$\alpha_2 = 1 + t_2^2 \neq 0.$$

Let $\tilde{a}_{21} = \tilde{a}_{31} = 0$, i.e.

$$\tilde{A} = \begin{bmatrix} \tilde{a}_{11} & \tilde{a}_{12} & \tilde{a}_{13} \\ 0 & \tilde{a}_{22} & \tilde{a}_{23} \\ 0 & \tilde{a}_{32} & \tilde{a}_{33} \end{bmatrix},$$

Then the parameter t_2 is determined by the following equations:

$$\begin{cases} b_{21} + b_{23}t_2 = 0, \\ b_{31} + (b_{33} - b_{11})t_2 - b_{13}t_2^2 = 0. \end{cases}$$

or

$$\begin{cases} b_{21}\hat{t}_2 + b_{23} = 0, \\ b_{31}\hat{t}_2^2 + (b_{33} - b_{11})\hat{t}_2 - b_{13} = 0, \end{cases} \tag{9}$$

where $\hat{t}_2 = t_2^{-1}$. We first suppose that $b_{21} \neq 0$. Then the equations (9) lead to the following condition:

$$\Delta = b_{21}(b_{13}b_{21} - b_{11}b_{23} + b_{33}b_{23}) - b_{31}b_{23}^2 = 0. \tag{10}$$

Substituting the equations (6) into condition (10), one may derive the following cubic equation

$$\lambda_3 \hat{t}_1^3 + \lambda_2 \hat{t}_1^2 + \lambda_1 \hat{t}_1 + \lambda_0 = 0, \tag{11}$$

where $\hat{t}_1 = t_1^{-1}$,

$$\begin{cases} \lambda_0 = a_{12}^2 a_{23} - a_{13}^2 a_{32} + (a_{22} - a_{33})a_{13}a_{12}, \\ \lambda_1 = \beta(a_{22} - a_{11}) - \alpha a_{12} + (2a_{23}a_{32} - a_{13}a_{31})a_{13}, \\ \lambda_2 = \alpha(a_{22} - a_{11}) - \beta a_{21} + (2a_{13}a_{31} - a_{23}a_{32})a_{23}, \\ \lambda_3 = a_{13}^2 a_{21} - a_{31}^2 a_{23} + (a_{33} - a_{11})a_{23}a_{21}, \end{cases} \tag{12}$$

and

$$\begin{cases} \alpha = a_{13}a_{21} - (a_{11} - a_{33})a_{23}, \\ \beta = -a_{12}a_{23} - (a_{22} - a_{33})a_{13}. \end{cases}$$

The above deduction is confirmed by use of Mathematica. If $\lambda_3 \neq 0$, there exists at least one real solution $\hat{t}_1$ of equation (11). If this solution happened to make $b_{21} = 0$, which we assumed to be non-zero in advance, then the first 3-D Jacobi rotator Q_1 acts as a 2-D Jacobi rotator to reduce a_{21} to be zero. In this case we only apply the first rotation. Thus we can conclude that either a 3-D or a 2-D Jacobi rotator may be applied to the cases of arbitrary matrices.

The only possibility of break down for this algorithm is that $\lambda_3 = 0$ and no real solution may be found from the equation:

$$\lambda_2 \hat{t}_1^2 + \lambda_1 \hat{t}_1 + \lambda_0 = 0,$$

We do no know how to avoid this break down, but the practical computation show that the 3-D Jacobi rotation presented in this paper is rarely broken down.

A similar formulation may be derived, if we assume $\tilde{a}_{31} = \tilde{a}_{32} = 0$. Then the rotation (8) produce another kind of upper Henssenberg matrix:

$$\tilde{A} = \begin{bmatrix} \tilde{a}_{11} & \tilde{a}_{12} & \tilde{a}_{13} \\ \tilde{a}_{21} & \tilde{a}_{22} & \tilde{a}_{23} \\ 0 & 0 & \tilde{a}_{33} . \end{bmatrix}$$

3   Computational Results

Example 1

In the first example, we input the desired eigenvalues in a diagonal matrix $D = \text{diag } \{\lambda_1, \lambda_2, \cdots, \lambda_n\}$. The matrix A is created by an arbitrary similar transformation,

$$A = Q^T D\, Q,$$

where Q is orthogonal. Table 1 shows the eigenvalues predicted by QR method (after 50 iterations), and 3-D Jacobi method (after 13 sweeps), and the computational time is the same.

Exact eigenvalues	QR method	3-D Jacobi(T)
20.000003	20.00000260	20.00000300
20.000002	20.00000228	20.00000200
20.000001	20.00000112	20.00000100
10.000003	10.00000299	10.00000300
10.000002	10.00000198	10.00000200
10.000001	10.00000103	10.00000100
5.000003	5.00000236	5.00000300
5.000002	5.00000192	5.00000200
5.000001	5.00000172	5.00000100
0.000003	0.00000301	0.00000300
0.000002	0.00000200	0.00000200
0.000001	0.00000100	0.00000100

Table 1 Computational results of symmetric matrix with clustered eigenvalues

The symbol (T) means that the technique of threshold is used to speed up the convergence of rotation.

Example 2

In the second example, we still input the desired eigenvalues, but take the similar transformation by an arbitrary non-singular matrix. Thus the resulting matrix A is non-symmetric. The following is an example:

$$
A = \begin{bmatrix}
4.298516 & 1.182469 & 0.410518 & -1.376300 & -0.165541 \\
0.721725 & 1.573222 & 0.605365 & -0.306129 & -0.235933 \\
0.533689 & -0.852280 & 1.641552 & 0.799227 & -0.588693 \\
4.476127 & 0.956653 & -1.322030 & 2.753537 & -1.545648 \\
-7.045039 & -1.665648 & -0.223594 & -3.847320 & 4.733174
\end{bmatrix}.
$$

The eigenvalues are 1, 2, 3, 4, 5. Both QR method and 3-D Jacobi rotation method will give the exact answer of this problem. Table 2 lists the performances of these methods.

	QR method	2-D Jacobi	3-D Jacobi	3-D Jacobi (T)
Number of sweeps	51	8	9	13
CPU time (sec.)	2.31	0.17	0.39	0.39

Table 2 Computational results of non-symmetric matrix

Example 3

The matrix is copied as below

$$
A = \begin{bmatrix}
0.066731 & 0.909534 & 0.932534 & 0.021070 & 0.202019 \\
0.204082 & 0.793114 & 0.267157 & 0.128374 & 0.737414 \\
0.915231 & 0.762058 & 0.471062 & 0.612560 & 0.498195 \\
0.041452 & 0.697869 & 0.916608 & 0.605423 & 0.142607 \\
0.877655 & 0.186270 & 0.634890 & 0.027592 & 0.156724
\end{bmatrix}.
$$

The eigenvalues of this matrix are unknown. The 2-D Jacobi method can not transfer this matrix to be Hessenberg matrix. The best result of 2-D Jacobi method is copied as below

$$
A = \begin{bmatrix}
-0.773238 & 0.374243 & 0.193314 & 0.028767 & 0.406433 \\
0.000000 & -0.083899 & -0.662002 & -0.526807 & 0.333437 \\
0.000000 & 0.000000 & 1.238849 & -0.556680 & 0.153480 \\
0.000000 & 0.000000 & 0.000000 & 0.283533 & 0.483691 \\
0.000000 & -0.228625 & 0.000000 & -0.009919 & 0.278166
\end{bmatrix}.
$$

The 3-D Jacobi rotation may successfully transfer this matrix to be reducible upper-Hessenberg after 5 sweeps. The following is the final result:

$$
A = \begin{bmatrix}
-0.773238 & -0.178285 & 0.102188 & 0.046248 & -0.344663 \\
0.000000 & 0.565115 & -0.639768 & -0.165686 & 0.339693 \\
0.000000 & 0.000000 & 1.238923 & 0.277370 & -0.192007 \\
0.000000 & 0.000000 & 0.000000 & 0.191187 & 0.349896 \\
0.000000 & 0.000000 & 0.000000 & -0.621104 & -0.279236
\end{bmatrix}.
$$

4   Conclusion

As shown above the 3-D Jacobi rotation method is probably workable for eigenvalue problems of both symmetric and non-symmetric matrices. The prove of convergence for the symmetric case will be reported in the paper in preparation. The theoretical prove of the convergence for the non-symmetric case is still unknown and the estimation of the convergence rate is expected too.

References

1. Bojanczyk, A.W., Lutoborski, A.: Computation of the Euler angles of a symmetric 3×3 matrix. SIAM J. Matrix Anal. Appl. **12**, (1991) 41-48
2. Eberlein, P.J.: A Jacob-like method for the automatic computation of eigenvalues and eigenvectors of an arbitrary matrix. SIAM J. **10**, (1962) 74-88
3. Forsythe, G.E., Henrici,P.: The cyclic Jacobi method for computing the principle values of a complex matrix. Tran. Amer. Math. Soc., **94**, (1960) 1-23
4. Hacon, D.: Jacobi method for skew-symmetric matrices. SIAM J. Matrix Anal. **14**, (1993) 619-628
5. Hansen, E.R.: On cyclic Jacobi method. SIAM J, **11**, (1963) 448-459
6. Rhee, N.H., Hari, V.: On the global and cubic convergence of a quasicyclic Jacobi method. Number. Math. **66**, (1993) 97-122
7. Shroof, G., Schreider, R.: On the convergence of cyclic Jacobi methods. Numerical linear algebra, digital signal processing and parallel algorithms. G.H. Golub and P. Van Dooren, editors, NATO ASI F Series **10**, Springer-Verlag, Berlin (1991), 597-604
8. Kerner, W.: Large–Scale complex eigenvalue problems, J. of Computational Physics, Vol. **85**, No. 1, (1989) 1–85
9. Golub, G.H., Van Loan, C.F.: Matrix Computation
10. Hari, V.: On sharp quadratic convergence bounds for the serial Jacobi methods, Numer. Math., **60**, (1991) 375-406
11. Watkins, D.S.: Fundamentals of Matrix Computations, John Wiley & Sons, (1991)

Perturbation Analyses for the Cholesky Factorization with Backward Rounding Errors

Xiao-Wen Chang

School of Computer Science, McGill University,
Montreal, Quebec, Canada, H3A 2A7

Abstract. This paper gives perturbation analyses of the Cholesky factorization with the form of perturbations we could expect from the equivalent backward error in A resulting from numerically stable computations. The analyses more accurately reflect the sensitivity of the problem than previous such results. Both numerical results and an analysis show the standard method of symmetric pivoting usually improves the condition of the problem. It follows that the computed R will usually have more accuracy when we use the standard symmetric pivoting strategy.

1 Introduction

Let $A \in \mathcal{R}^{n \times n}$ be a symmetric positive definite matrix. Then A has a unique Cholesky factorization of the form $A = R^T R$, where R is upper triangular with positive diagonal elements and is called the Cholesky factor of A.

Suppose there is a perturbation ΔA in A and $A + \Delta A$ is also symmetric positive definite, then $A + \Delta A$ has a unique Cholesky factorization

$$A + \Delta A = (R + \Delta R)^T (R + \Delta R).$$

The goal of the perturbation analysis is to give a bound on $\|\Delta R\|$ (or $|\Delta R|$) in terms of $\|\Delta A\|$ (or $|\Delta A|$). There have been several papers dealing with the perturbation analysis for the Cholesky factor for a general perturbation ΔA, see [2], [8], and [10] for normwise analyses, and [11,12] for componentwise analyses. Recently perturbation results of a different flavor were presented in [4]. For ΔA corresponding to the backward rounding error in A resulting from standard numerical stable computations, a nice norm-based perturbation result was presented in [4]. Note [11,12] also included component perturbation bounds for a different and somewhat complicated form of the backward rounding error in A. The purpose of this paper is to establish new first-order perturbation bounds for ΔA corresponding to the backward rounding error, which are sharper than the corresponding first-order result in [4].

Now we introduce notation. To simplify the presentation, for any $n \times n$ matrix X, we define the upper triangular matrix

$$\text{up}(X) \equiv \begin{bmatrix} \frac{1}{2}x_{11} & x_{12} & \cdots & x_{1n} \\ 0 & \frac{1}{2}x_{22} & \cdots & x_{2n} \\ \vdots & \vdots & \ddots & \vdots \\ 0 & 0 & \cdots & \frac{1}{2}x_{nn} \end{bmatrix}, \tag{1}$$

and the diagonal matrix

$$\text{diag}(X) \equiv \text{diag}(x_{11}, x_{22}, \ldots, x_{nn}) \equiv \text{diag}(x_{ii}).$$

For any matrix $C \equiv (c_{ij}) \equiv [c_1, \ldots, c_n] \in \mathcal{R}^{n \times n}$, denote by $c_j^{(i)}$ the vector of the leading i elements of c_j. Using 'u' to denote 'upper', we define

$$\text{uvec}(C) \equiv [c_1^{(1)^T}, c_2^{(2)^T}, \cdots, c_n^{(n)^T}]^T, \tag{2}$$

the vector formed by stacking the columns of the upper triangular part of C into one long vector. Note that for any upper triangular X, $\|\text{uvec}(X)\|_2 = \|X\|_F$. To help our norm analysis, for any matrix $C \in \mathcal{R}^{n \times n}$ we define

$$\text{duvec}(C) \equiv \text{diag}\,(1, \underbrace{\sqrt{2}, 1}_{2}, \ldots, \underbrace{\sqrt{2}, \sqrt{2}, \ldots, \sqrt{2}, 1}_{n})\,\text{uvec}(C). \tag{3}$$

Thus for any symmetric matrix X we have $\|\text{duvec}(X)\|_2 = \|X\|_F$. Finally let $\mathcal{D}_n$ be the set of all $n \times n$ real positive-definite diagonal matrices.

At the end of this section we summarize the strongest norm-based results for a general perturbation ΔA presented in [2], as those results will be compared with the new results for ΔA having the form of backward rounding error.

Theorem 1. (see [2]) *Let $A \in \mathcal{R}^{n \times n}$ be symmetric positive definite with the Cholesky factorization $A = R^T R$. Let ΔA be symmetric. If $\epsilon \equiv \|\Delta A\|_F / \|A\|_2$ satisfies $\kappa_2(A)\epsilon < 1$, where $\kappa_2(A) \equiv \|A\|_2 \|A^{-1}\|_2$, then $A + \Delta A$ has the Cholesky factorization*

$$A + \Delta A = (R + \Delta R)^T (R + \Delta R),$$

such that

$$\frac{\|\Delta R\|_F}{\|R\|_2} \leq \kappa_C(A)\epsilon + O(\epsilon^2), \tag{4}$$

$$\tfrac{1}{2}\kappa_2^{1/2}(A) \leq \kappa_C(A) \leq \kappa_C'(A) \leq \kappa_2(A), \tag{5}$$

where

$$\kappa_C(A) \equiv \|W_R^{-1}\|_2 \|A\|_2^{1/2}, \tag{6}$$

$$\kappa_C'(A) \equiv \inf_{D \in \mathcal{D}_n} \kappa_C'(A, D), \qquad \kappa_C'(A, D) \equiv \kappa_2(R)\kappa_2(D^{-1}R), \tag{7}$$

with $W_R \in \mathcal{R}^{\frac{n(n+1)}{2} \times \frac{n(n+1)}{2}}$ defined by

$$W_R = \begin{bmatrix} \begin{matrix} 2r_{11} \\ \sqrt{2}r_{12} & \sqrt{2}r_{11} \\ & 2r_{12} & 2r_{22} \\ & & \ddots & \ddots \\ \sqrt{2}r_{1n} & & & \\ & \sqrt{2}r_{1n} & \sqrt{2}r_{2n} & \end{matrix} & \begin{matrix} \\ \\ \\ \\ \sqrt{2}r_{11} \\ \sqrt{2}r_{12} & \sqrt{2}r_{22} \\ & \ddots & \ddots \\ 2r_{1n} & 2r_{2n} & \cdot & 2r_{nn} \end{matrix} \end{bmatrix}. \tag{8}$$

If standard symmetric pivoting is used in computing the Cholesky factorization $PAP^T = R^T R$, *we have*

$$\frac{1}{2}\kappa_2^{1/2}(A) \leq \kappa_C(PAP^T) \leq \kappa_C'(PAP^T) \leq \kappa_2^{1/2}(A)\sqrt{2n(n+1)(4^n+6n-1)/6}.$$

In [2] we showed the bound (4) is attainable to first order in ϵ. So $\kappa_C(A)$ is the condition number for the Cholesky factor R. It is expensive to estimate or evaluate $\kappa_C(A)$ directly by using the usual methods. Fortunately we can use $\kappa_C'(A)$ as an approximation of $\kappa_C(A)$. According to the well-known result of van de Sluis [13], if D is chosen to equilibrate the rows of R, i.e., all rows of $D^{-1}R$ have unit length, then $\kappa_C'(A, D)$ with such a D is at most a factor of $\sqrt{n}$ off the infimum $\kappa_C'(A)$. Notice that $\kappa_C'(A, D)$ can be estimated by a standard condition estimator in $O(n^2)$ flops.

2   Main results

In this section we first derive a tight perturbation bound, leading to the condition number $\chi_C(A)$ for perturbation having the form of equivalent backward rounding error. Also we derive a practical estimate $\chi_C'(A)$ of $\chi_C(A)$. Then we compare $\chi_C(A)$ with $\kappa_C(A)$, and $\chi_C'(A)$ with $\kappa_C'(A)$. Finally we show how standard pivoting improves the condition number $\chi_C(A)$.

Before proceeding we introduce the following result presented in [3], see also [6, Theorems 10.5 and 10.7].

Lemma 2. (see [3]) *Let* $A \equiv D_c H D_c \in \mathcal{R}^{n \times n}$ *be a symmetric positive definite floating point matrix, where* $D_c \equiv \text{diag}(A)^{1/2}$. *If*

$$n\epsilon \|H^{-1}\|_2 < 1, \tag{9}$$

where $\epsilon \equiv (n+1)u/(1-2(n+1)u)$ *with* u *being the unit round-off, then Cholesky factorization applied to* A *succeeds (barring underflow and overflow) and produces a nonsingular* $\tilde{R}$, *which satisfies*

$$A + \Delta A = \tilde{R}^T \tilde{R}, \qquad |\Delta A| \leq \epsilon dd^T, \tag{10}$$

where $d_i = a_{ii}^{1/2}$.

This lemma is applicable to any standard algorithms for the Cholesky factorization (see, for example, [5] for the standard algorithms). Based on this, we establish the following result.

Theorem 3. *Suppose all of the assumptions in Lemma 2 hold. Let* $A = R^T R$ *be the Cholesky factorization of* A. *Set* $\hat{R} \equiv RD_c^{-1}$. *Then for the perturbation* ΔA *and result* $\tilde{R}$ *in (10) we have with* $\Delta R \equiv \tilde{R} - R$ *that*

$$\frac{\|\Delta R\|_F}{\|R\|_2} \leq \chi_C(A)\epsilon + O(\epsilon^2), \tag{11}$$

$$\chi_C(A) \leq \chi_C'(A) \leq n\|H^{-1}\|_2, \tag{12}$$

$$\chi_C(A) \leq \frac{1}{\sqrt{2}}\|H^{-1}\|_2, \tag{13}$$

where

$$\chi_C(A) \equiv n\|\mathcal{D}_c W_{\hat{R}}^{-1}\|_2/\|R\|_2,$$
$$\chi_C'(A) \equiv \inf_{D \in \mathcal{D}_n} \chi_C'(A, D),$$
$$\chi_C'(A, D) \equiv n\|\hat{R}^{-1}\|_2 \|\hat{R}^{-1}D\|_2 \|D^{-1}R\|_2/\|R\|_2,$$

with

$$\mathcal{D}_c \equiv \text{diag}\,(a_{11}^{1/2}, \underbrace{a_{22}^{1/2}, a_{22}^{1/2}}_{2}, \ldots, \underbrace{a_{nn}^{1/2}, a_{nn}^{1/2}, \ldots, a_{nn}^{1/2}}_{n}) \in \mathcal{R}^{\frac{n(n+1)}{2} \times \frac{n(n+1)}{2}}, \quad (14)$$

and $W_{\hat{R}}$ being just W_R in (8) with each entry r_{ij} replaced by $\hat{r}_{ij}$. The bound in (11) is approximately attainable to first-order in ϵ.

Proof. Let $G \equiv \Delta A/\epsilon$. By (10) and (9) it is easy to show $A + tG$ is symmetric positive definite for all $t \in [0, \epsilon]$, and so it has the Cholesky factorization

$$A + tG = R^T(t)R(t), \qquad |t| \leq \epsilon, \quad (15)$$

with $R(0) = R$ and $R(\epsilon) = \tilde{R} \equiv R + \Delta R$. Differentiating (15) and setting $t = 0$ in the result gives

$$G = R^T \dot{R}(0) + \dot{R}(0)^T R, \quad (16)$$

which, combined with $R = \hat{R}D_c$, gives

$$\hat{R}^T \dot{R}(0)D_c^{-1} + D_c^{-1}\dot{R}^T(0)\hat{R} = D_c^{-1}GD_c^{-1}, \quad (17)$$

The upper and lower triangular parts of (17) contain identical information, and it is easy to show that by using (2) and (3) the upper triangular part of (17) can be rewritten as the following matrix-vector equation form

$$W_{\hat{R}} \text{ uvec}(\dot{R}(0)D_c^{-1}) = \text{duvec}(D_c^{-1}GD_c^{-1}). \quad (18)$$

Notice from (2) that $\text{uvec}(\dot{R}(0)D_c^{-1}) = \mathcal{D}_c^{-1} \text{uvec}(\dot{R}(0))$ with $\mathcal{D}_c$ as in (14), then from (18) we have

$$\text{uvec}(\dot{R}(0)) = \mathcal{D}_c W_{\hat{R}}^{-1} \text{duvec}(D_c^{-1}GD_c^{-1}), \quad (19)$$

which with $G = \Delta A/\epsilon$ and (10) gives

$$\|\dot{R}(0)\|_F \leq \|\mathcal{D}_c W_{\hat{R}}^{-1}\|_2 \|D_c^{-1}dd^T D_c^{-1}\|_F = n\|\mathcal{D}_c W_{\hat{R}}^{-1}\|_2. \quad (20)$$

The Taylor expansion of $R(t)$ about $t = 0$ gives at $t = \epsilon$

$$R + \Delta R \equiv \tilde{R} = R(\epsilon) = R(0) + \epsilon\dot{R}(0) + O(\epsilon^2), \quad (21)$$

which, combined with (20), gives (11).

Obviously there exists a symmetric matrix $F \in \mathcal{R}^{n \times n}$ such that

$$\|\mathcal{D}_c W_{\hat{R}}^{-1}\text{duvec}(D_c^{-1}FD_c^{-1})\|_2 = \|\mathcal{D}_c W_{\hat{R}}^{-1}\|_2 \|D_c^{-1}FD_c^{-1}\|_F.$$

Then by taking $G = (\min_{f_{ij} \neq 0} d_i d_j / |f_{ij}|) F$, we have $|\Delta A| \leq \epsilon \, dd^T$ and from (19) that

$$\|\dot{R}(0)\|_F = (\min_{f_{ij} \neq 0} d_i d_j / |f_{ij}|) \|\mathcal{D}_c W_{\hat{R}}^{-1}\|_2 \|D_c^{-1} F D_c^{-1}\|_F \geq \|\mathcal{D}_c W_{\hat{R}}^{-1}\|_2,$$

which shows the first-order bound in (11) is approximately attained for such G.

It remains to prove (12) and (13). From (16) with $R = \hat{R} D_c$, it follows that

$$\dot{R}(0) R^{-1} + R^{-T} \dot{R}^T(0) = \hat{R}^{-T} D_c^{-1} G D_c^{-1} \hat{R}^{-1}.$$

From this we see with notation 'up' (see (1)) that

$$\dot{R}(0) = \text{up}(\hat{R}^{-T} D_c^{-1} G D_c^{-1} \hat{R}^{-1}) R, \tag{22}$$

so with (19) and the fact that $\text{up}(XD) = \text{up}(X)D$ and $\|\text{up}(X)\|_F \leq \|X\|_F$ for any $X \in \mathcal{R}^{n \times n}$ and $D \in \mathcal{D}_n$ we have

$$\|\mathcal{D}_c W_{\hat{R}}^{-1} \text{duvec}(D_c^{-1} G D_c^{-1})\|_2 = \|\text{up}(\hat{R}^{-T} D_c^{-1} G D_c^{-1} \hat{R}^{-1}) R\|_F$$
$$\leq \|\hat{R}^{-1}\|_2 \|D_c^{-1} G D_c^{-1}\|_F \|\hat{R}^{-1} D\|_2 \|D^{-1} R\|_2.$$

Actually this holds for any symmetric $G \in \mathcal{R}^{n \times n}$ since it was essentially obtained from the matrix equation $R^T X + X^T R = G$ with X triangular. Notice $\|\text{duvec}(D_c^{-1} G D_c^{-1})\|_2 = \|D_c^{-1} G D_c^{-1}\|_F$, thus we must have

$$\|\mathcal{D}_c W_{\hat{R}}^{-1}\|_2 \leq \|\hat{R}^{-1}\|_2 \|\hat{R}^{-1} D\|_2 \|D^{-1} R\|_2.$$

Since this is true for any $D \in \mathcal{D}_n$, we see that the first inequality in (12) holds. The second inequality in (12) follows from $\chi_c'(A, I) = n \|\hat{R}^{-1}\|_2^2 = n \|H^{-1}\|_2$.

Similarly we can prove (13) by using the fact that $\|\text{up}(X)\|_F \leq \frac{1}{\sqrt{2}} \|X\|_F$ for any symmetric $X \in \mathcal{R}^{n \times n}$. $\qquad \square$

Since the first-order bound (11) is approximately attainable, $\chi_c(A)$ can be regarded as the *condition number* for the Cholesky factorization with the form of perturbation error satisfying (10).

Recently Drmač et al. [4] gave a perturbation analysis with the same form of perturbation error. Their asymptotic perturbation bound can be presented as follows:

$$\frac{\|\Delta R\|_2}{\|R\|_2} \leq (\tfrac{1}{2} + \lceil \log n \rceil) n \|H^{-1}\|_2 \epsilon + O(\epsilon^2). \tag{23}$$

This suggests that $\|H^{-1}\|_2$ can be regarded as a conditioning measure. From (12) we see our bound (11) is sharper than (23). In fact it can be much sharper. As an illustration suppose $A = \begin{bmatrix} 1 & \delta \\ \delta & \delta^2 + \delta^4 \end{bmatrix}$, $R = \begin{bmatrix} 1 & \delta \\ 0 & \delta^2 \end{bmatrix}$. Then it is easy to show for small $\delta > 0$, $\chi_c(A) \leq \chi_c'(A) = O(1/\delta)$, $\|H^{-1}\|_2 = O(1/\delta^2)$. Thus (23) may overestimate the true relative error of the computed Cholesky factor, and the approximation $\chi_c'(A)$ to the condition number $\chi_c(A)$ is a significant improvement to $\|H^{-1}\|_2$. Furthermore it is easy to see $\|H^{-1}\|_2$ is invariant if pivoting is used in computing the Cholesky factorization of A, whereas $\chi_C(A)$ and $\chi'_C(A)$ depend on any pivoting. Thus $\chi_C(A)$ and $\chi'_C(A)$ more closely reflects the true sensitivity of the Cholesky factorization than $\|H^{-1}\|_2$.

As far as we can see, it is expensive to compute or approximate $\chi_C(A)$ directly by the usual approach. Fortunately $\chi'_C(A)$ is quite easy to estimate. By the result of van der Sluis [13], $\|\hat{R}^{-1}D\|_2 \|D^{-1}R\|_2$ will be nearly minimum when the rows of $D^{-1}R$ are equilibrated. Then a procedure for obtaining an $O(n^2)$ condition estimator for the Cholesky factorization with backward rounding errors is to choose $D = D_r \equiv \operatorname{diag}(\|R(i,:)\|_2)$, and then use a standard norm estimator to estimate all factors in $\chi'_C(A, D)$. Numerical experiments suggest usually $\chi'_C(A)$ is a reasonable approximation of $\chi_C(A)$. But $\chi'_C(A)$ can still be very larger than $\chi_C(A)$, even though it can be much smaller than $n\|H^{-1}\|_2$. For the example above we have $\chi_C(A) = O(1)$.

Numerical tests suggest that usually $\chi_C(A)$ is smaller or much smaller than $\kappa_C(A)$, defined by (6). This is not surprising since (10) provides more information about the perturbation in the data. In [1] we proved the following results.

Theorem 4.

$$\frac{1}{n}\chi_C(A) \le \kappa_C(A) \le \frac{\max_i a_{ii}}{\min_i a_{ii}}\chi_C(A), \qquad (24)$$
$$\frac{1}{n}\chi'_C(A) \le \kappa'_C(A) \le \frac{\max_i a_{ii}}{\min_i a_{ii}}\chi'_C(A).$$

The first inequality in (24) is attainable, since equality will hold by taking $A = cI$ with $c > 0$. The second inequality is at least nearly attainable. In fact taking $A = R^T R$, where $R = \operatorname{diag}(\delta^{n-1}, \delta^{n-2}, \ldots, \delta, 1) + e_1 e_n^T$ with small $\delta > 0$, we easily obtain

$$\kappa_2(A) = O(\frac{1}{\delta^{2n-2}}), \qquad \frac{\max_i a_{ii}}{\min_i a_{ii}}\chi_C(A) = \frac{2}{\delta^{2n-2}}O(1) = O(\frac{1}{\delta^{2n-2}}).$$

This example also suggests that possibly $\kappa_C(A)$ is much larger than $\chi_C(A)$ if the maximum element is much larger than the minimum one on the diagonal of A.

The standard symmetric pivoting strategy can usually improve $\chi_C(A)$, just as it can usually improve $\kappa_C(A)$. In [1] we showed the following theorem.

Theorem 5. *Let $A \in \mathcal{R}^{n \times n}$ be symmetric positive definite with the Cholesky factorization $PAP^T = R^T R$ when the standard pivoting strategy is used. Then*

$$\chi_C(PAP^T) \le \chi'_C(PAP^T) \le \|H^{-1}\|_2^{1/2} n\sqrt{2n(n+1)(4^n + 6n - 1)}/6, \qquad (25)$$

where $PAP^T \equiv D_c H D_c$ with $D_c \equiv \operatorname{diag}(PAP^T)^{1/2}$.

One may not be impressed by the 4^n factor in the upper bound, and may wonder if it can significantly be improved. In fact we can prove for any n the upper bound can nearly be approximated by a parametrized family of matrices (cf. [1, pp.25–26]). But such examples are rare in practice.

Table 1. Results for Pascal matrices without pivoting.

n	$\chi_C(A)$	$\chi_{C1}(A, D_r)$	$\kappa_C(A)$	$\kappa_{C1}(A, D_r)$	$\frac{\sqrt{2}}{2} n \|H^{-1}\|_2$
1	5.0e−01	1.0e+00	5.0e−01	1.0e+00	7.1e−01
2	2.2e+00	6.0e+00	2.1e+00	6.3e+00	4.8e+00
3	8.9e+00	3.6e+01	9.7e+00	5.0e+01	3.6e+01
4	3.9e+01	2.3e+02	5.5e+01	4.8e+02	3.0e+02
5	1.7e+02	1.4e+03	3.5e+02	4.9e+03	2.5e+03
6	7.8e+02	9.2e+03	2.5e+03	5.2e+04	2.2e+04
7	3.5e+03	5.8e+04	1.9e+04	5.7e+05	2.0e+05
8	1.6e+04	3.7e+05	1.5e+05	6.3e+06	1.7e+06
9	7.7e+04	2.3e+06	1.3e+06	7.0e+07	1.5e+07
10	3.6e+05	1.5e+07	1.1e+07	7.9e+08	1.4e+08
11	1.7e+06	9.1e+07	9.8e+07	9.0e+09	1.2e+09
12	8.4e+06	5.6e+08	8.7e+08	1.0e+11	1.1e+10
13	4.1e+07	3.5e+09	7.8e+09	1.2e+12	9.8e+10
14	2.0e+08	2.2e+10	7.1e+10	1.4e+13	8.8e+11
15	9.7e+08	1.3e+11	6.5e+11	1.6e+14	7.9e+12

Our numerical experiments confirms that $\chi_C(PAP^T)$ is usually (much) smaller than $\chi_C(A)$. Thus Cholesky factorization with standard symmetric pivoting will usually give more accurate $\tilde{R}$.

By following the approach of [7], it is straightforward, but detailed and lengthy, to extend the first-order results to provide rigorous perturbation bounds. For such results, see [1].

3   Numerical experiments

In section 2 we presented a new first-order perturbation bound for the Cholesky factor with the change caused by backward rounding errors, defined $\chi_C(A) \equiv n\|\mathcal{D}_c W_{\hat{R}}^{-1}\|_2/\|A\|_2^{1/2}$ as the condition number, and suggested $\chi_C(A)$ could be estimated in practice by

$$\chi'_C(A, D_r) \equiv n\|\hat{R}^{-1}\|_2 \|\hat{R}^{-1} D_r\|_2 \|D_r^{-1} R\|_2/\|R\|_2$$

with $D_r = \text{diag}(\|R(i,:)\|_2)$, which can be estimated by standard norm estimators in $O(n^2)$ flops. Our new first-order result is (much) better than the previous corresponding result. Also we compare $\chi_C(A)$ with $\kappa_C(A)$, the condition number for general perturbation ΔA, and compare the corresponding estimates $\chi'_C(A)$ with $\kappa'_C(A)$ as well.

Now we give a set of examples to show our findings. The matrices are $n \times n$ Pascal matrices, (with elements $a_{1j} = a_{i1} = 1$, $a_{ij} = a_{i,j-1} + a_{i-1,j}$), $n = 1, 2, \ldots, n$. The results are shown in Table 1 without pivoting and in Table 2 with pivoting.

Note in Tables 1 and 2 how $\frac{1}{\sqrt{2}} n\|H^{-1}\|_2$ can be worse than $\chi_C(A)$. In Table 2 pivoting is seen to give a significant improvement to $\chi_C(A)$. Also we observe from both the tables that $\chi'_C(A)$ is a reasonable approximation of $\chi_C(A)$. We see $\chi_C(A)$ is smaller than $\kappa_C(A)$ for $n > 2$.

Table 2. Results for Pascal matrices with pivoting, $\tilde{A} \equiv PAP^T$.

n	$\chi_C(\tilde{A})$	$\chi_{C1}(\tilde{A}, D_r)$	$\kappa_C(\tilde{A})$	$\kappa_{C1}(\tilde{A}, D_r)$	$\frac{\sqrt{2}}{2}n\|H^{-1}\|_2$
1	5.0e–01	1.0e+00	5.0e–01	1.0e+00	7.1e–01
2	1.6e+00	4.9e+00	1.5e+00	4.2e+00	4.8e+00
3	4.1e+00	1.8e+01	5.1e+00	1.6e+01	3.6e+01
4	1.3e+01	6.1e+01	2.2e+01	8.0e+01	3.0e+02
5	3.6e+01	2.1e+02	8.3e+01	3.3e+02	2.5e+03
6	7.7e+01	6.7e+02	2.5e+02	1.3e+03	2.2e+04
7	1.8e+02	2.2e+03	9.4e+02	5.1e+03	2.0e+05
8	4.8e+02	7.8e+03	4.0e+03	2.4e+04	1.7e+06
9	1.2e+03	2.7e+04	1.6e+04	1.0e+05	1.5e+07
10	3.6e+03	9.0e+04	7.6e+04	4.7e+05	1.4e+08
11	7.5e+03	2.7e+05	2.4e+05	1.8e+06	1.2e+09
12	1.8e+04	9.2e+05	8.3e+05	8.2e+06	1.1e+10
13	3.9e+04	2.9e+06	3.2e+06	3.1e+07	9.8e+10
14	9.4e+04	8.5e+06	1.3e+07	1.2e+08	8.8e+11
15	2.2e+05	2.8e+07	5.4e+07	4.9e+08	7.9e+12

Acknowledgement.

I would like to thank Chris Paige for his valuable comments and suggestions.

References

1. Chang, X.-W.: Perturbation analysis of some matrix factorizations. Ph.D thesis, McGill University, February 1997
2. Chang, X.-W., Paige, C. C., Stewart, G. W.: New perturbation analyses for the Cholesky factorization. *IMA J. Numer. Anal.*, **16** (1996) 457–484
3. Demmel, J. W.: On floating point errors in Cholesky. Technical Report CS-89-87, Department of Computer Science, University of Tennessee, 1989. LAPACK Working Note 14
4. Drmač, Z., Omladič, M., Veselić, K.: On the perturbation of the Cholesky factorization. *SIAM J. Matrix Anal. Appl.*, **15** (1994) 1319–1332
5. Golub, G. H., Van Loan, C. F.: *Matrix computations*. Third Edition. The Johns Hopkins University Press, Baltimore, Maryland, 1996
6. Higham, N. J.: Accuracy and Stability of Numerical Algorithms. *Society for Industrial and Applied Mathematics*, Philadelphia, PA, 1996
7. Stewart, G. W.: Error and perturbation bounds for subspaces associated with certain eigenvalue problems. *SIAM Rev.*, **15** (1973) 727–764
8. Stewart, G. W.: Perturbation bounds for the QR factorization of a matrix. *SIAM J. Numer. Anal.*, **14** (1977) 509–518
9. Stewart, G. W.: On the perturbation of LU, Cholesky, and QR factorizations. *SIAM J. Matrix Anal. Appl.*, **4** (1993) 1141–1145
10. Sun, J.-G.: Perturbation bounds for the Cholesky and QR factorization. *BIT*, **31** (1991) 341–352
11. Sun, J.-G.: Rounding-error and perturbation bounds for the Cholesky and LDL^T factorizations. *Linear Algebra and Appl.*, **173** (1992) 77–97
12. Sun, J.-G.: Componentwise perturbation bounds for some matrix decompositions. *BIT*, **32** (1992) 702–714
13. van der Sluis, A.: Condition numbers and equilibration of matrices. *Numerische Mathematik*, **14** (1966) 14–23

An Efficient Finite Element Method for Interface Problems

Zhiming Chen[1] * and Jun Zou[2] **

[1] Institute of Mathematics, Academia Sinica, Beijing 100080, P.R. China
[2] Department of Mathematics, The Chinese University of Hong Kong, Hong Kong.

Abstract. In this talk an efficient finite element method is proposed for solving elliptic and parabolic interface problems over convex polygonal domains. Nearly the same optimal convergences as for regular problems are achieved when the interfaces are of arbitrary shape but are smooth. Numerical experiments are reported which are consistent with the theoretical results.

1 Introduction

Solving elliptic and parabolic problems with discontinuous coefficients are often encountered in material sciences and fluid dynamics. It is the case when two distinct materials or fluids with different conductivities or densities or diffusions are involved. When the interface is smooth enough, the solution of the interface problem is also very smooth in individual regions occupied by materials or fluids, but the global regularity is usually very low. Due to this and the irregular geometry of the interface, achieving the high order of accuracy seems difficult with finite element methods (cf. Babuska [1]), when elements not fitting with the interface of general shape.

One can find a few works on finite element methods for elliptic interface problems in the literature, see Babuska [1], Xu [5]. But there are much more works on finite difference methods for elliptic and parabolic interface problems, we refer to LeVeque and Li [3,4] and the references therein.

In this talk, we propose an efficient finite element method for solving elliptic and parabolic interface problems of second order and report some nearly optimal convergence and numerical results. The interface here is allowed to be of arbitrary shape but is smooth. Different from the existing finite element methods, the calculations of the stiffness matrix and the interface integral related to the jumps of normal derivatives are much simpler and more practical here.

2 Elliptic interface problems

We now formulate the interface problem. Let Ω be a convex polygon in R^2 and $\Omega_1 \subset \Omega$ be an open domain with C^2 boundary $\Gamma = \partial\Omega_1 \subset \Omega$. Let $\Omega_2 = \Omega \setminus \Omega_1$.

* The work of this author was partially supported by China NSF.
** The work of this author was partially supported by the Direct Grant of CUHK and Hong Kong RGC Grant No. CUHK 338/96E.

We consider the following elliptic interface problem

$$-\nabla \cdot (\beta \nabla u) = f \quad \text{in} \quad \Omega; \quad u = 0 \quad \text{on} \quad \partial \Omega \tag{1}$$

with jump conditions on the interface

$$[u] = 0, \quad \left[\beta \frac{\partial u}{\partial \mathbf{n}}\right] = g \quad \text{across} \quad \Gamma, \tag{2}$$

where $[v]$ is the jump of a quantity v across the interface Γ and $\mathbf{n}$ the unit outward normal to the boundary $\partial \Omega_1$. Without loss of generality, we consider only the case

$$\beta(x) = \beta_1 \quad \text{for} \quad x \in \Omega_1; \quad \beta(x) = \beta_2 \quad \text{for} \quad x \in \Omega_2.$$

The associated bilinear form with the problem (1)-(2) is

$$a(u,v) = \int_\Omega \beta(x) \nabla u \cdot \nabla v \, dx, \quad \forall u, v \in H^1(\Omega),$$

then its variational formulation reads as follows:

Problem (P). Find $u \in H_0^1(\Omega)$ such that

$$a(u,v) = (f,v) + \langle g, v \rangle, \quad \forall v \in H_0^1(\Omega). \tag{3}$$

Here $(\cdot, \cdot)$ and $\langle \cdot, \cdot \rangle$ denote the scalar products in the $L^2(\Omega)$ space and the interface space $L^2(\Gamma)$, resp.

We now formulate the finite element method. To triangulate the domain Ω, we first approximate Ω_1 by a domain Ω_1^h with a polygonal boundary Γ_h whose vertices all lie on the interface Γ. Let Ω_2^h stand for the domain with $\partial \Omega$ and Γ_h as its exterior and interior boundaries, resp. Then we triangulate Ω by a finite set of closed triangles $\mathcal{T}_h = \{K\}$ which satisfies the following conditions:

(**A1**) $\bar{\Omega} = \cup_{K \in \mathcal{T}_h} K$,
(**A2**)   if $K_1, K_2 \in \mathcal{T}_h$ and $K_1 \neq K_2$, then either $K_1 \cap K_2 = \emptyset$ or $K_1 \cap K_2$ is a common vertex or edge of both triangles,
(**A3**)   each $K \in \mathcal{T}_h$ is either in Ω_1^h or Ω_2^h, and has at most two vertices lying on Γ_h.

The triangles with one or two vertices on Γ_h are called interface triangles, the set of all interface triangles is denoted by $\mathcal{T}_h^*$ and we let $\Omega^* = \cup_{K \in \mathcal{T}_h^*} K$.

For each triangle $K \in \mathcal{T}_h$, we use h_K for its diameter, ρ_K and $\bar{\rho}_K$ for the diameters of its inscribed and circumscribed circles, respectively. Let $h = \max_{K \in \mathcal{T}_h} h_K$. We assume that the family of triangulations $\{\mathcal{T}_h\}_{h>0}$, is quasi-uniform, i.e. there are two positive constants C_0 and C_1 independent of h such that

$$C_0 \rho_K \leq h \leq C_1 \bar{\rho}_K, \quad \forall K \in \mathcal{T}_h, \quad \forall h. \tag{4}$$

Now we define V_h to be the piecewise linear finite element space over the triangulation $\mathcal{T}_h$ and V_h^0 the subspace of V_h with its functions vanishing on the boundary $\partial \Omega$. For the coefficient function $\beta(x)$, we define its approximation $\beta_h(x)$ as follows: for each triangle $K \in \mathcal{T}_h$, let $\beta_K(x) = \beta_i$ if $K \subset \Omega_h^i$, $i = 1$ or 2. Then β_h is defined by

$$\beta_h(x) = \beta_K(x), \quad \forall K \in \mathcal{T}_h.$$

It is easy to verify that

$$\text{supp}(\beta - \beta_h) \cap K = \tilde{K}, \quad \forall K \in \mathcal{T}_h^*.$$

Corresponding to the bilinear form $a(\cdot, \cdot) : H^1(\Omega) \times H^1(\Omega) \mapsto R^1$ defined previously, we introduce its discrete form $a_h(\cdot, \cdot) : H^1(\Omega) \times H^1(\Omega) \mapsto R^1$ by

$$a_h(u, v) = \sum_{K \in \mathcal{T}_h} \int_K \beta_K(x) \nabla u \cdot \nabla v \, dx, \quad \forall u, v \in H^1(\Omega).$$

Furthermore, we need an approximation g_h to the interface function g on Γ. Let $\{P_j\}_{j=1}^{m_h}$ be the set of all nodes of the triangulation $\mathcal{T}_h$ lying on the interface Γ, and $\{\phi_j^h\}_{j=1}^{m_h}$ the set of standard nodal basis functions corresponding to $\{P_j\}_{j=1}^{m_h}$ in the space V_h. Assume that $g \in C(\Gamma)$. Then we define $g_h \in V_h$ by

$$g_h = \sum_{j=1}^{m_h} g(P_j) \phi_j.$$

Then the finite element approximation to **Problem (P)** is defined as follows:
Problem (P_h). Find $u_h \in V_h^0$ such that

$$a_h(u_h, v_h) = (f, v_h) + \langle g_h, v_h \rangle_h, \quad \forall v_h \in V_h^0. \tag{5}$$

Here $\langle \cdot, \cdot \rangle_h$ denotes the scalar product in the space $L^2(\Gamma_h)$. It is easy to see that the finite element problem (5) has a unique solution u_h.

Our main results are the following convergence theorem:

Theorem 1. *Let u and u_h be the solutions to* **Problem (P)** *and* **Problem (P_h),** *respectively. Then, for $0 < h < h_0$, we have*

$$\|\nabla(u - u_h)\|_{L^2(\Omega)} \leq C h \,|\log h|^{1/2} \left(\|f\|_{L^2(\Omega)} + \|g\|_{H^2(\Gamma)} \right), \tag{6}$$
$$\|u - u_h\|_{L^2(\Omega)} \leq C h^2 \,|\log h| \left(\|f\|_{L^2(\Omega)} + \|g\|_{H^2(\Gamma)} \right). \tag{7}$$

Remark 2. The regularity of g in Theorem 1 can be weakened to be piecewise smooth. The error estimates (6)-(7) are optimal up to the factor $|\log h|$. In fact, the logorithm in H^1-error estimate can be removed.

The proof of Theorem 1 needs the following two lemmas. Let $\Pi_h : C(\bar{\Omega}) \mapsto V_h$ be the piecewise linear interpolant corresponding to the space V_h. As the solutions concerned are only in $H^1(\Omega)$ globally, one can not apply the standard interpolation theory directly. But we can still show

Lemma 3. *For the linear interpolation operator* $\Pi_h : C(\bar{\Omega}) \mapsto V_h$, *we have*

$$\|v - \Pi_h v\|_{L^2(\Omega)} + h \|\nabla(v - \Pi_h v)\|_{L^2(\Omega)} \leq C h^2 |\log h|^{1/2} \|v\|_X, \quad \forall v \in X. \quad (8)$$

Proof. The use of the discrete Hölder inequality and some Sobolev embedding theorem plays an important role in the proof, see [2] for details.

The second lemma is on the approximation property of g_h to the interface function g:

Lemma 4. *Assume that* $g \in H^2(\Gamma)$. *Then we have*

$$\left| \int_\Gamma g \, v_h \, ds - \int_{\Gamma_h} g_h \, v_h \, ds \right| \leq C h^{3/2} \|g\|_{H^2(\Gamma)} \|v_h\|_{H^1(\Omega^*)}, \quad \forall v_h \in V_h, \quad (9)$$

where $\Omega^* = \cup_{K \in \mathcal{T}_h^*} K$.

3   Parabolic interface problems

Let Ω be a convex polygon in R^2 and $\Omega_1 \subset \Omega$ be an open domain with C^2 boundary $\Gamma = \partial \Omega_1 \subset \Omega$, $\Omega_2 = \Omega \setminus \Omega_1$ and $Q_T = \Omega \times (0, T)$. In this section, we consider the following parabolic interface problem

$$\frac{\partial u}{\partial t} - \nabla \cdot (\beta \nabla u) = f(x, t) \quad \text{in} \quad Q_T \qquad (10)$$

with the initial and boundary conditions

$$u(x, 0) = u_0(x) \quad \text{in} \quad \Omega; \quad u = 0 \quad \text{on} \quad \partial\Omega \times (0, T) \qquad (11)$$

and jump conditions on the interface

$$[u] = 0, \quad \left[\beta \frac{\partial u}{\partial \mathbf{n}} \right] = g(x, t) \quad \text{across} \quad \Gamma \times (0, T), \qquad (12)$$

where $[v]$ and $\mathbf{n}$ are specified as in Section 2. For simplicity, we next consider only the case that the coefficient β is positive and piecewise constant, i.e.

$$\beta(x) = \beta_1 \quad \text{for} \quad x \in \Omega_1; \quad \beta(x) = \beta_2 \quad \text{for} \quad x \in \Omega_2.$$

We now propose the finite element approximation to the problem (10)-(12). We first divide the time interval $(0, T)$ into M equally-spaced subintervals by the following points:

$$0 = t^0 < t^1 < \cdots < t^M = T$$

with $t^n = n\tau$, $\tau = T/M$ the time step size. Let $I^n = (t^{n-1}, t^n]$ be the n-th subinterval. For a given sequence $\{w^n\}_{n=0}^M \subset L^2(\Omega)$, we introduce the backward difference quotient:

$$\partial_\tau w^n = \frac{w^n - w^{n-1}}{\tau}.$$

The piecewise linear finite element spaces V_h and V_h^0, and all other notation used in this section relevant to finite element discretizations are the same as in Section 2. To approximate the interface function $g(x,t)$, for $n = 1, 2, \cdots, M$, we define $\bar{g}_h^n \in V_h$ as

$$\bar{g}_h^n = \sum_{j=1}^{m_h} \bar{g}^n(P_j)\phi_j, \qquad \bar{g}^n(\cdot) = \tau^{-1} \int_{I^n} g(\cdot, t)\, dt.$$

With these notations, we can introduce the fully-discrete finite element approximation to the problem (10)-(12):

Problem ($P_{h,\tau}$). Let $u_h^0 = \Pi_h u_0$. For $n = 1, 2, \cdots, M$, find $u_h^n \in V_h^0$ such that

$$(\partial_\tau u_h^n, v_h) + a_h(u_h^n, v_h) = (f^n, v_h) + \langle \bar{g}_h^n, v_h \rangle_h, \quad \forall\, v_h \in V_h^0. \tag{13}$$

Now define a piecewise constant function $u_{h,\tau}$ in time by

$$u_{h,\tau}(x, t) = u_h^n(x), \quad \forall\, t \in (t^{n-1}, t^n], \quad n = 1, 2, \cdots, M,$$

then we can state our main convergence results as follows:

Theorem 5. *Let u and $u_{h,\tau}$ be the solutions of the problem (10)-(12) and Problem ($P_{h,\tau}$), respectively. Assume that $u_0 \in X$, $f \in H^1(0, T; L^2(\Omega))$ and $g \in L^2(0, T; H^2(\Gamma))$. Then, for $0 < h < h_0$, we have*

$$\|u - u_{h,\tau}\|_{L^2(Q_T)} \leq B(u, g, f)(\tau + h^2 |\log h|),$$

$$\|u - u_{h,\tau}\|_{L^2(0,T;H^1(\Omega))} \leq B(u, g, f)(\tau + h |\log h|^{1/2}).$$

The proof of Theorem 5 is based on the parabolic dual arguments, the stability estimates of the discrete dual solutions and the use of the discrete energy- and L^2-projections play important roles there (cf. [2]).

4   Numerical experiments

We now report some experiments on the numerical methods discussed in the talk. We solve the interface problem

$$\nabla \cdot (\beta(x, y)\nabla u) = f(x, y) \quad \text{in} \quad \Omega$$

with the true solution

$$u(x, y) = \frac{11}{4} - 8(x^2 + y^2) \quad \text{in} \quad \Omega^-; \quad u(x, y) = 1 - (x^2 + y^2) \quad \text{in} \quad \Omega^+$$

and the coefficient $\beta(x, y)$ is piecewise constant, i.e.

$$\beta(x, y) = \beta^- \quad \text{in} \quad \Omega^-; \quad \beta(x, y) = \beta^+ \quad \text{in} \quad \Omega^+$$

Here $\Omega = (-1, 1) \times (-1, 1)$, $\Omega^- = \{(x, y); \ x^2 + y^2 \leq 1/4\}$ and $\Omega^+ = \Omega \setminus \Omega^-$. The interface is the circle $\Gamma = \{(x, y); \ x^2 + y^2 = 1/4\}$. The source and interface functions $f(x,y)$ and $g(x,y)$ are computed by using the true solution $u(x,y)$. We use three triangulations with mesh sizes $h_1 = 1/2$, $h_2 = 1/4$ and $h_3 = 1/8$ (cf. Fig. 1, Fig. 2 and Fig. 3). We report the computed error order in L^2-norm. To see the order, let $E(h_1)$ and $E(h_2)$ be the L^2-norm error, then

$$p = \log(E(h_1)/E(h_2))/\log(h_1/h_2)$$

will be the computed error order. We also use the ratio

$$\rho = E(h_1)/E(h_2)$$

to measure the computed order. If the method is of order $O(h^2)$, then p and ρ will converge to 2 and 4 resp. We compute for the following coefficients:

Case A : $\beta^+ = 1$, $\beta^- = 10$;   Case B : $\beta^+ = 1$, $\beta^- = 10^2$

Case C : $\beta^+ = 10^4$, $\beta^- = 1$;   Case E : $\beta^+ = 10^5$, $\beta^- = 1$.

From the following tables we can see that the finite element method converges indeed with second order which is consistent with our theoretical analysis.

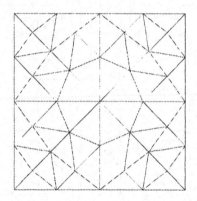

Fig. 1. Finite element triangulation with mesh size $h = 1/2$

Case A	$E(h)$	p	ρ
h_1	3.7138	–	–
h_2	0.8209	2.18	4.52
h_3	0.1996	2.04	4.11

Case B	$E(h)$	p	ρ
h_1	37.4943	–	–
h_2	8.3495	2.17	4.49
h_3	2.0540	2.02	4.06

Case C	$E(h)$	p	ρ
h_1	0.1555	–	–
h_2	0.0388	2.00	4.01
h_3	0.0100	1.95	3.86

Case D	$E(h)$	p	ρ
h_1	0.1544	–	–
h_2	0.0388	2.00	4.01
h_3	0.0100	1.95	3.86

Acknowledgements. The authors thank Prof. Guoping Liang for letting us use his finite element software FEPG V2.0 and Mr. K.C. Kung for conducting the numerical experiments.

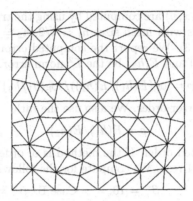

Fig. 2. Finite element triangulation with mesh size $h = 1/4$

Fig. 3. Finite element triangulation with mesh size $h = 1/8$

References

1. I. Babuška. The finite element method for elliptic equations with discontinuous coefficients. *Computing*, 5:207–213, 1970.
2. Z. Chen and J. Zou. Finite element methods and their convergence for elliptic and parabolic interface problems. Technical Report 96-25 (99), Dept of Math, The Chinese University of Hong Kong, 1996.
3. R. LeVeque and Z. Li. Immersed interface method for Stokes flow with elastic boundaries or surface tension. *SIAM J. Sci. Stat. Compt.* To appear.
4. Z. Li, H. Zhao, and S. Osher. A hybrid method for moving interface problems with application to the Hele-Shaw flow. Technical Report CAM Report 96-9, Department of Mathematics, University of California at Los Angeles, April 1996.
5. J. Xu. Error estimates of the finite element method for the 2nd order elliptic equations with discontinuous coefficients. *J. Xiangtan University*, (1):1–5, 1982.

Finite Element Methods Based on Two-Scale Analysis *

J.Z. Cui,[1] T.M. Shih,[2] F.G. Shin,[3] and Y.L. Wang[4]

[1] The Institute of Comp. Math.& Sci.-Engg. Comp., Academia Sinica, China
[2] Dept. of Applied Math., The Hong Kong Polytechnic Univ., Hong Kong
[3] Dept. of Applied Phy., The Hong Kong Polytechnic Univ.,Hong Kong
[4] Dept. of Math., Zhengzhou Univ., Zhengzhou, Henan, China

Abstract. In this paper the finite element method based on two-scale analysis for problems of composite materials and structures with small periodic configurations in two and three dimensional cases are presented. Some results of numerical experiments are shown.

1 Two-Scale Analysis

In structural engineering and in the design and manufacture of new industrial products, the analysis problems for the structures with periodicity are often encountered. The whole structure or its main part is made from composite materials with same cells, or composed of same basic configurations.

For brevity, the summation convention is applied, i.e. the index that appears twice, no matter upper or lower, will imply a summation with respect to it. Without further claim, the ranges of indices are supposed to be:

$$i, j, h, k, m, p, q = 1, \cdots, n; \; l = 0, 1, \cdots; \; \alpha_r = 1, \cdots, n \, (r = 1, \cdots, l)$$

where $n = 2$ or 3 is the dimension. The analysis problem can be simplified to the following problem with material constants changing sharply and periodically,

$$\begin{cases} \dfrac{\partial}{\partial x_j}(a^\varepsilon_{ijhk}(x)\dfrac{1}{2}(\dfrac{\partial u_h}{\partial x_k} + \dfrac{\partial u_k}{\partial x_h})) = f_i(x), & i = 1, \cdots, n, \quad x \in \Omega, \\ u(x) = \overline{u}, & x \in \partial\Omega, \end{cases} \tag{1}$$

where Ω denotes the structure to be analyzed, $x = (x_1, \cdots, x_n)^{\mathrm{T}}$ is the coordinates of a point, and $\partial\Omega$ denotes the boundary of Ω. At a point x, the body force in Ω, the given displacements on $\partial\Omega$ and the displacements of the structure are denoted by $f(x) = (f_1(x), \cdots, f_n(x))^{\mathrm{T}}$, $\overline{u}(x) = (\overline{u}_1(x), \cdots, \overline{u}_n(x))^{\mathrm{T}}$, and $u(x) = (u_1(x), \cdots, u_n(x))^{\mathrm{T}}$ respectively. $\{a^\varepsilon_{ijhk}(x)\}$ is the tensor of material parameters which are periodic functions with period length ε, i.e.

$$a^\varepsilon_{ijhk}(x + \varepsilon e_m) = a^\varepsilon_{ijhk}(x) \tag{2}$$

---

* This work is supported by The Hong Kong Polytechnic University Research Grant No. 353/073 and National Natural Science Foundation of China

and satisfies the following conditions

$$a_{ijhk}^\varepsilon = a_{ijkh}^\varepsilon = a_{jihk}^\varepsilon, \text{ and } \mu_1 \eta_{ih} \eta_{ih} \le a_{ijhk}^\varepsilon \eta_{ih} \eta_{jk} \le \mu_2 \eta_{ih} \eta_{ih}, \tag{3}$$

where e_m is the base, $\{\eta_{ih}\}$ is a symmetric matrix, μ_1, μ_2 are constants and $\mu_2 > \mu_1 > 0$. Thus, from Korn's inequality and Lax-Milgram lemma problem (1) has a unique solution $u(x)$.

The Two-Scale Analysis(TSA) method (see [5]) has been discussed for the problem with structure shown in Fig.1(a). Let Q denote a $1-$square, εQ an $\varepsilon-$square, and Q_c a basic configuration with prescribed composition of materials shown in Fig.1(b). From (2),$\{a_{ijhk}^\varepsilon(x)\}$ can be expressed as

$$a_{ijhk}^\varepsilon(x) = a_{ijhk}(\xi), \text{ and } \xi = x/\varepsilon, \tag{4}$$

where $\{a_{ijhk}(\xi)\}$ are now $1-$periodic functions.

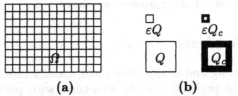

(a)   (b)
Fig.1 Structure with basic configurations

The displacements and stresses of the structures with periodic configuration depend on both the mechanical behaviour in macroscopic and the detailed configurations, so the displacements can be expressed as $u(x) = \tilde{u}(x, \xi)$, where x represents the global coordinates of the structure as well as the mechanical behaviour of the whole structure, and ξ is the local coordinates and embodies the effects of basic configurations. If Ω is a convex domain with piecewise smooth boundaries, $a_{ijhk}(\xi) \in C^1(Q)$ and $f(x) \in C^\infty(\Omega)$, then from [5] the solution $u(x)$ of the problem (1) can be formally expressed in two-scale variables x and ξ as follows

$$u(x) = u_0(x) + \sum_{l=1}^\infty \varepsilon^l \sum_{\alpha \in V_l} N_{m\alpha}(\xi) \frac{\partial^l u_{m0}}{\partial x_{\alpha_1} \cdots \partial x_{\alpha_l}}, \tag{5}$$

where

$$N_\alpha(\xi) = [N_{ij\alpha}(\xi)] \qquad V_l = \{\alpha = (\alpha_1, \alpha_2, \cdots, \alpha_l) \mid \alpha_r = 1, \cdots, n\}, \tag{6}$$

Every $N_{ij\alpha}(\xi)$ $(\alpha \in V_l)$ is an $1-$periodic function, and $u_0(x)$ is a vector valued function defined on whole Ω, and they can be determined by solving following problem (7), (8), (10) and (11)

$$\begin{cases} \frac{\partial}{\partial \xi_j} [a_{ijhk}(\xi) \frac{1}{2} (\frac{\partial N_{hm\alpha_1}}{\partial \xi_k} + \frac{\partial N_{km\alpha_1}}{\partial \xi_h})] = -\frac{\partial a_{ijm\alpha_1}}{\partial \xi_j}, \ \xi \in Q, \\ N_{m\alpha_1}(\xi) = 0, \quad \xi \in \partial Q, \end{cases} \tag{7}$$

$$\begin{cases} \widehat{a}_{ijhk}\dfrac{\partial}{\partial x_j}\dfrac{1}{2}\Big(\dfrac{\partial u_{h0}}{\partial x_k}+\dfrac{\partial u_{k0}}{\partial x_h}\Big)=f_i(x), & x\in\Omega,\\ u_0(x)=\overline{u}(x), & x\in\partial\Omega, \end{cases} \tag{8}$$

where

$$\widehat{a}_{ijhk}=\frac{1}{|Q|}\int_Q[a_{ijhk}(\xi)+a_{ijpq}(\xi)\frac{\partial N_{phk}}{\partial\xi_q}]d\xi, \tag{9}$$

in general $\{\widehat{a}_{ijhk}\}$ is called the homogenization constants of materials,

$$\begin{cases} \dfrac{\partial}{\partial\xi_j}[a_{ijhk}(\xi)\dfrac{1}{2}\Big(\dfrac{\partial N_{hm\alpha_1\alpha_2}}{\partial\xi_k}+\dfrac{\partial N_{km\alpha_1\alpha_2}}{\partial\xi_h}\Big)]=\widehat{a}_{i\alpha_2 m\alpha_1}-a_{i\alpha_2 m\alpha_1}(\xi)\\ -a_{i\alpha_2hk}(\xi)\dfrac{\partial N_{hm\alpha_1}}{\partial\xi_k}-\dfrac{\partial}{\partial\xi_j}(a_{ijh\alpha_2}(\xi)N_{hm\alpha_1}(\xi)), & \xi\in Q,\\ N_{m\alpha_1\alpha_2}(\xi)=0, & \xi\in\partial Q, \end{cases} \tag{10}$$

and

$$\begin{cases} \dfrac{\partial}{\partial\xi_j}[a_{ijhk}(\xi)\dfrac{1}{2}\Big(\dfrac{\partial N_{hm\alpha_1\cdots\alpha_l}}{\partial\xi_k}+\dfrac{\partial N_{km\alpha_1\cdots\alpha_l}}{\partial\xi_h}\Big)]=-a_{i\alpha_l hk}(\xi)\dfrac{\partial N_{hm\alpha_1\cdots\alpha_{l-1}}}{\partial\xi_k}\\ -\dfrac{\partial}{\partial\xi_j}(a_{ijh\alpha_l}(\xi)N_{hm\alpha_1\cdots\alpha_{l-1}}(\xi))-a_{i\alpha_l h\alpha_{l-1}}(\xi)N_{hm\alpha_1\cdots\alpha_{l-2}}(\xi), & \xi\in Q,\\ N_{m\alpha_1\cdots\alpha_l}(\xi)=0, & \xi\in\partial Q,\quad l=3,4,\cdots. \end{cases} \tag{11}$$

Let

$$u^{(M)}(x)=u_0(x)+\sum_{l=1}^{M}\varepsilon^l\sum_{\alpha\in V_l}N_{m\alpha}(\xi)\frac{\partial^l u_{m0}}{\partial x_{\alpha_1}\cdots\partial x_{\alpha_l}},\quad M=1,2,3,\cdots \tag{12}$$

The following theorem can be proved:

Theorem 1 *Suppose that Ω is a convex domain with piecewise smooth boundaries and $f(x)\in C^M(\Omega)$, then*

$$\|u^{(M)}(x)-u(x)\|_{H_0^1(\Omega)}\le(An\varepsilon)^{M-1}C, \tag{13}$$

A, and C are constants independent of ε.

2 Finite Element (FE) Computation of Homogenization Constants

In (8), in order to obtain the homogenization solution $u_0(x)$, one needs to solve (7) to obtain $N_{m\alpha_1}(\xi)$ first, and then evaluate the homogenization constants $\{\widehat{a}_{ijhk}\}$ in (9). Let

$$\varepsilon_{ij}(u)=\frac{1}{2}\Big(\frac{\partial u_i}{\partial x_j}+\frac{\partial u_j}{\partial x_i}\Big), \tag{14}$$

$$\int_Q \varepsilon_{ij}(v)a_{ijhk}(\xi)\varepsilon_{hk}(N_{m\alpha_1})d\xi = \int_Q \varepsilon_{ij}(v)a_{ijm\alpha_1}(\xi)d\xi, \qquad \forall v \in H_0^1(Q), \quad (15)$$

where ε_{ij} denote the strain tensors. It is well known that the virtual work equation (15) is equivalent to (7), and (15) can be solved by FE method. Let $N_{m\alpha_1}^{ho}(\xi)$ be the FE solution of (15) and

$$\widehat{a}_{ijhk}^{ho} = \int_Q [a_{ijhk}(\xi) + a_{ijpq}(\xi)\varepsilon_{pq}(N_{hk}^{ho})]d\xi, \tag{16}$$

where h_0 denotes the maximum diameter of the elements.

Since in practical engineering problems, $\{a_{ijhk}(\xi)\}$ change very sharply, $N_{m\alpha_1}(\xi) \notin H^k(Q)$ as $k > 2$. So in general linear elements on triangles and bilinear elements on quadrilaterals are used to evaluate $N_{m\alpha_1}(\xi)$, and then obtain $N_{m\alpha}^{ho}(\xi)$. In this case the following theorem holds through FE analysis.

Theorem 2 *In solving $N_{m\alpha_1}^{ho}(\xi)$, the approximate homogenization constants $\{\widehat{a}_{ijhk}^{ho}\}$ evaluated from (16) converges to the true ones as $h_0 \to 0$, where h_0 is the maximum diameter and*

$$|\widehat{a}_{ijhk} - \widehat{a}_{ijhk}^{ho}| \le Ch_0 \max_{1 \le h,k \le n} \{ \| \frac{\partial a_{jhk}^*}{\partial \xi_j} \|_{L^2(Q)} \}, \tag{17}$$

where $a_{jhk}^*(\xi) = (a_{1jhk}(\xi), \cdots, a_{njhk}(\xi))$ and C is a constant independent of h.

3   FE Algorithm of TSA

The FE algorithm for TSA is as follows:

1. Set up the mechanical and mathematical model
   · Form and verify the geometry of the structure, the material properties of components, the loading condition and constraints.
   · Form and verify the composition of basic configurations for every components with periodicity, the matrix, reinforce and their interfaces.

2. Set up the finite element model
   · Partition the structure into the set of elements according to its composition. Let $\overline{h}$ denotes the maximum diameter of elements in Ω.
   · Partition the 1−square domain into another set of elements (triangles or quadrilaterals) according to the composition of basic configuration. Let h_0 denotes the maximum diameter of elements in the 1−square Q.

3. According to the material properties of basic configurations, compute $N_{m\alpha_1}^{ho}(\xi)$ on Q by FE analysis program, and then compute

$$\varepsilon_{pq}(N_{m\alpha_1}^{ho}) = \frac{\partial N_{pm\alpha_1}^{ho}}{\partial \xi_q} + \frac{\partial N_{qm\alpha_1}^{ho}}{\partial \xi_p}. \tag{18}$$

Since the left sides of all equations satisfied by $N_{m\alpha_1}$ are the same as those by $N_{m\alpha_1\cdots\alpha_l}(\xi)$ $(l = 2, 3, \cdots)$, it needs to compute the stiffness matrix A_N of the FE equations for $N_{m\alpha_1\cdots\alpha_l}(\xi)$ and decompose A_N into LDL^{T} only once and save L and D.

4. Compute the homogenization constants by the following formula

$$\widehat{a}_{ijhk}^{ho} = \int_Q [a_{ijhk}(\xi) + a_{ijpq}(\xi)\varepsilon_{pq}(N_{hk}^{ho})]d\xi. \tag{19}$$

5. Based on the homogenization constants obtained in step 4, compute finite element solution $u_0^{\overline{h}}(x)$ on the whole structure by the $k-$order finite elements. Then evaluate

$$\varepsilon_{ij}(u_0^{\overline{h}}) = \frac{\partial u_{i0}^{\overline{h}}}{\partial x_j} + \frac{\partial u_{j0}^{\overline{h}}}{\partial x_i}. \tag{20}$$

6. Compute the following approximations of higher-order derivatives using the data process technique with $\overline{k}-$order approximation

$$\frac{\partial^l u_{m0}^{\overline{h}}}{\partial x_{\alpha_1}\cdots\partial x_{\alpha_l}} \quad (l = 1, \cdots, M).$$

7. Compute $N_{m\alpha_1\cdots\alpha_l}^{ho}(\xi)$ $(l = 1, \cdots, M)$ recurrently from previous results $N_{m\alpha_1\cdots\alpha_{l-1}}^{ho}(\xi)$ and $N_{m\alpha_1\cdots\alpha_{l-2}}^{ho}(\xi)$, and then compute

$$\varepsilon_{hk}(N_{m\alpha_1\cdots\alpha_l}^{ho}) = \frac{\partial N_{hm\alpha_1\cdots\alpha_l}^{ho}}{\partial \xi_k} + \frac{\partial N_{km\alpha_1\cdots\alpha_l}^{ho}}{\partial \xi_h}. \tag{21}$$

8. Compute the final results by following formulas

$$u_i^{h(M)}(x) = u_{i0}^{\overline{h}}(x) + \sum_{l=1}^{M} \varepsilon^l \sum_{\alpha\in V_l} N_{im\alpha_1\cdots\alpha_l}^{ho}(\frac{x}{\varepsilon})\frac{\partial^l u_{m0}^{\overline{h}}}{\partial x_{\alpha_1}\cdots\partial x_{\alpha_l}}, \tag{22}$$

$$\varepsilon_{hk}^{(M)}(x) = \frac{\partial u_{h0}^{\overline{h}}}{\partial x_k} + \frac{\partial u_{k0}^{\overline{h}}}{\partial x_h} + \sum_{l=1}^{M} \varepsilon^{l-1} \sum_{\alpha\in V_l} \varepsilon_{hk}(N_{m\alpha_1\cdots\alpha_l}^{ho}(\frac{x}{\varepsilon}))\frac{\partial^l u_{m0}^{\overline{h}}}{\partial x_{\alpha_1}\cdots\partial x_{\alpha_l}}$$

$$+\sum_{l=1}^{M} \varepsilon^l \sum_{\alpha\in V_l}[N_{hm\alpha_1\cdots\alpha_l}^{ho}(\frac{x}{\varepsilon})\frac{\partial^{l+1} u_{m0}^{\overline{h}}}{\partial x_{\alpha_1}\cdots\partial x_{\alpha_l}\partial x_k} + N_{km\alpha_1\cdots\alpha_l}^{ho}(\frac{x}{\varepsilon})\frac{\partial^{l+1} u_{m0}^{\overline{h}}}{\partial x_{\alpha_1}\cdots\partial x_{\alpha_l}\partial x_h}], \tag{23}$$

and

$$\sigma_{ij}^{(M)}(x) = a_{ijhk}(\frac{x}{\varepsilon})\varepsilon_{hk}^{(M)}(x), \tag{24}$$

where $\{\sigma_{ij}^{(M)}(x)\}$ denote the stress tensor.

It can be proved that the difference between the approximate solution $u^{h(M)}$ obtained from (22) and the true solution $u(x)$ can be estimated as follows

$$\| u - u^{h(M)} \|_{H^1(\Omega)} \le C_1(An\varepsilon)^{M-1} + C_2(h_0 + \overline{h}^k + h^{\overline{k}})F, \tag{25}$$

where ε is such that $An\varepsilon < 1$, A and F are constants independent of ε, h_0, $\overline{h}$, k, $\overline{k}$ and M, C_1 and C_2 are independent of ε, h_0, $\overline{h}$, k and $\overline{k}$.

4   Numerical results

We have coded the computing program of previous algorithm using COREP, which is a FE software developed by us, and some numerical experiments are evaluated to verify the effectiveness of the FE method based on TSA.

In order to test the FE computation of homogenization constants making use of two kinds of isotropic materials, we designed three kinds of basic configurations for periodic structures shown in Fig.2. The basic material constants corresponding to the shaded and white areas respectively are as follows:

$$
\begin{pmatrix} 3125000 & 625000 & 0 \\ 625000 & 3125000 & 0 \\ 0 & 0 & 1250000 \end{pmatrix} \quad \text{and} \quad \begin{pmatrix} 178571 & 71428 & 0 \\ 71428 & 178571 & 0 \\ 0 & 0 & 53571 \end{pmatrix}
$$

In the three basic configuration, though the areas of two materials are equal the corresponding homogenization constants are very different, not only in magnitude but also in the number of independent constants shown in Fig.2.

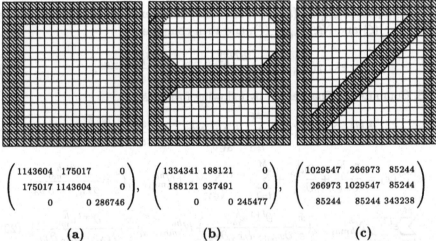

$$
\begin{pmatrix} 1143604 & 175017 & 0 \\ 175017 & 1143604 & 0 \\ 0 & 0 & 286746 \end{pmatrix}, \quad \begin{pmatrix} 1334341 & 188121 & 0 \\ 188121 & 937491 & 0 \\ 0 & 0 & 245477 \end{pmatrix}, \quad \begin{pmatrix} 1029547 & 266973 & 85244 \\ 266973 & 1029547 & 85244 \\ 85244 & 85244 & 343238 \end{pmatrix}
$$

(a)   (b)   (c)

Fig.2 Three basic periodic configuration and their homogenization constants.

In order to demonstrate the accuracy and efficiency of FE algorithm based on TSA, we computed the displacements and stresses of the periodic structure in Fig.3. For $u_0^{\bar{h}}(x)$ the whole structure is partitioned into 38×94 rectangles, and for $N_{m\alpha}^{ho}(\xi)$ 1-square Q is divided into 40×40 rectangles. The results of detailed stress distributions inside every basic configuration are obtained. The stress results on the dotted cell in Fig.3 with the basic configuration in Fig.2(b) are shown in Fig.4. For conventional FE method, much more refined rectangular meshes might be needed to achieve the same accuracy. Therefore the FE method based on TSA is very efficient for problems raised from composite materials and structures with small period.

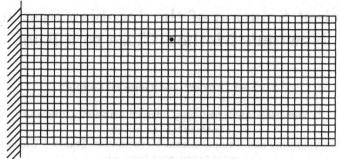

Fig.3 Structure with periodic configuration.

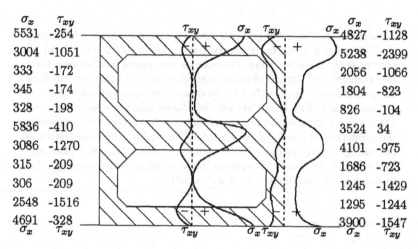

σ_x	τ_{xy}	σ_x	τ_{xy}
5531	-254	4827	-1128
3004	-1051	5238	-2399
333	-172	2056	-1066
345	-174	1804	-823
328	-198	826	-104
5836	-410	3524	34
3086	-1270	4101	-975
315	-209	1686	-723
306	-209	1245	-1429
2548	-1516	1295	-1244
4691	-328	3900	-1547

Fig.4 Stress distribution on the dashed line in the dotted cell in Fig. 3.

References

1. Aboudi, J.: Mechanics of Composite Materials - A Unified Micro-mechanical Approach. Elsevier, Amsterdam - Oxford - New York -Tokyo. (1991)
2. Bensoussan,A., Lions,J.L. and Papanicolaou,G.: Asymptotic Analysis for Periodic Structures. Amsterdam, North – Holland. (1978).
3. Cui, J.Z.: The Two-scale Analysis Methods for Woven Composite Material and Structures with Small Period. The Advances of Computational Mechanics, Intern. Academic Publisher. (1996)
4. Cui, J.Z., and Yang, H.Y.: A Dual Coupled Method of Boundary Value Problems of PDE with Coefficients of Small Period. Intern. J. Comp. Math. **14**. (1996) 159-174.
5. Cui, J.Z., Shih, T.M. and Wang, Y.L.: Two-Scale Analysis Method for Bodies with Small Periodic Configuration. invited paper in CASCM-97, Feb.11-14, 1997, Sydney, Australia.
6. Oleinik, O.A., Shamaev,A.S. and Yosifian,G.A.: Mathematical Problems in Elasticity and Homogenization. Amsterdam, North – Holland. (1992)

Efficient Solution of the Jacobian System in Newton's Method Close to a Root

Michael Drexler[1] and Gene H. Golub[2]

[1] Numerical Analysis Group, Oxford University, England.
namd@comlab.ox.ac.uk
[2] Scientific Computing and Computational Mathematics, Stanford University, USA.
golub@sccm.stanford.edu

Abstract. Newton's Method constitutes a nested iteration scheme with the Newton step as the outer iteration and a linear solver of the Jacobian system as the inner iteration. We examine the interaction between these two schemes and derive solution techniques for the linear system from the properties of the outer Newton iteration. Contrary to inexact Newton methods, our techniques do not rely on relaxed tolerances for an iterative linear solve, but rather on computational speedup achieved by exploiting the properties of the Jacobian update. This update shows a pattern of increasing sparsity in the solution vicinity for many practical problems. In this paper, we specify the sparsity pattern and present derived solution techniques for both direct and iterative solvers.

1   Introduction

Newton's method and its variants are powerful and popular algorithms to solve systems of non-linear equations $\mathbf{f}(\mathbf{x}) = \mathbf{0}$ via local linearisation. As an iterative method, its main loop consists in an update of the iterate vector $\mathbf{x}^{(k)}$

$$\mathbf{x}^{(k+1)} = \mathbf{x}^{(k)} + \mathbf{s}^{(k)}, \tag{1}$$

where $\mathbf{s}^{(k)}$ is the Newton shift vector. In each Newton iteration, a linear system of the form

$$\mathbf{J}^{(k)}\mathbf{s}^{(k)} = -\mathbf{f}\left(\mathbf{x}^{(k)}\right) \tag{2}$$

is solved to obtain the current shift vector for the Newton update. $\mathbf{J}^{(k)}$ is the Jacobian matrix of $\mathbf{f}$ evaluated at the kth iterate. We are interested in ways to solve this system of linear equations efficiently, using the fact that it arises from applying Newton's method with its well-defined convergence properties.

Matrix algorithms to solve (2) can be considered as the inner iteration and the Newton step (1) as the outer iteration of a nested iteration scheme. The outer iteration viewpoint has been helpful to establish variants of Newton's method with favourable global convergence properties ([1], [4], [11], [12]) and the numerical effect on the outer iteration depending on the tolerances of the inner iteration has been thoroughly analysed for inexact Newton Methods ([3], [7]). However, little has been said about the influence of the outer iteration on the properties of the inner iteration matrix. Such an approach leads to a matrix-based analysis of Newton's method and consequently to linear solvers that are specifically adapted to match the properties of the outer iteration.

2 Elementwise Convergence of the Jacobian

We denote the element at position (i, j) of the Jacobian $\mathbf{J} \in \mathbb{R}^{n \times n}$ in Newton iteration k by $\left[\mathbf{J}^{(k)}\right]_{ij} = \left.\frac{\partial f_i}{\partial x_j}\right|_{\mathbf{x}^{(k)}}$. For terminological clarity, we furthermore define

Definition 1. The non-linearity of a function $f(\mathbf{x}) : \mathbb{R}^n \mapsto \mathbb{R}$ in a closed domain $\mathcal{D}$ is determined by the size of the Lipschitz constant γ_f of its vector of first derivatives $\left[\frac{\partial f}{\partial \mathbf{x}}\right]$ on that domain.

This is in accordance with the geometrical notion that the first derivative of a non-linear function will vary across a wide range even in a small interval. On the other hand, the first derivative of a linear function stays constant on any interval. Although definition 1 is stated for a vector of variables, we employ it in the same meaning for functions of a scalar.

As a crucial preliminary result, we state the classical local convergence theorem for Newton's method after [5].

Theorem 2. *Considering Newton iterates inside a convergence radius r of the root $\mathbf{x}_*$, we demand that the Jacobian is Lipschitz continuous with constant γ. Furthermore, the condition $\left\| (\mathbf{J}_*)^{-1} \right\| \leq \beta$ is imposed on the inverse Jacobian at the root. Then, for each iteration k*

$$\|\mathbf{x}^{(k+1)} - \mathbf{x}_*\| \leq \beta \gamma \|\mathbf{x}^{(k)} - \mathbf{x}_*\|^2 \to 0 \quad as \quad k \to \infty. \tag{3}$$

For the proof and a more geometrically comprehensive interpretation of theorem 2, we refer to [5] or [6]. From the fact that $\mathbf{J}$ is the local linearisation of $\mathbf{f}$ at a point, it is obvious that $\mathbf{J}_*$ is a constant matrix. In theory and to infinite precision, the Newton iterations would continue infinitely with increasingly smaller changes to the iterates and the Jacobian. However, in the context of finite arithmetic, we can state that for a $k_f > 0$ specific to $\mathbf{f}$ and the starting iterate $\mathbf{x}^{(0)}$, $\left\{\mathbf{x}^{(k)} = \mathbf{x}_*\right\} \wedge \left\{\mathbf{J}^{(k)} = \mathbf{J}_*\right\}$ $\forall k > k_f$, i.e. the Jacobian converges towards a constant.

We are now interested in the specific way the Jacobian converges towards a constant matrix as the Newton iterates approach the root. To examine this, we define

Definition 3. We consider a connected region $\mathcal{D}^{(k)}$ with $\left\{\mathbf{x}^{(k)}, \mathbf{x}_*\right\} \in \mathcal{D}^{(k)}$. The elementwise Lipschitz constant $\gamma_{ij}^{(k)}$ in the kth Newton iteration is the smallest constant satisfying

$$\left| \left(\left. \frac{\partial f_i}{\partial x_j} \right|_{\mathbf{x}} - \left. \frac{\partial f_i}{\partial x_j} \right|_{\mathbf{x_*}} \right) \right| \leq \gamma_{ij}^{(k)} \cdot \| \mathbf{x} - \mathbf{x_*} \|, \ \forall \mathbf{x} \in \mathcal{D}^{(k)}.$$

From the fact that the $\mathcal{D}^{(k)}$ shrink with (3) as k grows, it follows that

$$\gamma_{ij}^{(k+1)} \leq \gamma_{ij}^{(k)}. \tag{4}$$

Combining this result with theorem 2, we obtain

Theorem 4. *The Jacobian of the Newton iterates $\mathbf{x}^{(k)}$ converges to the constant Jacobian at the root $\mathbf{x_*}$ in an elementwise fashion, such that for every $\epsilon > 0$, an iteration number k_{ij} can be found with*

$$\left| [\mathbf{J}^{(k)}]_{ij} - [\mathbf{J_*}]_{ij} \right| < \epsilon \qquad \forall k > k_{ij}$$

The convergence is determined by the non-linearity of f_i with respect to x_j.

The proof, following [6], is obtained from writing the Lipschitz condition for $\mathbf{J}^{(k)}]_{ij} - [\mathbf{J_*}]_{ij}$, which restates definition 3 in different notation. As the elementwise Lipschitz constant decreases monotonically with k according to (4) and the distance of the iterates to the root according to (3), k_{ij} can be determined by $\epsilon \geq \gamma_{ij}^{(k_{ij})} \cdot \| \mathbf{x}^{(k_{ij})} - \mathbf{x_*} \|$, proving the existence of k_{ij}. Within one specific iteration, the vector norm stays equal for all Jacobian entries, and the elementwise Lipschitz constant changes. Hence, convergence (via the size of k_{ij}) is determined by the size of the elementwise Lipschitz constant and therefore, employing definition 1, by the non-linearity of f_i with respect to x_j.

3   Implications for the Inner Iteration Scheme

As the results in the previous section have established, the elements of the Jacobian converge towards a constant at individual rates, determined by the non-linearity of the relevant function with respect to the variable determined by the element's position. In many practical problems, the $\gamma_{ij}^{(k)}$ will vary in magnitude, and therefore only a certain number of elements wil change significantly in the late Newton steps. This leads to an increasingly sparse update of the Jacobian between iterations, which is also of decreasing magnitude. We will examine the consequences of this result on linear solvers and propose modifications to established algorithms that speed up the computation.

3.1   Direct Solvers

We consider variants of the LU decomposition of a matrix. The factors are computed from the original elements in one sweep. Analysing the flow of information, we find that a change in a single element $[\mathbf{J}]_{ij}$ generally affects a region as

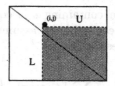

Fig. 1. Fan-Out of element $[\mathbf{J}]_{ij}$ in an in-place LU factorisation

depicted in Fig. 1 where the factors are assumed to be stored in place. It is there-fore obvious that a reordering of the Jacobian is preferable if only few elements change. The update of the factors can then be confined to the lower right corner with the associated decrease in computational work. As the factorisation of the reordered matrix changes, it is desirable that the correct ordering is established as early as possible. This could be achieved by monitoring the change of elements in the early iterations and reordering the matrix once Newton convergence sets in, using a drop-tolerance scheme.

We point out that similar approaches of matrix splittings into a linear and non-linear part have successfully been considered in the context of least-squares problems [10]. Also, the concept of low-rank modifications is discussed in many places (e.g. [8]).

The approach described so far is particularly beneficial if only a few highly non-linear elements change considerably. For certain problems, it is however also possible to incorporate a sparse update without the need for reordering. To demonstrate this, we will examine the important case of a diagonal update in a tridiagonal matrix. For simplicity, pivoting is omitted. Considering Crout's al-gorithm, and considering the bandwidth-preserving property of the factorisation [9], we obtain for the factors of a matrix $\mathbf{A} \in \mathbb{R}^{n \times n}$,

$$u_{(j-1),j} = a_{(j-1),j}, \qquad l_{(j+1),j} = \frac{a_{(j+1),j}}{u_{j,j}},$$
$$u_{j,j} = a_{j,j} - \frac{a_{(j-1),j} \cdot a_{j,(j-1)}}{u_{(j-1),(j-1)}}, \tag{5}$$

with $j = 1, \ldots, n$. Therefore, to describe the influence of a perturbation in $\mathbf{A}$, it is sufficient to analyse the recursion of the diagonal elements $u_{j,j}$. Motivated by practical applications (e.g. finite difference discretisations of second derivatives), we examine a perturbation $0 < \epsilon \ll 1$ of a diagonal element $a_{j,j}$. For notational simplicity, (5) is rewritten as

$$u_0 = a_0 - \frac{A_0}{u_1} = a_0 - \frac{A_0}{a_1 - \dfrac{A_1}{u_2}} = \cdots, \tag{6}$$

where $\quad a_k = a_{j-k,j-k}, \quad A_k = a_{(j-k-1),j-k} \cdot a_{j-k,(j-k-1)}, \quad u_k = u_{j-k,j-k}.$ (7)

The perturbation $a_1 \mapsto a_1 + \epsilon$ is now substituted into (6) to yield

$$u_0 = a_0 - \frac{A_0}{\left(a_1 - \dfrac{A_1}{u_2} \right) \left(1 + \dfrac{\epsilon}{a_1 - \dfrac{A_1}{u_2}} \right)}. \tag{8}$$

The denominator of the term containing ϵ is equivalent to u_1, and for well-conditioned problems, it is reasonable to assume $u_1 \gg 0$. We can therefore expand, using the Taylor series for $(1+x)^{-1}$,

$$u_0 = a_0 - \frac{A_0}{a_1 - \frac{A_1}{u_2}} \cdot \sum_{k=1}^{\infty} \left[(-1)^{k-1} \cdot \left(\frac{\epsilon}{a_1 - \frac{A_1}{u_2}} \right)^{k-1} \right]. \tag{9}$$

In the first-order approximation, this simplifies to

$$u_0 = a_0 - \frac{A_0}{a_1 - \frac{A_1}{u_2}} + \frac{A_0}{\left(a_1 - \frac{A_1}{u_2}\right)^2} \cdot \epsilon. \tag{10}$$

Hence, by dissipating down the diagonal by one element, the perturbation has become

$$\epsilon \mapsto \frac{A_0}{(u_1)^2} \cdot \epsilon. \tag{11}$$

Applying the above analysis recursively, we can therefore state for the influence of a perturbation $(a_{(j-k),(j-k)} + \epsilon)$ on the element $u_{j,j}$, using the notation (7)

$$\epsilon^{(k)} = \epsilon \cdot \prod_{i=1}^{k} \frac{A_{i-1}}{(u_i)^2}. \tag{12}$$

As the u_i and A_i are known from a previous decomposition or from the original matrix, the update on the diagonal can be done cheaply. Furthermore, if $|A_i| < (u_i)^2$, the perturbation will decay when being dissipated down the diagonal and therefore only affect the nearest neighbours significantly. Using this result, the direct method can be reformulated accordingly to account for perturbations in a quick update.

Despite being more complicated, a similar analysis can be carried out for the general LU factorisation. Research in this area is currently undertaken.

3.2   Iterative Solvers

We examine iterative solvers of the conjugate-gradient type. In general, these rely upon the iteration $\mathbf{x}^{(k+1)} = \mathbf{A}\mathbf{x}^{(k)}$ for solving the linear system $\mathbf{A}\mathbf{x} = \mathbf{b}$ from a starting guess $\mathbf{x}^{(0)}$. From writing down the iteration in the vector components, it can be seen that for the change of one element a_{ij} and the same starting iterate, only the jth component of the first iterate changes. However, in the next iterations, this change is propagated all over $\mathbf{x}^{(k)}$ and the complete matrix-vector multiplication has to be evaluated. Therefore, a sparse matrix update cannot be easily accounted for in a pure iterative method. The above argument also holds for matrix-splitting methods, as the change in $\mathbf{A}$ spills into the splitting matrices with the same effect.

In practice, pure iterative solvers are barely used and preconditioned solvers are preferred. In this context, a sparse matrix update can be accounted for easily. For most practical problems, as discussed in [6], it is a viable approach to keep the preconditioner unchanged in the last Newton steps as the Jacobian elements change only by a small magnitude. If the update is also sparse and an incomplete factorisation is used for preconditioning, the results on direct solvers can be incorporated, leading to considerable speedup in computing the preconditioner.

4   Numerical Experiments

To test the proposed algorithms, we consider the non-linear second-order ODE

$$-y''(x) = \tfrac{1}{2}y^5(x) \qquad \text{on } x \in [0, 1] \tag{13}$$

with boundary conditions $y(0) = 0, y(1) = 25$. The problem is discretised using central differences and stepsize h. For the numerical experiments, $h = 0.01$, leading to a 100×100 tridiagonal matrix. The iterates were converged to a point close to the true solution, from where Newton's Method was started with different inner iteration schemes.

As inner iteration schemes, three direct and three iterative variants were compared. The direct solvers were a sparse LU factorisation accounting for the matrix structure, the diagonal update algorithm discussed in section 3.1, and an algorithm keeping the LU factors constant and only substituting a new right-hand side in each Newton step. The latter variant is often used as a heuristic by practitioners to save computation time. The iterative schemes were preconditioned Conjugate Gradient solvers. The first variant (CG-p) updated the diagonal preconditioner [2] in each step, the second (CG-1) kept the preconditioner constant in the last three steps before convergence and the third (CG-2) did so over the last two steps before convergence.

The convergence histories are depicted in Fig. 2. As expected, only the non-updating LU shows strictly linear convergence [3]. Timings are shown in Table 1.

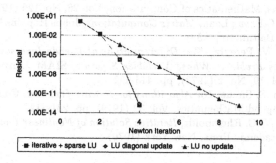

Fig. 2. Convergence for different Inner Iteration Schemes

Method	LU sparse	LU diagonal	LU no update	CG-p	CG-1	CG-2
Time [sec]	83.2	43.9	78.7	35.6	35.2	42.7

Table 1. Computational Cost for Different Inner Iteration Schemes

They were averaged over a number of runs on a Sun Sparc 4 platform, and state the time needed for 1000 repeated convergence runs.

The results confirm that the proposed algorithms are efficient within their respective class. The performance of the direct diagonal update is particularly worth noting. Within the preconditioned iterative methods, the proposed modification is suffering from the very cheap diagonal preconditioner. For more expensive preconditioners, much greater benefits are to be expected. Also, it is confirmed that the heuristic approach of not updating the Jacobian is no alternative to more elaborate linear solvers.

M.Drexler acknowledges the financial support of the DAAD through the programme HSPII/AUFE.

References

1. R.E. Bank, D.J. Rose, *Global Approximate Newton Methods*, Numerische Mathematik, Vol.37 (1981), pp. 279-295.
2. R. Barrett, T.F. Chan, et. al., *Templates for the Solution of Linear Systems: Building Blocks for Iterative Methods*, 2nd edition, SIAM, Philadelphia (1994).
3. R.S. Dembo, S.C. Eisenstat, T. Steihaug, *Inexact Newton Methods*, SIAM Journal on Numerical Analysis, Vol. 19, No. 2 (1982), pp. 400-408.
4. J.E. Dennis, J.J. Moré, *Quasi-Newton Methods, Motivation and Theory*, SIAM Review, Vol. 19, No. 1 (1977), pp. 46-89.
5. J.E. Dennis, R.B. Schnabel, *Numerical Methods for Unconstrained Optimization and Nonlinear Equations*, SIAM, Philadelphia (1996).
6. M. Drexler, G.H. Golub, *Convergence Properties of Newton's Method in the Proximity of a Solution*, submitted to SIAM Journal on Numerical Analysis.
7. S.C. Eisenstat, H.F. Walker, *Choosing the Forcing Terms in an Inexact Newton Method*, SIAM Journal on Scientific Computing, Vol. 17, No.1 (1996), pp. 16-32.
8. P.E. Gill, G.H. Golub, W. Murray, M.A. Saunders, *Methods for Modifying Matrix Factorizations*, Mathematics of Computation, Vol. 28, No. 126 (1974), pp. 505-535.
9. G.H. Golub, C.F. van Loan, *Matrix Computations*, Johns Hopkins University Press, 2nd edition, Baltimore (1989).
10. G.H. Golub, V. Pereyra, *The Differentiation of Pseudo-Inverses and Non-linear Least Squares Problems Whose Variables Separate*, SIAM Journal on Numerical Analysis, Vol. 10, No. 2 (1973), pp. 413-432.
11. M.W. Hirsch, S. Smale, *On Algorithms for Solving $f(x) = 0$*, Communications on Pure and Applied Mathematics, Vol. 32 (1979), pp. 281-312.
12. J.M. Ortega, W.C. Rheinboldt, *Iterative Solution of Nonlinear Equations in Several Variables*, Academic Press, New York (1970).

Computing Multivariate Normal Probabilities Using Rank-1 Lattice Sequences *

Fred J. Hickernell and Hee Sun Hong

Department of Mathematics, Hong Kong Baptist University
Kowloon Tong, Hong Kong

Abstract. Multivariate normal probabilities, which are used for statistical inference, must be computed numerically. This article describes a new rank-1 lattice quadrature rule and its application to computing multivariate normal probabilities. In contrast to existing lattice rules the number of integrand evaluations need not be specified in advance. When compared to existing algorithms for computing multivariate normal probabilities the new algorithm is more efficient when high accuracy is required and/or the number of variables is large.

1 Introduction

The most important probability distribution is the Gaussian or normal probability distribution. Normal probabilities are used to perform statistical inference and construct confidence intervals. The definition of the normal probability distribution involves an integral which cannot be evaluated in terms of elementary functions. Therefore, numerical methods are needed. Many software packages contain routines for evaluating *univariate* normal probabilities, however, the evaluation of *multivariate* normal probabilities is still an area of research. In particular, one would like to evaluate the s-dimensional integral

$$I(\mathbf{a}, \mathbf{b}, \boldsymbol{\Sigma}) = \frac{1}{\sqrt{|\boldsymbol{\Sigma}|(2\pi)^s}} \int_{a_1}^{b_1} \cdots \int_{a_s}^{b_s} e^{-\frac{1}{2}\boldsymbol{\theta}' \boldsymbol{\Sigma}^{-1} \boldsymbol{\theta}} \, d\boldsymbol{\theta}. \tag{1}$$

Here $\mathbf{a}$ and $\mathbf{b}$ are known s-dimensional vectors that define the interval of integration, and $\boldsymbol{\Sigma}$ is a given $s \times s$ positive definite covariance matrix. One or more of the components of $\mathbf{a}$ and $\mathbf{b}$ may be infinite.

Alan Genz [4] has compared several methods for computing multivariate normal probabilities. The best methods he found involved transforming the above integral to one over the unit cube, and then applying either an adaptive algorithm [1] or a Korobov rank-1 lattice rule [2,8]. One disadvantage of the adaptive algorithm is that it requires low order polynomials to be integrated exactly and so the number of integrand evaluations increases substantially as s increases. One disadvantage of existing lattice rules is that the number of integrand evaluations must be fixed in advance in order to generate the lattice. In this paper a new type of lattice rule is introduced that overcomes this disadvantage.

---

* This research was supported by HKBU Faculty Research Grant 95-96/II-01.

The following section outlines Genz's transformation of the s-dimensional integral in (1) into an $s - 1$-dimensional integral over the unit cube. Section 3 introduces a new rank-1 lattice sequence. Numerical experiments comparing the different methods for computing multivariate normal probabilities are given in Section 4. The last section contains a discussion of the results.

2   Transforming the Integrand

The integral appearing in (1) is not well-suited for numerical quadrature because the integration domain may be infinite, and the integrand may be exponentially small over a large part of the domain. Alan Genz [3] has proposed a transformation of variables that results in $I(\mathbf{a}, \mathbf{b}, \boldsymbol{\Sigma})$ being written as an integral over an $s - 1$-dimensional unit cube in the new integration variable $\mathbf{w} = (w_1, \ldots, w_{s-1})'$.

Let $\boldsymbol{\Sigma} = \mathbf{C}\mathbf{C}'$ be the Cholesky decomposition of the covariance matrix. Furthermore, let

$$a_1' = a_1/c_{11}, \qquad b_1' = b_1/c_{11}, \qquad d_1 = \varPhi(a'), \qquad e_1 = \varPhi(b'),$$

where $\varPhi$ is the univariate standard normal probability distribution function. For $i = 2, \ldots, s$ recursively define

$$y_{i-1}(w_1, \ldots, w_{i-1}) = \varPhi^{-1}(d_{i-1} + w_{i-1}(e_{i-1} - d_{i-1})),$$

$$a_i'(w_1, \ldots, w_{i-1}) = \frac{a_i - \sum_{j=1}^{i-1} c_{ij} y_j}{c_{ii}}, \qquad b_i'(w_1, \ldots, w_{i-1}) = \frac{b_i - \sum_{j=1}^{i-1} c_{ij} y_j}{c_{ii}},$$

$$d_i(w_1, \ldots, w_{i-1}) = \varPhi(a_i'), \qquad e_i(w_1, \ldots, w_{i-1}) = \varPhi(b_i').$$

Then the multivariate normal probability in (1) may be written as

$$I(\mathbf{a}, \mathbf{b}, \boldsymbol{\Sigma}) = (e_1 - d_1) \int_0^1 (e_2 - d_2) \int_0^1 \cdots \int_0^1 (e_s - d_s) \, d\mathbf{w}. \qquad (2)$$

3   New Infinite Lattice Sequences

The integral in (2) is a special case of $\int_{[0,1)^s} f(\mathbf{x}) \, d\mathbf{x}$, the integral over the unit cube of a general integrand, $f(\mathbf{x})$. (Here the dimension of the unit cube is considered to be s instead of $s - 1$ for convenience.) This integral may be evaluated numerically by a variety of techniques, including Monte Carlo, quasi-Monte Carlo and adaptive methods. Monte Carlo and Quasi-Monte Carlo methods approximate the integral by the sample mean on a set, P, of N points:

$$Q(f) \equiv \frac{1}{N} \sum_{\mathbf{z} \in P} f(\mathbf{z}). \qquad (3)$$

Monte Carlo methods choose P to be a random sample of points on the unit cube and attain an accuracy of $O(N^{-1/2})$. Quasi-Monte Carlo methods attain higher accuracy by choosing P to be a deterministic set of points spread evenly over the unit cube. These points are called low discrepancy points.

There are two major families of low discrepancy points — integration lattices [7,10,11] and (t, s)-sequences [10, Chapter 4]. An advantage of integration lattices is that they are very accurate for periodic integrands. (A non-periodic integrand may be made periodic by a suitable transformation.) An advantage of (t, s)-sequences is that one need not specify the number of points needed in advance. In this section we show how this advantage may be extended to lattice rules as well.

The formula for a rank-1 lattice is

$$\{\{i\mathbf{h}/N\} : i = 0, \ldots, N - 1\} \tag{4}$$

where N is the number of points, $\mathbf{h}$ is an s-dimensional integer generating vector that depends on N, and $\{\mathbf{x}\}$ denotes the fractional part of a vector $\mathbf{x}$. Unfortunately, rank-1 lattices require the number of points, N, and the generating vector, $\mathbf{h}$, to be chosen in advance. To define a rank-1 lattice sequence definition (4) is rewritten so that it does not depend on N.

This is done using the Van der Corput sequence in base b, which is denoted $\{\phi_b(i)\}_{i \geq 0}$. For any integer i, let $(\ldots i_3 i_2 i_1)_b$ be its b-nary representation. The Van der Corput sequence is defined by

$$\phi_b(i) = (0.i_1 i_2 i_3 \ldots)_b.$$

This sequence has the property that for any non-negative integer m the terms $\phi_b(0), \ldots, \phi_b(b^m - 1)$ take on the values $0/b^m, \ldots, (b^m - 1)/b^m$, although in a different order. If $N = b^m$, we may replace $i/N = i/b^m$ in (4) by $\phi_b(i)$, which does not depend on N.

The s-dimensional generating vector $\mathbf{h}$ in (4), also usually depends on N. It may be expressed in its b-nary form as:

$$\mathbf{h} = (\ldots h_{12} h_{11}, \ldots h_{22} h_{21}, \ldots, \ldots h_{s2} h_{s1})_b, \tag{5}$$

where $h_{jk} \in \{0, \ldots, b - 1\}$ are digits. In principle, one may think of the digits h_{jk} being defined for all positive k. However, in practice, the h_{jk} with $k \geq m$ do not affect the definition of a rank-1 lattice with $N = b^m$ points since they only contribute integers to the product $i\mathbf{h}/b^m$.

An infinite rank-1 lattice sequence is obtained by allowing for a generating vector with an infinite number of digits, as in (5), and removing the upper limit on the value of i, that is,

$$\{\{\phi_b(i)\mathbf{h}\} : i = 0, 1, \ldots\}. \tag{6}$$

The first b^m terms are a rank-1 lattice. In fact, succeeding runs of b^m terms are *shifted* copies of this initial rank-1 lattice.

Although (6) defines an infinite lattice sequence there are many theoretical and practical problems regarding the selection of a *good* generating vector $\mathbf{h}$.

A relatively simple and pragmatic approach was adopted for numerical experiments presented in the following section. The base was chosen to be $b = 2$ for convenience. A popular figure of merit for integration lattices, $D(P)$, was chosen as the objective function (see [5,6,11]):

$$[D(P)]^2 \equiv -1 + \frac{1}{N} \sum_{z \in P} \prod_{j=1}^{s} \left[1 + \frac{\beta^2}{2} B_2(\{z_j\}) \right],$$

where $B_2(x)$ is the quadratic Bernoulli polynomial. The value of β^2 was chosen to be 6. The first component of the generating vector, $h_1 = \ldots h_{13}h_{12}h_{11}$, was set to $\ldots 001 = 1$ without loss of generality. The digits h_{j1}, $j = 2, 3, \ldots$ were also chosen to be 1 so that the components of the generating vector would be relatively prime with respect to the base, 2. The digits of the second component, h_{2k}, $k = 2, 3, \ldots$ were chosen a few at a time so that the corresponding two-dimensional rank-1 lattice would minimize $D(P)$. Next, the digits h_{3k}, $k = 2, 3, \ldots$ were chosen in the same way assuming that the digits h_{1k} and h_{2k} to be known, and so on.

For any integers $M > 0$ and $m \geq 0$ let $Q(f)$ be the rank-1 lattice quadrature rule defined by (3), where $P = \{\{\phi_b(0)\mathbf{h}\}, \ldots, \{\phi_b(M2^m - 1)\mathbf{h}\}\}$. One may write $Q(f) = [Q_0(f) + \cdots + Q_{M-1}(f)]/M$, where $Q_k(f)$ is the rank-1 lattice rule based on

$$P_k = \{\{\phi_b(k2^m)\mathbf{h}\}, \ldots, \{\phi_b((k+1)2^m - 1)\mathbf{h}\}\}.$$

A heuristic estimate for the error of $Q(f)$ based on the M embedded rules is

$$E \equiv \left| \int_{C^s} f(\mathbf{x})\, d\mathbf{x} - Q(f) \right| \lessapprox c \sqrt{\frac{1}{M} \sum_{k=0}^{M-1} [Q(f) - Q_k(f)]^2}, \qquad (7)$$

where c is a fudge factor. The numerical experiments reported here used M between 4 and 7 and $c = 1$. Practical error estimation for this new lattice quadrature rule is an area of ongoing research.

4   Numerical Experiments

The test problems considered are the same as those in [4]. The first kind can be described as follows:

$$a_1 = \cdots = a_s = -\infty, \qquad b_j \text{ i.i.d. uniformly on } [0, \sqrt{s}] \qquad (8a)$$

$$\Sigma = (\sigma_{ij}), \text{ where } \sigma_{ij} = \begin{cases} 1 & i = j \\ \sigma & i \neq j \end{cases}, \qquad (8b)$$

$$\sigma \text{ distributed uniformly on } [0, 1] \qquad (8c)$$

Because all the off-diagonal elements of the covariance matrix are the same and the lower limits of integration are $-\infty$, the normal probability may be written as (see [12])

$$I(-\infty, \mathbf{b}, \boldsymbol{\Sigma}) = \frac{1}{\sqrt{2\pi}} \int_{-\infty}^{+\infty} e^{-\frac{1}{2}t^2} \prod_{j=1}^{s} \Phi((b_j + \sqrt{\sigma}t)/\sqrt{1-\sigma}) \, dt. \qquad (9)$$

This integral can be evaluated by standard univariate quadrature techniques.

Numerical comparisons were made using the adaptive algorithm DCHURE [1], the Korobov rank-1 lattice rule [2,8], and the new rank-1 lattice rule proposed above. For the two lattice rule algorithms the periodizing transformation $w_j = |2x_j - 1|$ was applied to the integral in (2). The computations were carried out in FORTRAN on a Unix workstation in double precision. Absolute error tolerances of $\epsilon = 10^{-4}$ and 10^{-6} were supplied to these routines. The true absolute errors, E, as defined in (7) were calculated using (9). Computation times were recorded to the nearest 0.01 second. For each s fifty test problems were generated randomly. Figure 1 shows the dependence of E/ϵ and the computation time on the method and on the dimension. In these box and whisker plots the boxes contain the middle half of the values and are divided by the median values. The whiskers show the full range of values except for the outliers, denoted by $*$.

The second kind of test problem considers the more general case of completely random covariance matrices $\boldsymbol{\Sigma}$.

$$a_1 = \cdots = a_s = -\infty, \qquad b_j \text{ i.i.d. uniformly on } [0, \sqrt{s}] \qquad (10a)$$
$$\boldsymbol{\Sigma} \text{ generated according to } [4,9]. \qquad (10b)$$

The absolute error tolerance was chosen to be $\epsilon = 10^{-4}$. The scaled absolute errors and computation times are given in Figure 2. Since the true value of the integral is unknown for this test problem, the value given by the Korobov algorithm with a tolerance of $\epsilon = 10^{-8}$ was used as the "exact" value for computing the error.

5   Discussion and Conclusion

For $\epsilon = 10^{-4}$ all three algorithms are relatively reliable, especially for lower dimensions. However, the adaptive algorithm seems to be unreliable when higher accuracy is required, since it consistently underestimates the error for problem (8) when $\epsilon = 10^{-6}$. In general problem (10) seems to be more difficult than problem (8) since there were more cases of errors being underestimated by all three algorithms. This suggests that further research needs to be done on reliable error estimation for multidimensional quadrature.

The adaptive algorithm is the most efficient for low dimensions. However, the two rank-1 lattice rule algorithms become more efficient as the dimension increases. The three algorithms have similar computation times for $s = 9$, but for $s = 15$ the adaptive algorithm is ten to several hundred times slower than the rank-1 lattice rule algorithms. For higher accuracy and harder problems, the new rank-1 lattice sequence algorithm is substantially faster than the Korobov algorithm.

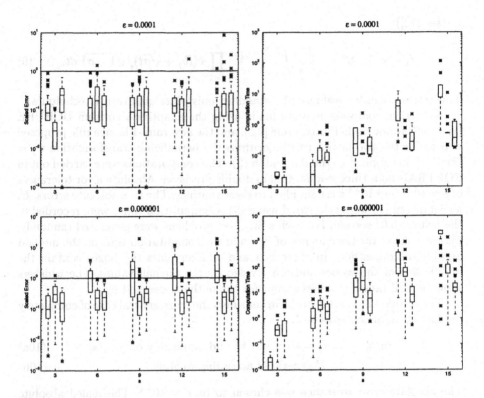

Fig. 1. Box and whisker plots of the scaled errors, E/ϵ, and computation times in seconds for 50 randomly chosen test problems (8). For each dimension s results from left to right correspond to the DCHURE algorithm, Korobov rank-1 lattice rules and the new rank-1 lattice sequence rules.

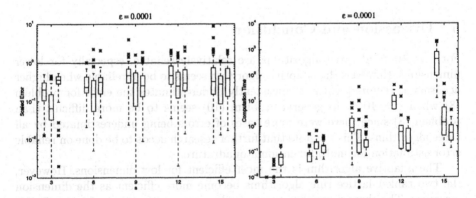

Fig. 2. Box and whisker plots of the scaled errors, E/ϵ, and computation times in seconds for 50 randomly chosen test problems (10). For each dimension s results from left to right correspond to the DCHURE algorithm, Korobov rank-1 lattice rules and the new rank-1 lattice sequence rules.

Although this paper considers the specific problem of multivariate normal probabilities, we believe that the results are indicative of what one may expect for multidimensional quadrature problems in general. Some important problems, such as those arising in finance, have even larger dimensions than those considered here. Thus, we conjecture that the new rank-1 lattice sequence rule may be effective for such problems as well.

Acknowledgments

The authors would like to thank Prof. Alan Genz for providing us his code to facilitate the numerical experiments and for answering our questions.

References

1. J. Bernsten, T. O. Espelid, and A. Genz, *An adaptive algorithm for the approximate calculation of multiple integrals*, ACM Trans. Math. Softw. **17** (1991), 437–451.
2. R. Cranley and T. N. L. Patterson, *Randomization of number theoretic methods for multiple integration*, SIAM J. Numer. Anal. **13** (1976), 904–914.
3. A. Genz, *Numerical computation of multivariate normal probabilities*, J. Comput. Graph. Statist. **1** (1992), 141–150.
4. _____, *Comparison of methods for the computation of multivariate normal probabilities*, Computing Science and Statistics **25** (1993), 400–405.
5. F. J. Hickernell, *Quadrature error bounds with applications to lattice rules*, SIAM J. Numer. Anal. **33** (1996), 1995–2016.
6. _____, *A generalized discrepancy and quadrature error bound*, Math. Comp. **66** (1997), to appear.
7. L. K. Hua and Y. Wang, *Applications of number theory to numerical analysis*, Springer-Verlag and Science Press, Berlin and Beijing, 1981.
8. P. Keast, *Optimal parameters for multidimensional integration*, SIAM J. Numer. Anal. **10** (1973), 831–838.
9. G. Marsaglia and I. Olkin, *Generating correlation matrices*, SIAM J. Sci. & Statist. Comput. **5** (1984), 470–475.
10. H. Niederreiter, *Random number generation and quasi-Monte Carlo methods*, SIAM, Philadelphia, 1992.
11. I. H. Sloan and S. Joe, *Lattice methods for multiple integration*, Oxford University Press, Oxford, 1994.
12. Y. L. Tong, *The multivariate normal distribution*, Springer-Verlag, New York, 1990.

Wavelets, Signal Processing and Textile Surfaces

Holger Kraus *

School of Design and Manufacture, De Montfort University, LE1 9BH, UK
(hka@dmu.ac.uk)

Abstract. In recent years, wavelet transform has been introduced as a mathematical tool to the field of signal and image processing and has attracted great attention [9]. The need for a true time–frequency representation of a signal leads to the introduction of the wavelet transform with its multiresolution capability. This paper then provides a summary on the use of the wavelet transform in combination with some signal processing methods that have been designed in an attempt to characterise and classify textile fabrics.

Introduction

This paper is divided in two parts. Part 1 gives a brief introduction of the theoretical background. This is followed by part 2, giving a review of five papers within the scope of textile fabric classification and defect detection, and one paper of a texture analysis approach to corrosion image classification.

1 Theoretical Considerations

A transformation is a mathematical algorithm applied to a signal to obtain further information that is not readily available in the original signal. It is often followed by thresholding and intends to select the most relevant features of a signal or image. It may then be possible to apply an inverse transform, for the purpose of reconstructing the geometry of the original signal or image, except with the desired features explicit or enhanced.

1.1 Fourier Transform

The Fourier transform (FT) has been used as a reliable tool in signal analysis for many years. Invented in the early 1800's by the French mathematician Jean-Baptiste Joseph Fourier, it has become a cornerstone of modern signal analysis and has proven incredibly versatile in applications ranging from pattern recognition to image processing. A function in the time domain is translated by the FT into a function in the frequency domain, where it can be analysed for its frequency content [2]. A limitation is, the FT works under the assumption that the

---

* The research was funded by this School and the Bentley Fund.

original time-domain function is periodic in nature. As a result, the FT has difficulty with functions having transient components, that is components localised in time. This is especially apparent when a signal has sharp transitions. Another problem is that the FT of a signal does not convey any information pertaining to translation of the signal in time or space. Until recently, applications that use the FT often worked around the first problem by windowing the input data so that the sampled values converge to zero at the endpoints. Attempts to solve the second problem have met with marginal success.

1.2   The Wavelet Transform

The wavelet transform (WT) is the result of work by a number of researchers. Initially, the French geophysicist Jean Morlet began searching in the 1980's to analyse the sound echoes used in oil prospecting. He experimented with equations that substituted individual simpler waves — or wavelets — for the endless series of sine and cosine curves used in Fourier analysis. Yves Meyer, a mathematician, recognised his work to be part of the field of harmonic analysis, and came up with a family of wavelets that he proved were most efficient for modelling complex phenomena [10]. This work was improved upon by two American researchers, Stéphane Mallat of New York University and Ingrid Daubechies of AT&T Bell Labs. Since 1988, there has been an explosion of activity in this area, as engineers and researchers apply the wavelet transform to applications ranging from image compression to fingerprint analysis. Wavelets, which are basically new families of orthonormal basis functions, lead to transforms which overcome the problems of the FT. The fundamental difference between the two is that the WT breaks up the signal into contributions from wave pulses (mother wavelets) which are dilated, translated and rotated [3].

2   Current Developments and Applications in Praxis

All the above mentioned advantages have brought forth new developments in the fields of signal analysis, image processing and data compression. For example, wavelets have been successfully applied by the FBI to the compression of electronic fingerprints. Another possible area of application could be the characterisation of image features such as edges, shapes or texture of surfaces within the textile industry; the advantage in speed makes wavelets very suitable for on-line inspection in a production environment to detect faults or defects.

2.1   Cloth Texture Classification Using the Wavelet Transform

"... it is shown that the WT is capable of providing features that may be used to discriminate between eight different cloth textures. The effectiveness of using WT features to classify texture was compared with that of a commonly used method that extracts features

from spatial grey-level dependency matrices called co-occurrence matrices. Texture features were input to a decision tree classification algorithm to discriminate different textures. The WT features correctly identified 86% of 64 cloth textures. The co-occurrence features correctly classified 76% of the cloth samples." [5]

Mary B. Henke-Reed is the first person (1992) who has published results of research into the application of WT for textile fabrics, be it for texture classification or defect detection. This paper explores the viability of using the WT for texture feature extraction for the purpose of image classification. It represents an initial investigation and is based on a study carried out on 8 different cloth textures with only 8 samples of each. This small sample size is unlikely to produce statistically significant results. However, it was merely intended to be a feasibility study of using the WT for texture classification, and as such it fulfils its aim demonstrating that feature extraction using the WT is an effective technique for texture classification.

2.2 Image Analysis of Mispicks in Woven Fabric

"Images of a woven fabric with missing picks are digitised, and three different image analysis techniques are compared: the Sobel edge operator, the fast Fourier transform (FFT) and the discrete WT. The WT, used as a multiresolution spectral filter, is able to give both spectral and frequency information about a fabric. For the samples tested, the WT can characterise defects due to missing picks and ends faster and more accurately than the other methods." [6]

This paper focuses on the real-time measurement of weaving defects (missing picks and missing ends) in plain cotton fabric. From an image processing perspective, identifying weaving defects is complicated by the texture of the fabric itself. Although mispicks look like dark or light horizontal edges, conventional edge detection algorithms are often fooled by the edges of the weave itself.

A short and excellent review of current defect detection approaches is presented, none of which is well suited for on-line inspection, either because of computational reasons or because the reliability and repeatability are not sufficient for industrial applications. Current methods are mainly based on comparisons between a template and the test image. A more suitable approach is one that directly measures the test image to determine whether or not a defect exists.

The use of WT has been described as a fast and robust fabric mispick detection system. The defective regions of the fabric are easily located in the horizontal edge component of the wavelet representation of the image. Since measurements are made directly on the image as opposed to comparisons with as template, the method offers significant advantages over current fabric inspection systems. Comparisons with other edge detection methods such as the Sobel edge operator and the FFT demonstrate that the wavelet method is more effective and reliable. While the Sobel operator produces many artefacts and the FFT is not able to distinguish between poor quality and first quality fabrics, the wavelet representation gives good localisation of the defect and a good discrimination between first quality and defective fabrics.

2.3   Real-time Fabric Defect Detection and Control in Weaving Processes

"...research between Georgia Institute of Technology and North Carolina State University addresses the monitoring requirements for performance assessment of a weaving machine under on-line real-time conditions and control of the machine parameters to minimise fabric defects. Novel ideas for fault detection and identification of woven textile structures are introduced and implemented. ... Fractal scanning, a new technique, is developed to scan the digitised image of textile fabrics. A fuzzified WT algorithm with adaptive noise rejection and on-line learning is used to extract features ... " [4]

This three year project commenced in March 1994 and has so far produced very successful and promising results. The method used for fabric defect detection is a combination of WT techniques and fuzzy inferencing methods. The resulting arrangement is called Fuzzy Wavelet Analysis (FWA). The algorithms provide the ability to analyse image or target signatures in space/frequency localised manner while accommodating uncertainty. The FWA provides on-line adaptability and robust pattern classification through learning.

Wavelets are used to generate a multiresolution analysis of a signal. The texture of woven fabrics can be described by periodic functions whereas a defect such as a missing yarn (a mispick) can be described by a high frequency event in one direction (across the weave direction) and a low frequency content along the weave direction. The WT has the property of giving both frequency and spatial information about an image. As detection of some faults, such as mispicks, requires information in both frequency and spatial domains, this property of wavelets is very important for the correct detection of such faults.

The fuzzy wavelet analysis procedure employs WT of different wavelet functions to process the output data streams obtained from the scanning operation. The wavelet transforms generate the appropriate coefficients that are non-linearly combined by a fuzzy inferencing mechanism. Since most features of interest produce signatures in a wide range of frequencies, a number of wavelet coefficients are buffered and subjected to a transformation in order to obtain the trend of the signature. Next, these coefficients are processed further, and finally evaluated by a fuzzy expert system. The aim is not only to detect the presence of a defect, but also to distinguish between various defects and to assess their severity. The technology being developed in this project is very promising and will be well-suited for applications in the textile industry.

2.4   Texture Characterisation and Defect Detection Using Adaptive Wavelets

"Many textures such as woven fabrics and composites have a regular and repeating texture. This paper presents a new method to capture the texture information using adaptive wavelet bases. ... Texture constraints are used to adapt the wavelets to better characterise specific textures. An adapted wavelet basis has very high sensitivity to the abrupt changes in the texture structure caused by defects. This paper demonstrates how adaptive wavelet basis can be used to locate defects in woven fabrics." [7]

This paper develops research results presented in [6] and [4] further. It introduces the relatively new idea of using image texture data for the formation of wavelet bases. A two-dimensional, linear, spatially invariant filter is designed for a given texture such that the filter gives a zero response to that texture. Disturbances in the texture due to noise and defects will produce a non-zero output from the filter. This piece of work attempts to address the issue of locating small differences in a particular texture by developing a texture-specific wavelet basis for a filter such that the filter is tuned to this particular texture. The response of the filter will be close to zero when the texture is presented to the filter; when the presented pattern is different (due to defects), the filter response will be significantly different, thereby enabling detection. An adaptation of the wavelet basis can be achieved by dynamically altering the wavelet coefficients. Least-squares techniques are used to construct adaptive wavelet bases. Therefore, this method develops new wavelets which are able to incorporate the texture information, employing numerical methods such as Newton–Raphson or Levenberg–Marquardt. Adaptive wavelet bases have been developed in order to apply them to an on-line loom-based fabric inspection system. A range of test images are described as superior to the use of Daubechies wavelets in terms of reducing the response of the filters to the texture. This appears to be the most effective method to characterise texture of fabric as well as to detect fabric defects with a single combination of algorithms.

2.5   An On-line Fabric Classification Technique Using a Wavelet-based Neural Network Approach

"A sewing system is described that classifies both the fabric type and number of plies encountered during apparel assembly, so that on-line adaptation of the sewing parameters to improve stitch formation and seam quality can occur. Needle penetration forces and presser foot forces are captured and decomposed using the WT. Salient features extracted using the WT of the needle penetration forces form the input to an artificial neural network, which classifies the fabric type and number of plies being sewn. A functionally linked wavelet neural network is trained on a moderate number of stitches for five fabrics, and can correctly classify both fabric type and number of plies being sewn with 97.6% accuracy. ... " [1]

The object of this research is to investigate the on-line identification of a fabric within a finite set of fabrics. The role of the fabric in affecting the dynamics of the sewing system is seen as very important. Information from the interaction of the sewing machine with the fabric is used to identify the kind of fabric being sewn. By applying mathematical transforms, particular aspects of the fabric/machine interaction are used to determine the fabric type.

Signal power and FT have been applied to the experimental data obtained, but the data clusters overlap significantly and do not allow for a definite separation of data clusters. To further distinguish between fabrics, more explicit information is required about the relationship between frequencies in the signal and the moment in time they occur within the signal. It does not seem clear how the important wavelet coefficients have been chosen, but according to the authors the WT is helpful in extracting features of the measured forces. The insights into the fabric/machine interaction can be taken advantage of by using neural networks to learn the mapping of significant wavelet coefficients of measured forces to the fabric/ply combination. This ability allows the sewing machine to use pre-defined sewing parameters to automatically adjust the sewing machine settings. This method has been tested with only five different fabric types.

2.6   A Texture Analysis Approach to Corrosion Image Classification

"A method is described for the classification of corrosion images using texture analysis methods. Two morphologies are considered: pit formation and cracking. The analysis is done by performing a wavelet decomposition of the images, from which energy feature sets are computed. A transform that turns the wavelet features into rotation invariant ones is introduced. The classification is performed with a Learning Vector Quantisation network and comparison is made with Gaussian and k-Nearest Neighbour (k-NN) classifiers. The effectivity of the method is shown by tests on a set of 398 images." [8]

Since images showing the same morphology can have very different appearances, the underlying processes that cause the corrosion are too complex for use in an automated recognition system. Therefore, only the images themselves define the classes. The task of a supervised classifier is to assign data to pre-defined classes. This is usually done by first segmenting artefacts (here: pits and cracks) automatically from the background, which was not possible because of the large variability of the extensive set of examples. It was suggested to use wavelets to generate a multiresolution representation to discriminate between the two morphologies. The extracted feature data are then processed by a neural network classifier. A LVQ network has been compared with two well-known statistical classifiers, namely a Gaussian Quadratic Classifier (GQC) and a k-NN. The novel concept is to introduce an energy feature vector from the subimages.

Larger discrepancies between training and test results show that the results should be treated with care, and especially generalisation becomes rather difficult. Nevertheless, a successful scheme, with a classification score of 86.2%.

Conclusions

All of the above articles provide prove that wavelet transform, combined with appropriate algorithms is better suitable than conventional methods to detect surface defects and to classify surface texture of textile fabrics, not only in terms of computational requirements. There is a lot of scope to extend the application of wavelets to other types of texture and engineering surfaces.

References

1. Barrett, G., Clapp, T., Titus, K.: *An On-line Fabric Classification Technique Using a Wavelet-based Neural Network Approach* Textile Research Journal, **66** (1996)
2. Bracewell, R.: *The Fourier Transform and Its Applications* McGraw-Hill (1986)

3. Daubechies, I.: *Ten Lectures on Wavelets* SIAM (1992)
4. Dorrity, J., Vachtsevanos, G., Jasper, W.: *Real-Time Fabric Defect Detection and Control in Weaving Processes* North Carolina State University (1995)
5. Henke-Reed, M., Cheng, S.: *Cloth Texture Classification Using the Wavelet Transform* Journal of Imaging Science and Technology **37** (1993) 610–614
6. Jasper, W., Potlapalli, H.: *Image Analysis of Mispicks in Woven Fabric* Textile Research Journal **65** (1995) 683–692
7. Jasper, W., Garnier, S., Potlapalli, H.: *Texture Characterisation and Defect Detection Using Adaptive Wavelets* Optical Engineering **35** (1996) 3140–3149
8. Livens, S., Scheunders, P., Van de Wouwer, G., Van Dyck, D.: *A Texture Analysis Approach to Corrosion Image Classification* Microscopy, Microanalysis, Microstructures **7** (1996) 143–152
9. Mallat, S.: *Wavelets for a Vision* Proceedings of the IEEE **84** (1994) 604–614
10. Meyer, Y.: *Wavelets, Algorithms & Applications* SIAM (1993)

Table 1. Summary of the signal processing techniques used in the articles

Paper	Algorithm	Compared with	Results / Comments
2.1 [5]	- 64 images of 8 textured cloth samples - Standard WT - Decision tree classifier	- Spatial grey-level dependency matrices (Co-occurrence matrices)	- WT - 86% success rate - Co-occurrence matrix - 76% success rate
2.2 [6]	- Plain weave cotton fabric - Scanning using CCD camera - WT	- Sobel edge operator - FFT	- No template used
2.3 [4]	- 2-D images of woven fabric - Fractal scanning - WT - Fuzzification algorithm - Fuzzy inferencing mechanism		- Process a 1-D signal - Use of template
2.4 [7]	- Image - Spatially invariant filter with a texture specific wavelet basis		- Process a 2-D signal
2.5 [1]	- Fabric data collection - WT - Energy analysis of wavelet coefficients - Inverse WT with wavelet coefficients that contain important details - Map the resulting waveforms to a decision variable - Sewing parameters for a particular fabric type	- Signal power - Fourier transform	
2.6 [8]	- 2-D corrosion image - Wavelet-based multi-resolution analysis - Neural Net. Classifier (LVQ)	Statistical classifier: - GQC - k-NN	- Make declaration about the best suitable method for corrosion images

Size-location patterns of different wavelets in signal processing

Mong-Shu Lee

Department of Computer and Information Sciences,
National Taiwan Ocean University
Keelung, Taiwan R.O.C.
E-mail: mslee@ntou66.ntou.edu.tw

Abstract. Wavelets are used as a cutting-edge tool in the signal processing area. One of the characteristics is that wavelets can be used to detect the local changes of a signal(transition), which is the traditional Fourier method can not do. When we implement the wavelet transform to the signal, the transition of signal will reflect upon the magnitude of the coefficients at that location. In this article, we experiment some signals with wavelets in different lengths, then we compare the size-location patterns with distinct wavelets. From these demonstrations, we conclude that the size-location of the wavelets coefficients remains constant from one wavelet to another.

1   Introduction

Fourier transform is the traditional method to analyze signals. We decompose signals into linear combinations of waves(sines and cosines). The Fourier coefficients of such a signal involve integrals over the full signal and each coefficient reflects the presence of some frequency somewhere. From the point of view of the uncertainty principle, it tells us that the value of signal f at a particular point are encoded in the global properties of $\hat{f}$. Recently, wavelet decompositions have emerged as useful alternatives for many applications. The success of wavelets comes from the fact that they are both local with respect to the "space" variable and to the "frequency" variable, the only limitation comes from uncertainty principle. Wavelet decompositions are better behaved than Fourier decompositions in the sense that in the standard Fourier analysis there is no relationship between the local behavior of a signal f and the size of its Fourier coefficients while the wavelet coefficients of f will provide a rich and deep information on this local behavior. The aim of this article is to study whether different wavelets affect this size-location relationship in signal processing.

2   Wavelets and Properties

The wavelet theory can be described in the framework of the multiresolution analysis. A chain of closed subspaces $\{V_j : j \in Z\}$ are called a multiresolution analysis of $L_2(R)$ if they satisfy the following conditions:

(a)$V_j \subset V_{j-1}, j \in Z$.

(b)$f(x) \in V_j \iff f(2x) \in V_{j-1}$.

(c)$f(x) \in V_0 \iff f(x+1) \in V_0$.

(d)$\overline{\cup_{j \in Z} V_j} = L_2(R)$ and $\cap_{j \in Z} V_j = \{0\}$.

(e)There is a function $\phi(x) \in V_0$ whose translate $\{\phi(x-l) : l \in Z\}$ forms an orthonormal basis of V_0.

Now, we conclude from $(a), (b)$, and (e) of the above conditions that there exists a sequence (h_l) such that

$$\phi(x) = \sqrt{2} \sum_{l \in Z} h_l \phi(2x - l) \tag{1}$$

where the $\sqrt{2}$ is for the normalization of ϕ, and the numbers h_l, called the *filter* coefficients of ϕ, can be finite. Take $(h_l)_{l=0}^{l=2N-1}$ as example(see[2]), then the function ϕ will have compact support on $[0, 2N - 1]$.

The equation (1), called the *two-scale equation*, can be generalized as follows:

$$\phi_{j,l}(x) = \sum_{k \in Z} h_{k-2l} \phi_{j-1,k}(x)$$

where $\phi_{j,l}(x) := 2^{-j/2} \phi(2^{-j} x - l)$.

If we define the orthogonal complementary subspace $W_j = V_{j-1} \ominus V_j$ so that $V_{j-1} = V_j \oplus W_j$, then we can telescope the union in the "increase" property (d) of multiresolution conditions to write the orthogonal wavelet decomposition of $L_2(R)$ as

$$L_2(R) = \oplus_{j \in Z} W_j. \tag{2}$$

The subspaces W_j are called orthonomal wavelet subspaces, and W_0 is generated by a function ψ satisfying the wavelet equation:

$$\psi(x) = \sqrt{2} \sum_{k \in Z} g_k \phi(2x - k), g_k = (-1)^k h_{1-k}$$

In other words, for any fixed j, $\{2^{-j/2} \psi(2^{-j/2} x - l) : l \in Z\}$ gives an orthonormal basis for W_j, and the functions $\{\psi_{j,l}(x) : j, l \in Z\}$ provide an orthonormal basis for $L_2(R)$.

The Haar function is a special case of wavelet, nevertheless we will use it here to start the illustration of size-location relationship of wavelet decomposition. Let $h(x) := \chi_{[0,1/2)} - \chi_{[1/2,1)}$ be the Haar function that takes value 1 on the left half of $[0, 1]$ and the value -1 on the right half. By translation and dilation, we form the functions $h_{j,l}(x)$ defined as $\phi_{j,l}$ and $\psi_{j,l}$. Then, $h_{j,l}(x)$ is supported at the dyadic interval $I_{j,l} := [l2^j, (l+1)2^j)$. It is well-known, for any $f \in L_2(R)$, that we have the decomposition

$$f(x) = \sum_{j \in Z} \sum_{l \in Z} < f, h_{j,l} > h_{j,l}(x). \tag{3}$$

Here $< \cdot, \cdot >$ denote the inner product in $L_2(R)$. In other words, the functions $\{h_{j,l}(x), j, l \in Z\}$ forms an orthonormal basis for $L_2(R)$.

Because each of the functions $h_{j,l}(x)$ is nonzero only on a relative small set, the Haar basis thus allow us to focus on important pieces of f. For example, if we are only interested in f for $1 \leq x \leq 1.6$, then we can discards all the terms in the decomposition except the ones with $h_{j,l}(x) \neq 0$ there. So we only need the terms with $0 < 2^{-j}x - l < 1$ for some $1 \leq x \leq 1.6$. For a given j, corresponding to the scale 2^j, this means that only the terms with l satisfying $2^{-j} \leq l \leq 2^{-j}1.6$ give contribution.

Unfortunately, the frequency spectrum of each of the functions is not nearly as nice. since each Haar function is built up from two "boxes", the Fourier transform decays very slowly. As a result, we get some complicated interactions on the frequency side between the different terms in the Haar decomposition of a given function.

The ideal expansion similar to the Haar series (3) would be the form :
Given $f \in L_2(R)$,

$$f(x) = \sum_{j=-\infty}^{\infty} \sum_{l=-\infty}^{\infty} d_{j,l}\psi_{j,l}(x) \tag{4}$$

where $\psi_{j,l}(x) = 2^{-j/2}\psi(2^{-j}x - l)$ and

$$d_{j,l} = < f(\cdot), \psi_{j,l}(\cdot) >,$$

also with a function ψ such that

$$\left\{ \begin{array}{l} \psi(x) \neq 0 \ \ \text{for} \ \ 0 \leq x < 1. \\ \hat{\psi}(\xi) \neq 0 \ \ \text{for} \ \ 1 \leq |\xi| < 2. \end{array} \right. \tag{5}$$

We would then know exactly where the functions $\psi_{j,l}(x)$ are nonzero, just as for the Haar functions. Furthermore, since the Fourier transform of $\psi_{j,l}$

$$\hat{\psi}_{j,l}(\xi) = 2^{-j/2}e^{i2^j l\xi}\hat{\psi}(2^j \xi),$$

the terms $\psi_{j_0,l}(x)$ would exactly catch the corresponding frequency component of $\hat{f}$ between 2^{-j_0} and 2^{-j_0+1} for any fixed j_0. We would then avoid all the interactions between the different frequency components (or scales). We could focus on an interesting piece of f and then obtain a frequency decomposition similar to the Fourier series.

The problem is that there are no functions ψ with the properties (5). Wavelet decompositions are representations of the form (4), but in terms of the function ψ which satisfies each of two properties (5) to different degrees. The Haar functions form one extreme case satisfying the first property exactly and the other one not at all. On the other hand, the Fourier basis of trigonometric system are just the opposite. Other famous wavelets, such as the "Daubechies wavelets" (see [2]) are better compromises and have some of both properties (5). In spite of the fact that there are no "ideal" wavelets with the properties (5), the way to understand the wavelet decomposition almost always uses the "ideal" wavelets as a guiding model.

For example, let us try to understand what information is contained in each coefficient

$$d_{j,l} = \langle f(\cdot), \psi_{j,l}(\cdot) \rangle = \int f(x) 2^{-j/2} \psi(2^{-j}x - l) dx$$

when we have picked a wavelet ψ. Using our model, we assume that ψ is the ideal wavelet. Since the function $\psi(2^{-j}x - l)$ is zero outside the interval $I_{j,l} := [l2^j, (l+1)2^j)$, only the behavior of f on $I_{j,l}$ influence the value of $d_{j,l}$. On the other hand, by Plancherel's formula,

$$d_{j,l} = \langle \hat{f}(\cdot), 2^{-j/2} e^{i2^j l\xi} \hat{\psi}(2^j \cdot) \rangle$$

Now $\hat{\psi}(2^j \xi)$ is zero unless $1 \le 2^j \xi < 2$, which means that only the behavior of $\hat{f}$ in the frequency range $2^{-j} \le \xi < 2^{-j+1}$ enter the calculation of d_{jl}. Putting everything together, we thus see that $d_{j,l}$ contains information about the frequency component $\xi \approx 2^{-j}$ at location $x \approx 2^j l$.

Let us consider two wavelets ψ^1 and ψ^2 which are two basis functions for our function f from L_2 with their coefficients $\{\langle f, \psi_{j,l}^1 \rangle\}$ and $\{\langle f, \psi_{i,k}^2 \rangle\}$ respectively. Now, we have

$$f = \sum_{j,l \in Z} \langle f, \psi_{j,l}^1 \rangle \psi_{j,l}^1$$

$$= \sum_{j,l \in Z} \langle f, \psi_{j,l}^1 \rangle \sum_{i,k \in Z} \langle \psi_{j,l}^1, \psi_{i,k}^2 \rangle \psi_{i,k}^2$$

where $\psi_{j,l}^1 = \sum_{i,k \in Z} \langle \psi_{j,l}^1, \psi_{i,k}^2 \rangle \psi_{i,k}^2$,
Equivalently, we get

$$f = \sum_{i,k \in Z} (\sum_{j,l \in Z} \langle f, \psi_{j,l}^1 \rangle \langle \psi_{j,l}^1, \psi_{i,k}^2 \rangle) \psi_{i,k}^2.$$

and

$$(\sum_{j,l \in Z} \langle f, \psi_{j,l}^1 \rangle \langle \psi_{j,l}^1, \psi_{i,k}^2 \rangle) = \langle f, \psi_{i,k}^2 \rangle.$$

It means that if f is first represented by the basis $\{\psi_{j,l}^1\}$ and we change basis to $\{\psi_{i,k}^2\}$, then the representation in this new basis is given by

$$\langle f, \psi_{i,k}^2 \rangle = \sum_{j,l \in Z} \langle f, \psi_{j,l}^1 \rangle \langle \psi_{j,l}^1, \psi_{i,k}^2 \rangle \qquad (6)$$

Suppose that $\{\psi_{j,l}^1\}$ and $\{\psi_{i,k}^2\}$ were identical, then we would have $\langle \psi_{j,l}^1, \psi_{i,k}^2 \rangle = \langle \psi_{j,l}^1, \psi_{i,k}^2 \rangle = \delta_{j,i} \delta_{l,k}$. In other words, $\{\langle \psi_{j,l}^1, \psi_{i,k}^2 \rangle\}$ would be a diagonal "matrix" and using (6) there would be a perfect correspondence between $\langle f, \psi_{j,l}^1 \rangle$ and $\langle f, \psi_{i,k}^2 \rangle$.

Next, when we choose different wavelets, they still have so much in common that the "matrix" $\{< \psi_{j,l}^1, \psi_{i,k}^2 >\}$ is almost diagonal. If $i = j$, with k and l far apart, say, $|k - l| \geq 50 \cdot 2^j$, then the matrix elements $\{< \psi_{j,l}^1, \psi_{i,k}^2 >\}$ is very small(in magnitude). Similarly, if $i \neq j$, then, again $\{< \psi_{j,l}^1, \psi_{i,k}^2 >\}$ is small.

3   Implementation

In this section, we apply the *Discrete wavelet transform(DWT)* algorithm from Stephane Mallat [6] to signal decomposition and reconstruction. More details about this algorithm can be found in [2,8].

Now, let the signal data "bumps" with 1024 sample points (see the top row of fig. 1). We compute the wavelet transform with two different filters to the signal "bumps", and set the resolution levels $j = 4$. These wavelets are the cases of Daubechies wavelets with $N = 4$ and 8.(see [2]). The results are displayed in Fig. 1.

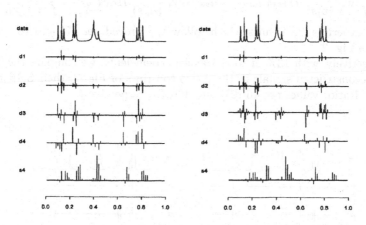

Fig. 1. : Wavelet transform coefficients of data "bumps". Left: W_4 wavelet(N=4). Right: W_8 wavelet(N=8).

We notice that the coefficients sizes and the corresponding locations of W_4 and W_8 wavelets are kept to be constant. The patterns of these figures support the property of wavelets as we mentioned in the last section. Also from Figure 1, we can see that almost all of the wavelet coefficients are nearly zero. For the further demonstration, we utilize the thresholding in signal compression. The idea behind thresholding is the removal of small wavelet coefficients considered to be neglected. So we choose a threshold value C(see [4]), if the j-scale coefficients $| < f, \psi_{j,l} > | \geq C$ keep; otherwise, throw out. The (j, l) here correspond to a certain location: $2^j l$. When we apply the *Inverse Discrete Wavelet Transform(IDWT)* to reconstruct the signal, we pull these coefficients from their locations to approximate our signal. The following figure is the overall process we are using.

$$f \overset{DWT}{\longrightarrow} \{< f, \psi_{j,l} >\} \overset{Thresholding}{\longrightarrow} \{ \text{ subset of } < f, \psi_{j,l} >\} \overset{IDWT}{\longrightarrow} \tilde{f} \approx f$$

where f and $\tilde{f}$ denote the original and reconstructed signal, respectively.

In order to verify the close relation between where coefficients of one wavelet $| < f, \psi^1_{j,l} > |$, and coefficients of another wavelet $| < f, \psi^2_{i,k} > |$, are large. In other words,

$$\{(j,l) : | < f, \psi^1_{j,l} > | \geqslant C\} \text{ corresponds to } \{(i,k) : | < f, \psi^2_{i,k} > | \geqslant C\}$$

We use the predominant coefficients of the second wavelet in the locations where the first wavelet coefficients are big. We should still be able to reconstruct a reasonable approximation of our signal. Here, let the first wavelet be W_8 and the second wavelet be W_4, then we have

$$f \overset{W_4}{\longrightarrow} \{d_l s\} \overset{Thresholding \ related \ to \ W_8}{\longrightarrow} \{d'_k s\} \overset{IDWT \ of \ W_4}{\longrightarrow} \tilde{f} \approx f(x). \qquad (7)$$

To interchange the roles of W_4 and W_8, we get

$$f \overset{W_8}{\longrightarrow} \{d_l s\} \overset{Thresholding \ related \ to \ W_4}{\longrightarrow} \{d'_k s\} \overset{IDWT \ of \ W_8}{\longrightarrow} \tilde{f} \approx f(x). \qquad (8)$$

The experimental results which follow the procedures (7) and (8) are displayed in Fig. 2.

Objectively, with 13% of total wavelets coefficients, we compare the quality of the reconstructed signals in these two top rows of Fig. 2, their SNR(Signal-to-Noise Ratio) values are 46(dB) and 40(dB), respectively.

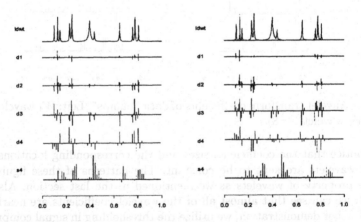

Fig. 2. : With resolution levels $j = 4$, left is the wavelet transform coefficients followed the procedure in (7), and right is the wavelet transform coefficients followed the procedure in (8). Both the top rows represent the reconstructed signals.

4   Conclusion

The theoretical as well as the experimental results have indicated the existence of size-location correlation concerning different wavelets. We, using the same scheme to test other signals, also find that choosing high order wavelets(with big N)as a thresholding criteria when applied to low order wavelets(with small N) produces better results. Some further explorations about how strong correlation of this relationship are needed.

References

1. C. K. Chui, (1992) *An Introduction to wavelets*, Academic Press.
2. I. Daubechies, (1993) *Ten Lectures on Wavelets*, SIAM, CBMB Reginal Conference in Applied Mathematics Series, No. 61.
3. R. Devore, B. Jawerth and B. Lucier, *Image compression through wavelets transform coding, IEEE Trans. on Information Theory*, vol. 38 , No. 2, pp.719-746, 1992.
4. D. L. Donoho and I. M. Johnstone, *Ideal spatial adaptation via wavelet shrinkage.* Technical report, Stanford University, 1992.
5. S. Mallat and S. Zhong *Wavelets transform maximum and multiscale edges*, In Ruskai M. et. al., editors, *Wavelets and their applications*, pp. 67-104, Jone and Bartlet publishers, 1992.
6. S. Mallat, *A theory for   multiresolution signal decomposition: the wavelet representation.* IEEE Transactions on Pattern Analysis and Machine Intelligence, 11(7): 674-693, 1989.
7. Y. Meyer (1992), *Wavelets and Operator*, Cambridge University Press.
8. O. Rioul and M.Vetterli, *Wavelets and signal processing.* IEEE Signal Proc. Mag. 14-38, October 1991.

Efficient Image Compression Scheme Using Tree Structure Wavelet Transforms

Jiaming Li and Jesse S. Jin

School of Computer Science and Engineering,
The University of New South Wales
Sydney 2052, AUSTRALIA.

Abstract. In this paper, we propose a novel wavelet transform (WT)-based image coding scheme. By applying the perceptual characteristics of human visual system (HVS) to WT domain before doing the embedded zerotree wavelet (EZW) coding, coding performance is enhanced based on SNR/WSNR measure. To further take the advantage of structure correlation in WT, perceptually important edge positions are detected and used to enhance the weighting. Image quality improvement has been observed subjectively with the compression rate ranging from 8 to 64 for image Lena, which has also been confirmed through the calculation of weighted-signal-to-noise-ratio (WSNR).

1 Introduction

Most recent research on applying wavelet transform (WT) in image coding has shown that WT-based methods generally outperform all other methods including discrete cosine transform (DCT) based approaches. In comparing with other transform-based image coding schemes, the most attractive feature of WT is that it offers a flexible image representation with multiple resolutions. Among various versions of WT-based coding algorithm [1–7], it has been claimed that the Embedded Zero-tree Wavelet (EZW) coding proposed by J. Shapiro [6] can consistently produce compression results that are competitive with virtually all known compression algorithms (by 1993) over a set of conventional test images. Much effort has been made to elevate the performance of tree-structure-related algorithm [7,8]. The main ideal is: the coefficient tree is encoded recursively if the absolute value of a WT coefficient is greater than a threshold, otherwise the tree is truncated and zero coefficients are implied. In the course of finding ways of improving the performance of EZW coding, we realize that the purpose of image compression is to achieve low bit rate representation of image with a minimum perceived loss of image quality, and that the final observer of the processed image is human being. We incorporate the characteristics of Human Visual System (HVS) into the coding process so that the algorithm is adapted to the sensitivity of the visual perception and produces subjectively less noise in images. The core of the perceptual weighting is that human viewer can hardly perceive details in high frequency region and it is desirable to use more bits to represent viewer sensitive part and less bits to represent insensitive part. During the developing of our early work on perceptual weighting [9], it was found that though the quality of the perceptually weighted and compressed image as a whole is subjectively improved comparing with no-perceptually weighted image, the distortion in areas of sharp edges is more visible due to more severe compression aliasing than that in flat area, especially at high compression rate. Because the degree of distortion perception is also different according to the edge height [10], we rectify the perceptual weighting along edges. The edge areas are detected with no further coding overhead, based on the unique character of WT, namely, there exists correlation between the subbands and the measure of activity in the lower band can predict the amplitude range (therefore the sharp areas) of the coefficients of upper bands. The following will present more details of our algorithm and some experimental results.

2   GENERAL DESCRIPTION OF OUR CODING ALGORITHM

Figure 1 is the block diagram of structure-related perceptual weighting embedded zerotree coding (SRPEZW) system. In the coding process, an original image is first undergone a WT. Secondly, in order to incorporate characteristics of HVS and correlations between subbands, each WT coefficient is weighted accordingly first and then the so weighted coefficients are encoded with EZW algorithm (SRPEZW) to produce symbol stream. Finally, the symbol streams are entropy-encoded [11], resulting compressed symbol sequence. The decoding process is the inverse operation of the encoding.

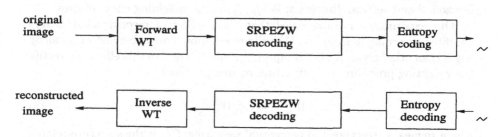

Fig. 1. The block diagram of our structure-related perceptual weighting coding system

In the coding process, the major error will be the quantization error introduced in the EZW coding. This error can be interpreted as a noise signal which will be interpolated and filtered by the synthesis filters and weighted by human eyes. So the key issue in system design is how to use HVS characteristic to minimize the noise sensitivity. HVS can be modelled [12] as a curve representing the function of eye's relative sensitivity $A(f)$ to spatial frequency f, as depicted in Figure 2. It can be seen that the visibility of noise is less in regions of high frequency than in regions of low frequency. It means that in the regions of high

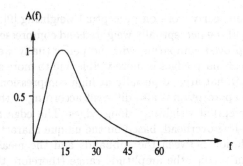

Fig. 2. The sensitivity of HVS to spatial frequency (cycles/degree)

frequency we can tolerate more distortion than the regions of low frequency. Therefore, more bits (bigger weighting) can be given to coefficients of low frequency, and less bits (smaller weighting) to those of high frequency. Extending to 2D WT domain, the perceptual weighting value for subband (j_x, j_y), $W^p_{j_x, j_y}$, can be represented as

$$W^p_{j_x, j_y} = \frac{r_x r_y}{N_{j_x, j_y}} \int \int_{D_x, D_y} W(f_x, f_y) |G_{j_x, j_y}(r_x f_x, r_y f_y)|^2 df_x df_y \qquad (1)$$

where $j_x(j_y) = 1, 2, \ldots, J_x(J_y)$ are the scales of subband decomposition, with $J_x(J_y)$ representing uppermost band; p means perceptual weighting; N_{j_x, j_y} is the product of the decimation factors of the subband (j_x, j_y) in 2D; G_{j_x, j_y} is the synthesis filter of the subband; r_x and r_y are the sampling intervals in the horizontal and vertical directions; $W(f_x, f_y)$ is the weighting curve of eyes.

Our experiments have revealed that visual effect of reconstructed images weighted according to Equation (1) will present more noticeable distortion along edges than other areas at high compression rate. To reduce this effect, we rectify the weighting procedure to a structure-related process as

$$W^{sp}_{j_x, j_y} = C^j_{xy} W^p_{j_x, j_y} \qquad (2)$$

where sp means structure and perceptual weighting; C^j_{xy} is the structure-related rectifying factor for coefficient at point (x,y) of the jth scale. The success of the rectifying is decided by the correctness of calculation of C^j_{xy}, which is based on the unique characteristic of subband decomposition, namely, if there exist sizable amplitude changes in the lower (coarser) band, there will be high amplitude values in the corresponding locations of the upper band. Therefore the locations corresponding to these high amplitude values can be considered as edges and be paid more attention. The factor C^j_{xy} can be calculated as

$$C^j_{xy} \begin{cases} = 1 \text{ when no sizeable amplitude changes} \\ \quad \text{are detected in lower band;} \\ > 1 \text{ otherwise.} \end{cases}$$

In our implementation, positions with 5% top values are considered as edge points. The amplitude changes are calculated in three ways depending on different spatial directions: for the low-filtered x and high-filtered y band, the amplitude difference between 3 points in the same row is used; for the high-filtered x and low-filtered y band, the amplitude difference between 3 points in the same column is used; for the high-filtered x and high-filtered y band, the amplitude difference of a 3*3 block is used.

Based on the assumption that the coefficients with large magnitudes are more important than smaller ones and the assumption that if the coefficients in low band are insignificant then the coefficients in high band with same orientation are also insignificant, the EZW coding has the unique embedding and adaptivity features. This makes it one of most successful WT-based coding schemes. Details of the algorithm can be found in [6].

3   EXPERIMENTS AND RESULTS

In experiments, we compared the performance of our SRPEZW with the performances of original EZW scheme (EZW) and the simple perceptual weighting EZW scheme (PEZW). In all simulations, a biorthogonal (9,7) filter pair [13] was used, which makes it possible to apply efficient symmetric image extension [14]. Figure 3 gives examples of image Lena (512*512) reconstructed respectively with the EZW (a), the PEZW (b), and our SRPEZW (c), all at 0.125 bits/pixel. It can easily be seen that the one reconstructed with our scheme presents less distortion on edges, such as shoulder and hat, and in smooth areas than the other two methods, resulting in subjectively more pleasing effects.

The most often used image quality measurement in digital image compression research is the mean square error (MSE) between the original image and the processed image. However, it has been empirically determined that the MSE and its variants do not correlate well with subjective (human) quality assessments, since they do not adequately "mimic" what the human visual system does in assessing image quality. To substantially compare the performance, we define the weighted-signal-to-noise ratio as

$$WSNR = 10\log(\frac{S^2}{\lambda_{q,w}^2}) \tag{3}$$

where S is the original image luminance value, $\lambda_{q,w}^2$ is the weighted noise power, which can be expressed as Equation 4.

$$\lambda_{q,w}^2 = \sum_{j_x=0}^{J_x-1}\sum_{j_y=0}^{J_y-1} W_{j_x,j_y}^{sp} \sigma_{j_x,j_y}^2 \tag{4}$$

(J_x, J_y) are the number of bands in the vertical and horizontal directions; σ_{j_x,j_y}^2 is the variance of the quantization error of the $(j_x,j_y)^{th}$ subband; W_{j_x,j_y}^{sp} is the structure perceptual weighting coeffieient for subband j_x,j_y.

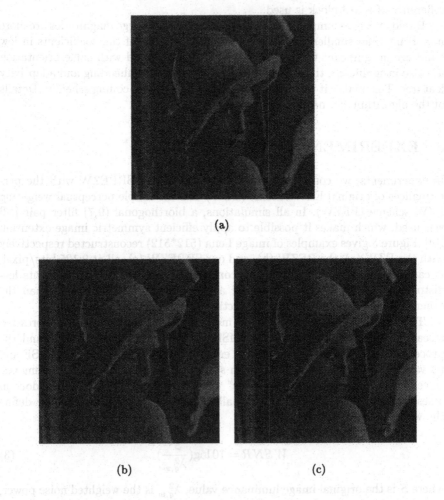

(a)

(b)   (c)

Fig. 3. Performance comparison of EZW,PEZW and SRPEZW. All images are reconstructed at 0.125 bits/pixel.

Table 1 presents the SNR and WSNR of the algorithms under different bit rates. It can be observed that though the SNRs of weighted images are slighted smaller that those of no-weighted, much bigger difference of WSNR can be seen between the weighted and the non-weighted. The WSNR improvement of our SRPEZW algorithm over others coincides with the subjective image quality improvement.

Table 1. Coding Performance Comparison of three algorithm for Image Lena(512x512)

bit rate	EZW		PEZW		SRPEZW	
bits/pixel	SNR	WSNR	SNR	WSNR	SNR	WSNR
1.0	31.3	57.3	29.1	64.0	29.5	64.9
0.5	29.1	52.7	27.3	60.3	28.0	61.5
0.25	26.4	47.7	24.0	55.2	24.1	55.7
0.125	22.3	41.1	21.3	45.2	21.9	46.7

References

1. I. Daubechies, "Orthonormal Bases of Compactly Supported Wavelets", *Communications on Pure and Applied Mathematics*, Vol.XLI, pp.909-996, 1988.
2. P. J. Burt and E. H. Adelson, "The Laplacian Pyramid as a Compact Image Code", *IEEE Transactions on Communications*, Vol.Com-31, No.4, pp.532-540, April 1983.
3. S. G. Mallat, "A Theory for Multiresolutuion Signal Decomposition:The Wavelet Representation", *IEEE Transactions on Pattern Analysis and Machine Intelligence*, Vol.11, No.7, pp.674-692, July 1989.
4. K. Ramchandran and M. Vetterli, "Best Wavelet Packet Bases in Rate-Distortion Sense", *IEEE Transactions on Image Processing*, Vol.2, pp.160-175, 1994.
5. A. S. Lewis and G. Knowles, "Image Compression Using the 2-D Wavelet Transform", *IEEE Transactions on Image Processing*, Vol.1, No.2, pp.244-250, April 1992.
6. J. M. Shapiro, "Embedded Image Coding Using Zerotrees of Wavelet Coefficients", *IEEE Transactions on Signal Processing*, Vol.41, No.12, pp.3445-3462, Dec. 1993.
7. J. Li and P. Y. Cheng and C. C. J. Kuo, "On the Improvements of Embedded Zerotree Wavelet (EZW) Coding", *SPIE, Visual Communications and Image Processing*, Vol.2501, pp.1590-1501, May 1995.
8. D. Taubman and A. Zakhor, "Multirate 3-D Subband Coding of Video", *IEEE Trans. on Image Processing*, Vol.3, No.5, pp.572-588, Sep. 1994.
9. J. Li and J. Jin, "An Image Compression Scheme Based on Wavelet Transform and the Human Visual System", *SPIE, Wavelet Applications in Signal and Image Processing IV*, Vol.2825, pp.879-889 , Aug. 1996.
10. A. N. Netravali and B. Prasada, "Adaptive Quantization of Picture Signals Using Spatial Masking", *Proceedings of IEEE*, Vol.65, No.4, pp.536-548, April. 1977.
11. R. C. Gonzalez and R. E. Woods, "Digital Image Processing", 1992.
12. J. L. Mannos and D. J. Sakrison, "The Effects of a Visual Fidelity Criterion on the Encoding of Images", *IEEE Transactions on Information Theory*, Vol.IT-20, No.4, pp.525-536, July 1974.
13. Gilbert Strang and Truong Nguyen, "Wavelets and Filter Banks", 1996.
14. M. J. T. Smith and S. L. Eddins, "Analysis/Synthesis Techniques for Subband Image Coding", *IEEE Trans. on Acoustics Speech and Signal Processing*, Vol.38, No.8, pp.1446-1456, Aug. 1990.

Parallel Optimization for TSP

Hai Xiang Lin and Kees Lemmens

Dept. of Applied Mathematics and Informatics, Delft University of Technology,
P.O. Box 356, 2600 AJ Delft, The Netherlands, e-mail: h.x.lin@twi.tudelft.nl

Abstract. The Traveling Salesman Problem (TSP) is a combinatorial
optimization problem which is NP-hard. So, in practice heuristics are
used to find good approximations of the optimum of a TSP. However,
for large problems even these heuristics still require a huge amount of
computer time to find a good approximation. We consider the use of
parallel computers to speed up the computation for large TSPs. We
describe a partitioning based parallel heuristic. Implementation issues
and computational results on a distributed memory parallel computer
are reported.

1 Introduction

The traveling salesman problem (TSP) is probably the most well-known member
among the combinatorial optimization problems. It tries to find a shortest (or
most economical) path visiting a number of cities exactly once with the same city
as the starting and ending point. The TSP has many practical applications in real
life, such as in vehicle routing, determining the best route for the pen movement
of a plotter, ordering-picking problem in warehouses, drilling of printed circuit
boards, etc.

The TSP is a very computation intensive (NP-hard) problem. The number
of feasible solutions grows exponentially as $\frac{(n-1)!}{2}$ (n is the number of cities).
Although problems of up to a few hundred cities can often be solved to optimality using the recent Branch-and-Cut algorithms and occasionally some large
problems with a few thousands cities ([8], [9]), it is however in general unrealistic
to find the optimum for large problems within a reasonable amount of computer
time. On the other hand, the determination of the optimum solutions of TSP
is seldom necessary in practical applications. Usually one is satisfied with solutions that is within a certain percentage of an optimum solution. So, in practice,
heuristics are used for finding a good approximation of the optimization problem.

However, for a TSP with a very large number of cities, the most heuristics
such as the well-known Lin-Kernighan (L-K) heuristic still require too much
computer time. One way to speed up the optimization for TSP is parallel computation. In this paper we report a promising parallel algorithm and some results
on finding good approximation for very large TSPs using parallel computers.

In general, the optimization techniques for TSP can be classified into the
following categories: 1) Branch-and-Bound; 2) Construction heuristics; 3) Improvement heuristics; 4) Analogies of nature processes.

1. Branch-and-Bound: this is the important exact algorithm for the TSP, recently the efficient branch-and-cut algorithm has been added to this family [3]. The branch-and-cut algorithm is a powerful method for finding the optimum of a TSP and it is of importance where the optimum is required (e.g. to answer some theoretical questions). Because the required computer time increases fastly for large problems (say $n > 1000$), it is less useful in practice when a good approximation of a large TSP often needs to be computed in short time.

2. Construction heuristics: This type of heuristics try to determine a Hamiltonian cycle according to some construction rules, e.g., selects the nearest neighbor as the next city to be visited; or selects a city not in the cylce and inserts it into the cycle such that the increase in the length is minimal; etc. This type of heuristics are usually very fast (e.g. with a time complexity of $O(n^2)$), but the approximations found are also of moderate quality.

3. Improvement heuristics: This type of heuristics starts from some initial Hamiltonian cycle, and tries to improve it by means of a number of modifications. Examples of such modification rules are: a *2-exchange* by eliminating two edges and reconnecting the two resulting paths to obtain a new cycle; or a *3-exchange* breaks a cycle into 3 parts instead of 2 and combines the resulting paths in the best possible way. A *k-opt* heuristic checks all possible *k-exchange* combinations, if a *k-exchange* improves the current cycle, then it is accepted as the new cylce, until all possible combinations are failed to improve. A direct implementation of the *2-opt* and the *3-opt* heuristic has a time complexity of $O(n^2)$ and $O(n^3)$ respectively. There exist efficient implementations which limit the number of *k-exchanges* through using *neighbor-lists* [9].

The well known Lin-Kernighan (L-K) heuristic performs a modification based on a number of subsequent *k-exchanges* which allows slight increase during some intermediate exchanges [6]. For every city, upto a predefined maximum number (e.g. 15) of subsequent *k-exchanges* are performed, if the cycle is improved then it is accepted as the new cycle, otherwise take the next city until it failed to improve for all cities. At the present, the L-K heuristic is consider to be the best which produces the best approximations.

4. Analogies of nature processes: This type of heuristics consists of methods which tries to simulate some nature's process in the optimization process. Two recently well known such analogies are the *simulated annealing* [5] and the *genetic algorithm*. Simulated annealing simulates the physical process of cooling down after annealing which leads to an orderly configuration of an ensemble of atoms/molecules. When modeling the length of the Hamiltonian of the TSP as the entropy, simulating the probabilistic changes in the Hamiltonian with a slowly cooling process can lead to a global minimum length (in theory). The genetic algorithm emulates the evolution process in nature. By relating the length of the Hamiltonian to some fitness function, and starting with a (random) population of feasible solutions (Hamiltonians), the algorithm consists of a process of regeneration with recombinations and selecting the fittest ones as the new generation of the population. These methods have the nice feature that they can compute the approximation to the optimality if the algorithm runs for a sufficient long time (i.e., very long!). However, experimental results show that they usually produce porer results and are much slower compared to the L-K heuristic.

2   Parallel optimization

One way to apply parallel processors for speeding up an improvement heuristic is to partition a tour into a number of segments and then using parallel processors to (locally) improve each of the segment. Let an initial tour be broken up into k segments of length $\frac{n}{k}$ where k is equal to a multiple of the number of processors available. Each segment becomes a subtour after adding an edge between its endpoints, and is handed to a processor. This subtour can be improved by applying a 2-Opt or L-K heuristic (the added edge between the endpoints cannot be removed during the local improvement). After all the k subtours are improved, they are combined together by removing the added edges and connect the endpoints in the same way as the original tour before the parallel local improvement. This procedure can be repeated by breaking up the improved tour again into k segments but now each segment takes half of the cities from each of the two adjacent old segments. Good parallel performance can be expected, but there is a considerable penalty in the quality of the results when compared to that of the L-K. This can be intuitively explained as follows. Although some global effect can be achieved by repeatedly shifting the endpoints of the segments, this is however a kind of 'one-dimensional' shifting, changes in the 'other' dimension (for a 2-D or 3-D TSP) are difficult to obtain.

Another parallel approach is described in [7], the so-called large-step Markov chains heuristic, which attempts to combine simulated annealing with local search heuristics such as the L-K. The algorithm repeatedly attempts to improve the currently local optimal tour by first introducing some local changes (e.g. a 4-exchange) followed by applying the L-K to this modified tour. After the L-K, if a shorter tour is obtained then this tour is accepted as the new local optimum, otherwise it is rejected. Parallelization is obtained by running p simultaneous Markov chains on p processors. After a certain number of steps (e.g. every 10 or 100 steps per processor), the best tour at that time is replicated to all processors. And the processors then continues the iterated L-K. The parallelization is basically running a number of L-K iterations on p processors. The authors report that the communication overhead is low especially when the number of steps between replicating the best tour is kept large (the price paid is then the increased possiblity that the effort of some processors are wasted).

3   A parallel algorithm based on geometric partitioning

We consider a divide and conquer method, which is based on the partitioning heuristic proposed by Karp [4]. We modify it to achieve more accurate approximation and introduce a parallel improvement heuristic at the combining phase

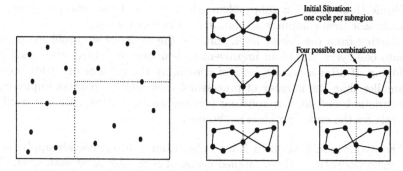

Fig. 1. 1. Illustration of a map divided into 4 subregions (left). 2. the four possible combinations of merging two cycles (right).

for efficient parallel optimization. First the Karp's heuristic is described in more detail.

1. Divide: The cities are clustered into a number of subregions each consisting of maximal m cities. Cities close to each other are clustered into the same subregion by recursively dividing a region into two subregions. The bisection is performed alternately along the x- and y-directions. Exactly one common city in the middle of two subregions is determined at each bisection (see Fig. 1).

2. Subregion solution: For the solution of a TSP in a subregion we can use the branch and bound algorithm. However, for a subregion with a moderate size m this algorithm takes a long time to compute. So, a L-K type algorithm is implemented as an alternative for approximating the subregion TSPs.

3. Combining: After a (sub)optimal cycle for each subregion has been computed, these cycles are pairwisely combined into larger cycles until a cycle for the whole region is formed. This combining process is illustrated in Fig. 1. The combination resulting in the shortest cycle is chosen for merging the two cycles.

Karp has shown some nice results based on theoretical analysis. The length of an approximated solution is shown to be within an $O(m^{-0.5})$ percentage with respect to the optimum, where m is the number of cities in a subregion which is solved to the optimum. But this is not very useful in practice. Because even for a 10% error we need to solve subproblems of size 100 to the optimum, which may require much computer time. So, in order to obtain an accurate approximation (say within a few percent of the optimum), we need to do something more. Furthermore, it can be observed that a cycle obtained with the above combination algorithm clearly resembles the local structure of the subregions and thus less optimal (see Fig. 2).

In the following we describe the improvements and modifications of the above algorithm for more accurate approximations and introducing balanced parallelism. The parallel algorithm is outlined below.

1. Divide the cities recursively into subproblems contained in subregions;

2. While there is still a subproblem left, assign it to an idle processor; The processor then computes a solution for the subproblem.

3. Starting from the lowest level, combine the cycles of subproblems pairwise into one cycle. Apply an improvement heuristic (e.g. L-K) after each combining. If after a certain level of combining the number of cycles becomes smaller than the number of processors, then apply a parallel improvement heuristic based on tour-splitting (see explanation below). Repeat until one cycle for the entire problem is obtained.

The major modification to Karp's algorithm is to recursively apply an improvement heuristic to the combined cycles of subproblems at each level. Computational results show that this is very effective for improving the quality of the final solution. At the beginning we applied the L-K improvement heuristic to the last cycle (i.e. the combined cycle for the whole problem) only. That gives a siginficant improvement to the cycle obtained from the basic Karp's algorithm, but plots of the solutions show that the final tour after applying the L-K heuristic often resembles many local structures of subregion cycles. The application of the L-K algorithm to the (sub)cycles at each combining level generally improves the final solution by 2 to 6%, which is very siginificant because we usually want the final solution to be within 2% of the optimal solution.

The same problem of bad local structure has also been pointed out in [1]. They have attempted to overcome this problem by first computing the glabal tour structure using the 'center of gravity' of each subproblem and using this to choose a pair of subproblems for combining. Furthermore a parallel tour-based partitioning improvement heuristic is used to improve the final cycle. However, as they mentioned their algorithm has the disadvantage that some of the local bad structures at the border of the subregions, which are not neighbor in the cycle, will never be improved. The application of L-K heuristic at each combining level in our algorithm solves this problem.

During the combining phase, the remaining number of Hamiltonian cycles to be merged is decreasing to eventually one single cycle. Moreover, the last few bigger cycles requires the most of the time for improvement (the computing time required by a L-K heuristic increases more than linearly with the number of cities). Therefore, for an efficient parallelization, the improvement heuristic after combining two Hamiltonian cycles into one cycle must be parallelized too.

A tour-based partitioning L-K heuristic is used for the improvement of each combined cycle. The tour is split into two segments each with half of the cities from one of the two regions. The two segments can then be improved in parallel using two processors. More parallelism can be introduced near the end of combining by splitting the tour into k $(k > 2)$ segments. This looks like the tour-based partitioning scheme for parallel optimization in Section 2, but there are essential differences among them. In the tour-based partitioning scheme improvement heuristic is applied to the global tour only. Here we apply the improvement at each level during the hierarchical combining process. In this way we eliminate the problem of 'one-dimensional' improvement drawback as is mentioned in Section 2. This has been confirmed by computational experiment (Table 1).

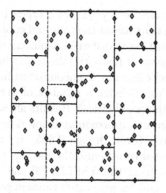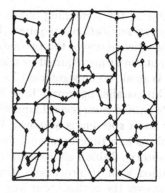

Fig. 2. A problem divided into a number of regions (left); A tour obtained by Karp's algorithm (without post improvement at combining step) (right).

4 Parallel implementation

We have implemented the hybrid scheme based on geometric partitioning on a distributed memory parallel computer, a Parsytec system with T805 transputers configured in a mesh topology.[1]

After the partitioning, each subregion is handed to a processor. These smaller TSPs in the subregions are then optimized in parallel without requiring communication. The basic unit of workload to be distributed across the processors is the subregion solution and combining two cycles followed by a L-K improvement. Because the workload for the subregion solutions may vary largely even if the two subregions consisting the same number of cities. So static load distribution gives a poor load balance. We apply a simple dynamic load distribution strategy: Each time when a processor has finished its current assigned work, the master will assign a new task of subregion solution or a task of combining two cycles.

Table 1. Computational results using 16 processors. The number in the name is the number of cities of the problem.

Problems	Average excess	Speedup
PR76	1.6%	11.5
RAND120	1.1%	13.8
RAND500	1.6%	14.5
FNL4461	2.0%	15.1
RAND5000	1.9%	15.1
D15112	2.2%	15.6

---

[1] The T805 has a speed of about 1.5 Mflops and is a slow processor for today's standard. However, it suffices for testing the parallel performance of our algorithm.

During the combining phase where two smaller cycles are merged into one cycle communication is required. The information about the tour and coordinates of the cities in one of the two subregions must be transferred. Compared to the computation time required by a local improvement heuristic, the communication cost is small, so no significant loss in parallel efficiency is encountered here.

Our parallel optimizer have been used to optimize a number of very large TSPs (varied from several hundreds to ten thousand cities). Results show that a near linear speedup is obtained. We can compute approximated solutions to the problems in a time between 1 and 20 minutes (even though T805s are not fast processors). Table 1 shows some computational results and the parallel speedup obtained with 16 processors. For problems from the TSPLIB [9] the average excess over the optimal tour length is given, and for generated random problems (name beginning with Rand) the average excess over the Held-Karp lower bound is reported. The size of the subregions has no noticable influence the quality of the final tour (in the original algorithm of Karp larger subregion size is required for more accuracy). The obtained results with our algorithm is better than that of [1], both in accuracy and parallel performance (they reported parallel efficiency of around 60%). Furthermore, for a solution with the same quality, our algorithm is much faster than the large-step Markov chains heuristic.

References

1. A. Bachem, B. Steckemetz and M. Wottawa, "An efficient parallel cluster-heuristic for large Traveling Salesman Problems", Tech. Report No. 94-150, University of Koln, Germany, 1994.
2. D.S. Johnson, "Local optimization and the traveling salesman problem", *Proc. 17th Colloquium on Automata, Languages and Programming*, Springer-Verlag, 1990, pp. 446-461.
3. M. Junger, G. Reinelt and G. Rinaldi, "The traveling salesman problem", in *Handbooks in Oper. Res. and Manag. Sci.*, Vol. 7, M.O. Ball, et. al. (eds.), 1995, pp. 225-330.
4. R.M. Karp, "Probabilistic analysis of partitioning algorithms for the traveling-salesman in the plane", *Math. Oper. Res.*, Vol. 2, 1977, pp. 209-224.
5. S. Kirkpatrick, C.D. Gelatt, Jr., and M.P. Vecchi, "Optimization by simulated annealing", *Science*, 220, 1983, pp. 671-680.
6. S. Lin and B. Kernighan, "An effective heuristic algorithm for the traveling salesman problem", *Oper. Res.*, Vol. 21, 1973, pp. 498-516.
7. O.C. Martin, and S.W. Otto, "Combining simulated annealing with local search heuristics", *Meta-Heuristics in Combinatorial Optimization*, G. Laporte and I.H. Osman (eds.), Annals of Oper. Res, Vol. 60, 1995.
8. M.W. Padberg and G. Rinaldi, "A branch and cut algorithm for the resolution of large-scale symmetric traveling salesman problems", *SIAM Review*, Vol. 33, 1991, pp. 60-100.
9. G. Reinelt, "TSPLIB - A traveling salesman problem library", *ORSA J. Computing*, 3, 1991, pp. 376-384. (data sets and results of many TSPs can be down loaded from http: //www.iwr.uni-heidelberg.de/iwr/comopt /soft/TSPLIB95/TSPLIB.html)

Numerical Analysis of the Bulging of Continuously Cast Slabs *

Zhouchen Lin and Tsimin Shih

Dept. of Applied Math., The Hong Kong Polytechnic Univ., Hong Kong

Abstract. To guarantee the quality of continuously cast steel, the solidifying process should be studied. Both theory and experiments have shown that excessive stress inside the steel strand, which results from thermal contraction, bending and bulging of the solidified shell, can cause internal and surface cracks. The bulging of the shell is in general a visco-elasto-plastic phenomenon coupled with phase change. In this paper, the theory of elasticity is applied to study numerically the bulging behavior of the solid shell between consecutive supporting rollers. The boundary conditions between the shell and the rollers are described by the physical elastic contact theory. By an adjusting algorithm, the boundary conditions can be established. Considering the geometry of the steel slab, both heat transfer and stress analysis can be simplified to two-dimensional problems. Furthermore, the heat transfer process can be treated independently of the deformation. With these simplifications, numerical results are obtained.

1 Introduction

Continuous casting is a new and prosperous technology in steel industry. In the casting process, molten steel is poured into a bottomless mold and leaves the mold with thin solidified shell, which is supported by rollers. Due to ferrostatic pressure, the shell will bulge. Fig.1 is a schematic illustration of a continuous thin slab caster and the bulging phenomenon. Under the mold exit, water is sprayed onto the surface of the strand till the solidification process completes. The rollers not only guide the strand, but support the solid shell against the ferrostatic pressure of the liquid core. Although in general the magnitude of bulging is small, it is not negligible, because the liquid core is susceptible to the deformation of shell, which will result in the internal defects, such as centerline segregation and especially segregated cracks. Misalignment of rollers and improper roller gaps can make the bulging deteriorate. Therefore, the designer of a continuous caster is concern about the bulging properties of the machine. Comparing with the experimental way to measure the bulging, mathematical analysis is more convenient, economic and hence preferable.

---

* This work is supported by the Research Grant (No. 350/369) of The Hong Kong Polytechnic University.

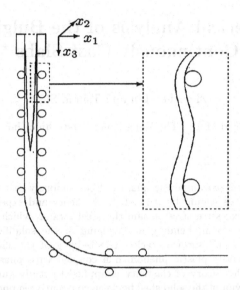

Fig. 1 The thin slab caster and the bulging of solidified shell.

The bulging has already been studied by other authors ([1] and the references therein). In those papers, elasticity, plasticity as well as creep were considered in the bulging model. However, all of them encountered the problem of contacting condition between the shell and the rollers ([5] and etc.). The most natural treatment is to constrain the displacement normal to the surface to be zero and neglect the friction. Apparently, since the stress around the contacting area is much greater than elsewhere, such a measure seems unconvincing. As for the contact problem between two bodies, both analytical and numerical approaches have been proposed. The analytical, or physical theory is rooted from Hertz's work. In numerical approach, boundary element method (BEM) has been widely adopted ([8]). The numerical examples in [8] showed that, when the contact area is small comparing to the scale of the bodies, the Hertz's theory works well. In this paper, we will concentrate on the elastic aspect of the bulging and apply Hertz's theory to depict the contacting condition between the shell and the rollers.

2   Governing Equations

Let σ_{ij}, ϵ_{ij} and u_i ($i = 1, 2, 3; j = 1, 2, 3$) be the stress tensor, strain tensor and the displacement vector, and T be the temperature. Following the conventional notation in elasticity theory, the general steady-state thermoelastic problem for the solid shell can be written as follows ([3]):

$$\sigma_{ij,j} = -F_i, \tag{1}$$

$$\sigma_{ij} = 2\mu\epsilon_{ij} + [\lambda\epsilon_{kk} - (3\lambda + 2\mu)\alpha_T(T - T_0)]\delta_{ij}, \tag{2}$$

$$\epsilon_{ij} = \frac{1}{2}(u_{i,j} + u_{j,i}), \tag{3}$$

where F_i are the componenets of body force, α_T is the coefficient of thermal expansion and λ, μ are the Lamé constants which are related to the Young's modulus E and the Possion's ratio ν as follows:

$$\lambda = \frac{\nu E}{(1+\nu)(1-2\nu)}, \qquad \mu = \frac{E}{2(1+\nu)}. \tag{4}$$

The system (1)–(3) can be reformulated in terms of temperature and the displacements:

$$\mu u_{i,kk} + (\lambda + \mu)u_{k,ki} - (3\lambda + 2\mu)\alpha_T T_{,i} = -F_i. \tag{5}$$

Neglecting the contribution of thermal stress, by the conservation law of energy, the thermal transition of the steel, including liquid, solid and the mushy region in between, is governed by the following equation:

$$\mathbf{v} \cdot \nabla H = K\Delta T. \tag{6}$$

Here $\mathbf{v}$ is the velocity field of particles inside the strand, K is the coefficient of thermal conductivity, H is the enthalpy which is related to T as follows:

$$H = H(T) = \int_{T_0}^{T} \rho(\xi)\left(C(\xi) + L\frac{d\chi}{d\xi}\right) d\xi, \tag{7}$$

where ρ is the density, C is the heat capacity, L is the latent heat and χ is the liquid fraction.

In addition, to avoid the Navier-Stokes equations, one may also assume that $\mathbf{v} = v\mathbf{x}_3$. This is widely accepted ([4],[6] and etc.).

In view of the facts that the width of the slabs is much larger than the thickness and the heat conduction in the direction of casting is by a factor 10^{-4} smaller than those in the width and thickness directions ([6]), we may reduce (5) and (6) to two dimentional:

$$(\lambda + 2\mu)\frac{\partial^2 u_1}{\partial x_1^2} + (\lambda + \mu)\frac{\partial^2 u_3}{\partial x_1 \partial x_3} + \mu\frac{\partial^2 u_1}{\partial x_3^2} - (3\lambda + 2\mu)\alpha_T\frac{\partial T}{\partial x_1} = 0, \tag{8}$$

$$\mu\frac{\partial^2 u_3}{\partial x_1^2} + (\lambda + \mu)\frac{\partial^2 u_1}{\partial x_1 \partial x_3} + (\lambda + 2\mu)\frac{\partial^2 u_3}{\partial x_3^2} - (3\lambda + 2\mu)\alpha_T\frac{\partial T}{\partial x_3} = -\rho g, \tag{9}$$

and

$$v\frac{\partial H}{\partial x_3} = K\left(\frac{\partial^2 T}{\partial x_1^2} + \frac{\partial^2 T}{\partial x_2^2}\right). \tag{10}$$

The boundary conditions for (10) are:

$$-K\frac{\partial T}{\partial n} = \alpha_{mold}(T - T_{mold}), \quad \text{if } x_3 \le L_{mold},$$

$$-K\frac{\partial T}{\partial n} = \alpha_{roll}(T - T_{roll}), \quad \text{if } x_3 > L_{mold} \text{ and the shell}$$

$$\text{touches the rollers,} \tag{11}$$

and

$$-K\frac{\partial T}{\partial n} = \alpha_{water}(T - T_{water}), \text{ otherwise.}$$

(10)–(11) can be solved by phase relaxation method ([9]) without iterations.

3   Boundary Conditions for Displacements

The computation is carried out only on the shell between L_{mold}, the mold exit, and L_{core}, the end of the liquid core. To specify, the solid shell is the part with $T \leq \frac{1}{2}(T_s + T_l)$, where T_s is the solidus temperature and T_l is the liquidus temperature. At both ends the displacements are prescribed to be 0. On the solid-liquid interface, the boundary condition is $\bar{p}\mathbf{n} = \sigma \cdot \mathbf{n}$, where $\bar{p}$ is the ferrostatic pressure. By virtue of (2) and (3), it is equivalent to:

$$\left[(\lambda+2\mu)\frac{\partial u_1}{\partial x_1}+\lambda\frac{\partial u_3}{\partial x_3}-(3\lambda+2\mu)\alpha_T(T-T_0)\right]n_1+\mu\left(\frac{\partial u_1}{\partial x_3}+\frac{\partial u_3}{\partial x_1}\right)n_3 =\bar{p}n_1,$$

and

$$\mu\left(\frac{\partial u_1}{\partial x_3}+\frac{\partial u_3}{\partial x_1}\right)n_1+\left[(\lambda+2\mu)\frac{\partial u_3}{\partial x_3}+\lambda\frac{\partial u_1}{\partial x_1}-(3\lambda+2\mu)\alpha_T(T-T_0)\right]n_3 =\bar{p}n_3.$$

On the outer surface where the shell does not contact the rollers, $\sigma \cdot \mathbf{n} = 0$. The determination of the boundary condition where the shell is tightly pressed against is not trivial and is derived as follows.

In Hertz's classical theory on the elastic contact of two cylinders pressed together with parallel axes ([2]), the contact patch is assumed to be small in comparison with R_1 and R_2 (Fig.2), which are the radii of the cylinders, and the section of the contact patch is assumed to be a parabola. Due to friction, the center of the parabola will shift by a small amount c.

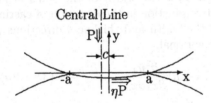

Fig. 2 Elastic contact of two cylinders with parallel axes.

Introduce

$$A = \frac{2(1 - \nu_1^2)}{E_1} + \frac{2(1 - \nu_2^2)}{E_2},$$

and

$$\beta = \frac{(1+\nu_1)(1-2\nu_1)}{AE_1} - \frac{(1+\nu_2)(1-2\nu_2)}{AE_2},$$

where E_1 and E_2 are the Young's modulus of the two bodies, ν_1 and ν_2 are the Possion's ratios. According to the theory, the half width a of the area and the pressure p distributed in the contact area will be:

$$a = \sqrt{\frac{PA}{2\pi m(1 - m)\phi}},$$

and

$$p(s) = \frac{\phi a}{A}(\sin m\pi)(1-s)^m(1+s)^{1-m},$$

where P is the total pressure, η is the coefficient of friction, and

$$m = \frac{1}{\pi}\arctan\left(\frac{1}{|\beta|\eta}\right), \quad s = \frac{x}{a},$$

$$c = 2m - 1, \qquad \phi = \frac{1}{R_1} + \frac{1}{R_2}.$$

Then, the boundary condition at the contact area will be $\sigma \cdot n = pn$.

Now the main problem boils down to determine the total pressure P. Denote the distance of rollers from the meniscus by Z_i, $i = 1, 2, \cdots, N$, where N is the last roller before the tip of the liquid core. To obtain the total pressure P_i at the ith roller, we first roughly estimate it by the pressure $\tilde{P}_i$ exerted by the liquid core between $Z_{start} = \frac{1}{2}(Z_i + Z_{i-1})$ (if $i = 1$, then $Z_{start} = L_{mold}$) and $Z_{end} = \frac{1}{2}(Z_i + Z_{i+1})$ (if $i = N$, then $Z_{end} = L_{core}$). Obviously,

$$\tilde{P}_i = \int_{Z_{start}}^{Z_{end}} \bar{p}\,dx_3.$$

Next, taking advantage of the fact that the rollers should maintain their positions, one may apply the following algorithm to adjust P_i:

1. Choose $\epsilon > 0$, $0 < \xi \ll 1$, and set $k = 0$, $\omega_i^{(k)} = 0$, $i = 1, \cdots, N$.
2. Letting $P_i = \omega_i^{(k)}\tilde{P}_i$, establish the boundary conditions for (8)–(9) and solve (8)–(9).
3. If $k = 0$, set $k = 1$, $\omega_i^{(k)} = 1$, $i = 1, \cdots, N$, goto 2; otherwise go to 4.
4. If at every $Z_i, i = 1, \cdots, N$, the horizontal displacements $|u_1^{(k)}| < \epsilon$, then stop; otherwise go to 5.
5. At Z_i, if $|u_1^{(k)}| < \epsilon$, then $\omega_i^{(k+1)} = \omega_i^{(k)}$; else if $|u_1^{(k-1)} - u_1^{(k)}| \geq \xi$, then set $\omega_i^{(k+1)} = \omega_i^{(k-1)} - (\omega_i^{(k)} - \omega_i^{(k-1)})u_1^{(k-1)}/(u_1^{(k)} - u_1^{(k-1)})$; otherwise set $\omega_i^{(k+1)} = \omega_i^{(k)}(1 + \frac{u_1^{(k)}E}{P_i})$.
6. Set $k = k + 1$, go to 2.

4   An Numerical Example

The numerical example is based on finite element method. (10)–(11) is solved beforehand. Then partition the solid part between L_{mold} and L_{core} and apply the above-mentioned algorithm to solve (8)–(9) iteratively. Professor Guoping Liang, Institute of Mathematics, Academia Sinica developed a software called Finite Element Program Generator (FEPG). The programs to solve (8)–(9) and (10)–(11) are automatically generated by this software.

Table 1 lists the distances of the rollers from the meniscus; in Table 2, part of the data used in computation are listed; Table 3 are the Young's modulus at some temperatures, the same as those in [1] except that at T_l the modulus is assumed to be zero. For temperatures not occuring in this table, interpolations by natural cubic spline are applied.

Table 1 Position of rollers

i	1	2	3	4	5	6	7	8	9	10	11
Z_i(m)	1.4	1.8	2.2	2.6	3.0	3.4	3.8	4.2	4.6	5.0	6.0

Table 2 Some data

T_l	1485°C	K	110kJ/(mh°C)	L	320kJ/kg
T_s	1377°C	C_s	0.674kJ/(kg°C)	ρ_s	7400kg/m³
T_0	25°C	C_l	0.82kJ/(kg°C)	ρ_l	7000kg/m³
T_{mold}	80°C	α_{mold}	6000kJ/m²h°C	L_{mold}	1m
T_{water}	27°C	α_{water}	2300kJ/m²h°C	L_{thick}	0.05m
T_{roll}	400°C	α_{roll}	2000kJ/m²h°C	L_{width}	1.2m
ν	0.3	α_T	2×10^{-5}/°C	v	250m/h

Table 3 Young's modulus

T(°C)	0	130	315	500	771	1000	1250	1485
$E(10^{10}\text{N/m}^2)$	21	20.5625	18.8125	15	8.4583	3.7917	3.0625	0

Fig. 3 is the contour of temperature on the half longitudinal cross section of the strand. The difference between adjacent contours is 47.2°C. The dotted line indicates the solid-liquid interface. Fig.4 is the profile of the outer surface of the deformed shell.

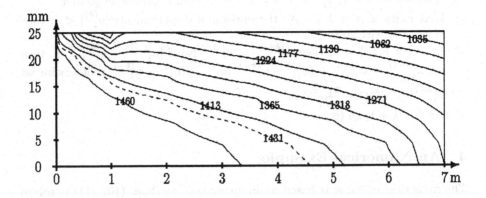

Fig. 3 Temperature field inside the steel strand.

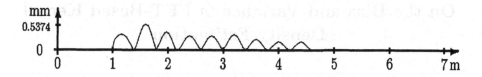

Fig. 4 Normal displacement of the shell between L_{mold} and L_{core}.

References

1. Binder, A.: On the numerical treatment of bulging of continuously cast slabs. Computing **49** (1992) 265–278
2. Hills, D.A., et. al.: Mechanics of elastic contacts. Butterworth Heinemann (1993)
3. Kovalenko, A.D.: Thermoelasticity. Wolters-Noordhoff (1969)
4. Laitinen, E., Neittaanmäki, P.: On numerical solution of the problem connected with the control of the secondary cooling in the continuous casting process. Control – Theory and Advanced Technology **4** (1988) 285–305
5. Lewis, B.A., et. al.: Boundary condition difficulties encountered in the simulation of bulging during the continuous casting of steel. Appl. Math. Modelling **7** (1983) 274–277
6. Rogberg, B.: Testing and application of a computer program for simulating the solidification process of a continuously cast strand. Scand. J. of Metallurgy **12** (1983) 13–21
7. Sheng, Y.P., et. al.: The computation of the bulging of continuously cast slabs (in Chinese). Iron and Steel **28** (1993) 20–25
8. Sun, H.C.: An automatic incremental techniques of BEM for 2D contact problems with friction (PhD. Thesis). University of Cincinnatti (1989)
9. Verdi, C., Visintin, A.: Error estimates for a semi-explicit numerical scheme for Stefan-type problem. Numerische Mathematik **52** (1988) 165–185

On the Bias and Variance of FFT-Based Kernel Density Estimation

Ramesh Natarajan

I.B.M. Thomas J. Watson Research Center,
P. O. Box 218,
Yorktown Hts., N.Y. 10598, U. S. A.

Abstract. An efficient computational procedure for kernel density estimation using the FFT algorithm has been given by Silverman ([5]). This procedure requires the empirical characteristic function to be interpolated on a regular mesh, and the high-frequency interpolation errors can result in a significant loss of accuracy when the kernel density estimates or its derivatives are used as a part of some larger statistical procedure. In this paper, we describe systematic finite element discretization procedures for improving the accuracy of the FFT-based algorithms. We derive the bias and variance of the FFT-based kernel density estimates, and suggest modifications to eliminate interpolation bias. Simulation studies that verify the results of the analysis are presented.

1 Introduction

An efficient computational procedure for kernel density estimation using the FFT algorithm has been given by Silverman ([5]), along with some extensions by Jones and Lotwick ([4]). This algorithm requires the empirical characteristic function to be interpolated on a uniform mesh, which introduces high-frequency approximation errors. In the case of density estimation, these high-frequency errors are subsequently damped by the smoothing effect of the kernel multiplier, but nevertheless, they can still be significant in applications where the density estimates and its derivatives are required as part of some larger statistical procedure. For example, Silverman ([6]) describes several applications of the use of density estimation in statistical procedures such as discriminant analysis, clustering, bump-hunting and projection pursuit.

Our specific interest in this error analysis arose in the context of using higher-order optimization methods in projection pursuit analysis. There, the usual low-order discretization methods for the FFT-based kernel density estimates lead to inaccurate and even inconsistent results in the evaluation of the gradient and Hessian for the optimization procedure.

Let $g_N(x)$ be the N-point sample kernel density estimate of the density function $g(x)$, so that

$$g_N(x) = \frac{1}{Nh} \sum_{i=1}^{N} K\left(\frac{x - X_i}{h}\right), \tag{1}$$

where $\{X_i\}_{i=1}^N$ denote *i.i.d.* random variables sampled from $g(x)$, and h denotes the kernel bandwidth parameter. In this paper, we are specifically concerned with symmetric, non-negative kernels, and a suitable choice when high-order differentiability is required is the Gaussian kernel $K(x) = (2\pi)^{-1/2}e^{-x^2/2}$.

The straightforward direct evaluation of $g_N(x)$ at M points from (1) requires $O(MN)$ operations. In contrast, for M points on a fixed regular grid, the FFT-based algorithm requires $O(M \log M + N)$ operations.

The Fourier transform of $g_N(x)$ in the case of the Gaussian kernel is

$$\widetilde{g_N}(s) = (2\pi)^{-1/2}e^{-h^2s^2/2}\sum_{i=1}^N \frac{e^{isX_i}}{N}. \tag{2}$$

Denoting the empirical characteristic function by

$$\gamma_N(s) = \sum_{i=1}^N \frac{e^{isX_i}}{N}, \tag{3}$$

it can be seen from (2) that the components of $\gamma_N(s)$ for $hs \gg 1$ are strongly damped in the evaluation of $\widetilde{g_N}(s)$. This is advantageously used in the FFT-based algorithms, by replacing the empirical characteristic function by another estimate with possible high-frequency errors, which are in any case damped by convolution with the Gaussian kernel.

2   Discretization and Computational Details

Consider the interval (a, b), which is chosen large enough to completely contain the sample data $\{X_i\}_{i=1}^N$. This interval is partitioned into $M = 2^m$ equal subintervals, where $\delta = (b-a)/M$ is the discretization parameter, and the grid points are given by $x_k = a + k\delta$, $k = 0, 1, \cdots, M - 1$, respectively.

A finite element interpolation of the empirical characteristic function on this grid yields

$$\gamma_N(s) = \frac{1}{N}\sum_{i=1}^N e^{isX_i} \approx \sum_{k=0}^{M-1} \xi_k e^{isx_k}, \quad \text{where} \quad \xi_k = \frac{1}{N}\sum_{i=1}^N \phi_k(X_i), \tag{4}$$

with $\phi_k(x)$ denoting the finite-element basis function associated with the grid point x_k. Suitable basis functions include, for example, the piecewise constant $\phi_k^{(0)}$ (Silverman [6]), piecewise linear $\phi_k^{(1)}$ (Jones and Lotwick [3]), and piecewise quadratic $\phi_k^{(2)}$ cases (considered in this paper for the first time). In these specific cases, denoting $\mu_k = (x - k\delta)/\delta$, we have in terms of the indicator function χ,

$$\phi_k^{(0)}(x) = \chi_{[x_{k-\frac{1}{2}}, x_{k+\frac{1}{2}}]}(x), \tag{5}$$

$$\phi_k^{(1)}(x) = (1 - |\mu_k|)\chi_{[x_{k-1}, x_{k+1}]}(x), \tag{6}$$

$$\phi_k^{(2)}(x) = \begin{cases} \frac{1}{2}(1 - |\mu_k|)(2 - |\mu_k|)\chi_{[x_{k-2}, x_{k+2}]}(x), & \text{even } k, \\ (1 - |\mu_k|)(1 + |\mu_k|)\chi_{[x_{k-1}, x_{k+1}]}(x), & \text{odd } k. \end{cases} \tag{7}$$

The values of ξ_k in (4) are computed by histogramming and interpolation. For example, in the piecewise quadratic case, we have the following procedure

1. Set $\xi_k = 0$ for all k.
2. For each $i = 1, N$
   If $X_i \in [x_{2m}, x_{2m+2}]$, then with $k = 2m$ and $\alpha = 1/2N\delta^2$,

$$\xi_k \longleftarrow \xi_k + \alpha(x_{k+1} - X_i)(x_{k+2} - X_i)$$
$$\xi_{k+1} \longleftarrow \xi_{k+1} + \alpha(x_{k+2} - X_i)(X_i - x_k)$$
$$\xi_{k+2} \longleftarrow \xi_{k+2} + \alpha(X_i - x_k)(X_i - x_{k+1})$$

Endif

Now consider the following modification to the approximation in (4),

$$\gamma_N^*(s) = e^{isa}\beta(s) \sum_{k=0}^{M-1} \xi_k e^{is\delta k}, \tag{8}$$

where the frequency-dependent factor $\beta(s)$ has been introduced to adjust for any interpolation bias (as discussed below), with the unadjusted case in (4) recovered by setting $\beta(s) = 1$.

We consider this sum for the discrete values of s given by

$$s_j = \frac{2\pi j}{(b-a)} = \frac{2\pi j}{M\delta}, \text{ for } j = 0, \ldots, M-1, \tag{9}$$

since, if we assume that the kernel density estimate can be periodically extended outside the interval (a, b), then these discrete values are sufficient to reconstruct the kernel density estimate on (a, b). In practice, this is hardly a restrictive assumption since (a, b) can be chosen sufficiently larger than the range of the sample values. Then, denoting $\beta_j \equiv \beta(s_j)$, we have

$$\gamma_N^*(s_j) = e^{is_j a}\beta_j \sum_{k=0}^{M-1} \xi_k e^{i\frac{2\pi jk}{M}} = e^{is_j a}\beta_j \hat{\xi}_j, \tag{10}$$

in which the sum is evaluated in terms of the DFT components $\{\hat{\xi}_j\}_{j=0}^{M-1}$ of the sequence $\{\xi_j\}_{j=0}^{M-1}$.

Now, the Fourier integral in (8) can be discretized by the trapezoidal rule,

$$g_N(x_k) \approx \frac{(2\pi)^{\frac{1}{2}}}{M\delta} \sum_{j=-M/2+1}^{M/2} \hat{g}_j e^{-is_j x_k} = \frac{(2\pi)^{\frac{1}{2}}}{M\delta} \sum_{j=-M/2+1}^{M/2} \hat{g}_j e^{-is_j a} e^{-i\frac{2\pi jk}{M}} = g_N^*(x_k), \tag{11}$$

in which all the modes upto the Nyquist frequency on the grid have been retained (although in practice, the contribution of the higher-order modes on the grid to this integral will be negligibly small). Finally, combining (10) and (11) yields in the specific case of the Gaussian kernel,

$$g_N^*(x_k) = \frac{1}{M\delta} \sum_{j=-M/2+1}^{M/2} \left[\beta_j \xi_j e^{-\frac{1}{2}\left(\frac{2\pi k}{M\delta}\right)^2 j^2} \right] e^{-i\frac{2\pi jk}{M}}. \tag{12}$$

This sum can be evaluated by an inverse DFT to obtain the required approximation to the kernel density estimates at the grid points x_k.

3   Analysis of Discretization Errors

The estimates $g_N(x)$ and $g_N^*(x)$ along with their derivatives are all functions of the $i.i.d$ random variables $\{X_i\}_{i=1}^N$. Therefore, letting $g_N^{(r)}(x)$ and $g_N^{*(r)}(x)$ denote the corresponding r'th derivatives, the integrated mean square discretization error $(I.M.S.D.E.)$ is given by

$$I.M.S.D.E. \equiv \mathcal{M}^{(r)} = E \int_{-\infty}^{\infty} (g_N^{(r)}(x) - g_N^{*(r)}(x))^2 dx$$

$$= \int_{-\infty}^{\infty} s^{2r} \widetilde{K}(hs)^2 E(|\gamma_N(s) - \gamma_N^*(s)|^2) ds, \tag{13}$$

with the second step following from Parseval's theorem. Now, we have the following standard results for the expectation and variance of the empirical characteristic function,

$$E(\gamma_N(s)) = \gamma(s), \quad E(|\gamma_N(s)|^2) = \frac{1}{N} + \frac{N-1}{N}|\gamma(s)|^2, \tag{14}$$

where $\gamma(s) = E(e^{isx})$ denotes the characteristic function of $g(x)$. Using these, yields after some manipulations (Lotwick and Jones [3])

$$E(|\gamma_N(s) - \gamma_N^*(s)|^2) = E(|\gamma_N(s)|^2) + E(|\gamma_N^*(s)|^2) - E(\gamma_N(s)\overline{\gamma_N^*(s)}) - E(\overline{\gamma_N(s)}\gamma_N^*(s)$$

$$= \frac{1}{N} + \frac{N-1}{N}|\gamma(s) - \beta(s) \sum_{k=0}^{M-1} E(\phi_k(x))e^{isx_k}(s)|^2$$

$$- \frac{2\beta(s)}{N} \sum_{k=0}^{M-1} E(\phi_k(x)\cos s(x - x_k)) + \frac{\beta(s)^2}{N} \sum_{k=0}^{M-1} E(\phi_k(x)^2)$$

$$+ \frac{2\beta(s)^2}{N} \sum_{k=0}^{M-1}\sum_{k<l}^{M-1} E(\phi_k(x)\phi_l(x))\cos s(x_k - x_l)). \tag{15}$$

The arrangement of the terms in (15) exposes a square bias term, whose evaluation is of independent interest below. The various terms in (15) can now be evaluated using the following direct consequence of the Euler-Maclaurin summation formula for periodic functions.

Lemma 1. *Let the density function $g(x) \in C^{\infty}(a, b)$ be extended periodically outside this interval, and let $g^{(n)}(x), n = 0, 1, \ldots$ denote its derivatives. Then, using the notation $g_k^{(n)} \equiv g^{(n)}(x_k), k = 0, 1, \ldots, M - 1$, we have*

$$(-is_j)^n \gamma_j = \sum_{k=0}^{M-1} \delta g_k^{(n)} e^{is_j x_k}, \quad \text{for } s_j = \frac{2\pi j}{b-a}, \quad j = 0, 1, \ldots, M-1, \quad (16)$$

where $\gamma_j \equiv \gamma(s_j) = \int_a^b g(x) e^{is_j x} dx$. Similarly, for even M,

$$\frac{1}{2}(-is_j)^n \gamma_j = \sum_{\substack{k=0 \\ k \text{ even}}}^{M-2} \delta g_k^{(n)} e^{is_j x_k} = \sum_{\substack{k=1 \\ k \text{ odd}}}^{M-1} \delta g_k^{(n)} e^{is_j x_k}. \quad (17)$$

Finally, $\gamma_0 \equiv \gamma(s_0) = 1$.

First, consider the evaluation of the bias terms in (15), which are given by the following lemma

Lemma 2. *For $s_j = 2\pi j/(b - a)$, $j = 0, 1, \ldots, M - 1$, let $\gamma_{N,j}^{[i]} \equiv \gamma_N^{[i]}(s_j), i = 0, 1, 2$ respectively denote the approximation to $\gamma_N(s_j)$ obtained by using piecewise constant, linear and quadratic basis functions as in (5)-(7). Then,*

$$E(\gamma_{N,j}^{[i]}) = \alpha_j^{[i]} \beta_j \gamma_j, \quad \text{for } i = 0, 1, 2, \quad (18)$$

where

$$\alpha_j^{[0]} = \frac{2 \sin \frac{1}{2} \delta s_j}{\delta s_j}, \quad \alpha_j^{[1]} = \frac{2(1 - \cos \delta s_j)}{(\delta s_j)^2}, \quad (19)$$

$$\alpha_j^{[2]} = \frac{(2 \sin \delta s_j - \sin 2 \delta s_j)}{(\delta s_j)^3} + \frac{(3 - 4 \cos \delta s_j + \cos 2 \delta s_j)}{2(\delta s_j)^2}. \quad (20)$$

The details of the proof will be given in the long version of this paper.

For the case $\beta_j = 1$ (i.e., without introducing a bias adjustment factor), the $E(\gamma_{N,j}^{[i]})$ are unbiased estimates of γ_j only for $\delta s_j \longrightarrow 0$. An unbiased estimator for all δs_j can be obtained by taking $\beta_j = (\alpha_j^{[i]})^{-1}$, but in practice it is often sufficient to use the following approximations for $\delta s_j \ll 1$,

$$\beta_j^{[0]} \approx 1 + \frac{1}{24}(\delta s_j)^2 + \frac{7}{5760}(\delta s_j)^4 + \frac{31}{96780}(\delta s_j)^6 + O((\delta s_j)^8), \quad (21)$$

$$\beta_j^{[1]} \approx 1 + \frac{1}{12}(\delta s_j)^2 + \frac{1}{240}(\delta s_j)^4 + \frac{1}{6048}(\delta s_j)^6 + O((\delta s_j)^8), \quad (22)$$

$$\beta_j^{[2]} \approx 1 + \frac{1}{60}(\delta s_j)^4 - \frac{13}{7560}(\delta s_j)^6 + O((\delta s_j)^8). \quad (23)$$

These approximations are adequate for the necessary bias adjustment in the low-to-moderate frequency regime of interest in kernel density estimation.

With these results in hand, the square bias term in (15) is then given by

$$|\gamma(s_j) - E(\gamma_N(s_j)|^2 = (1 - \beta_j^{[i]}\alpha_j^{[i]})^2|\gamma_j|^2, \text{ for } i = 0, 1, 2. \quad (24)$$

The final results can be summarized in the following two lemmas

Lemma 3. *For* $s_j = 2\pi j/(b-a), j = 0, 1, \ldots, M - 1$, *the mean square discretization error of the characteristic function is given by*

$$E(|\gamma_{N,j} - \gamma_{N,j}^{[i]}|^2) = \frac{1}{N} + \frac{N-1}{N}(1 - \beta_j^{[i]}\alpha_j^{[i]})^2|\gamma_j|^2 - \frac{2}{N}\beta_j^{[i]}\alpha_j^{[i]} + \frac{1}{N}(\beta_j^{[i]})^2\psi_j^{[i]}, \quad (25)$$

for $i = 0, 1, 2$, *corresponding to the case of piecewise constant, piecewise linear and piecewise quadratic basis functions respectively. In (25), we have denoted*

$$\psi_j^{[0]} = 1, \quad \psi_j^{[1]} = \frac{1}{3}(2 + \cos \delta s_j), \quad \psi_j^{[2]} = \frac{1}{15}(12 + 4\cos \delta s_j - \cos 2\delta s_j). \quad (26)$$

Lemma 4. *For* $\delta s_j \longrightarrow 0$, *from (25) and (26) we have the following asymptotic estimates*

$$E(|\gamma_{N,j} - \gamma_{N,j}^{[0]}|^2) = \frac{1}{12N}(\delta s_j)^2 + O((\delta s_j)^4), \text{ with } \beta_j^{[0]} = 1, \quad (27)$$

$$E(|\gamma_{N,j} - \gamma_{N,j}^{[1]}|^2) = [\frac{1}{120N} + \frac{N-1}{144N}|\gamma_j|^2](\delta s_j)^4 + O((\delta s_j)^6), \text{ with } \beta_j^{[1]} = 1, \quad (28)$$

$$E(|\gamma_{N,j} - \gamma_{N,j}^{[2]}|^2) = \frac{2}{945N}(\delta s_j)^6 + O((\delta s_j)^8), \text{ with } \beta_j^{[0]} = 1. \quad (29)$$

The leading order term in (28) depends on the underlying density being estimated via the term involving $|\gamma_j|^2$. *Since this dependence arises from the square bias term in (24), we can obtain the alternative estimate*

$$E(|\gamma_{N,j} - \gamma_{N,j}^{[1]}|^2) = \frac{1}{720N}(\delta s_j)^4 + O((\delta s_j)^6), \text{ with } \beta_j^{[1]} = 1 + \frac{1}{12}(\delta s_j)^2. \quad (30)$$

With these results in hand, we now note from (13) that

$$\mathcal{M}^{(r)} \approx \frac{(2\pi)^{\frac{1}{2}}}{M\delta} \sum_{j=-M/2+1}^{M/2} s_j^{2r}\widetilde{K}(hs_j)^2 E(|\gamma_N(s_j) - \gamma_N^*(s_j)|^2), \quad (31)$$

where again all the modes upto the Nyquist frequency on the grid have been retained although $\widetilde{K}(hs_j)$ will be heavily damped for the higher-order modes. The evaluation of the summation above can be carried in the case $\delta s_j \ll 1$, if we write

$$E(|\gamma_N(s_j) - \gamma_N^*(s_j)|^2) = \lambda(\delta s_j)^{2n} + O((\delta s_j)^{2n+1}). \quad (32)$$

Here, the leading-order coefficient λ is assumed to be independent of underlying characteristic function $\gamma(s)$. From (24) and (32), we note that any dependence of λ on $\gamma(s)$ can only arise through the square bias terms, and it is therefore important to choose the value of β_j via (22) to at least cancel the bias in the leading order terms for small δs_j. The required values of λ and n for the various discretization procedures are given above in (27), (29) and (30). Thus finally, we have

$$
\mathcal{M}^{(r)} \approx \lambda \delta^{2n} \left\{ \frac{(2\pi)^{\frac{1}{2}}}{M\delta} \sum_{j=-M/2+1}^{M/2} s_j^{2(n+r)} \widetilde{K}(hs_j)^2 \right\}
$$

$$
\approx \lambda \delta^{2n} \int_{-\infty}^{\infty} s^{2(n+r)} \widetilde{K}(hs)^2 \, ds = \lambda \delta^{2n} \mathcal{A}_{n+r}. \tag{33}
$$

Here, the integrals $\mathcal{A}_j$ depend only on the kernel function, and in the particular case of the Gaussian kernel can be evaluated as

$$
\mathcal{A}_j = \frac{1.3 \ldots (2j-1)}{2^{j+1} \sqrt{\pi} h^{2j+1}}. \tag{34}
$$

We therefore obtain the main theoretical result of this section,

Lemma 5. *The Integrated Mean Square Discretization Error (I.M.S.D.E.) of the approximations to the sample density estimates and its derivatives (as defined in (13)) is given asymptotically for $\delta \longrightarrow 0$ by*

1. *Piecewise constant:* $\mathcal{M}^{(r)} \approx \frac{1}{12N} \delta^2 \mathcal{A}_{1+r}$.
2. *Piecewise linear:* $\mathcal{M}^{(r)} \approx \frac{1}{720N} \delta^4 \mathcal{A}_{2+r}$, *when the bias correction factor is used as in Lemma 4.*
3. *Piecewise quadratic:* $\mathcal{M}^{(r)} \approx \frac{2}{945N} \delta^6 \mathcal{A}_{3+r}$.

These theoretical results are in excellent agreement with numerous simulations that have been performed. Figure (1) shows theory and simulation results for the constant factor in the asymptotic convergence rate (for samples drawn from a standard gaussian, a well separated 50-50 gaussian mixture and a uniform distribution). The simulations used a fixed kernel bandwidth $h = 0.1$, with each input data realization consisting of $N = 100$ randomly-sampled points (with the results averaged over 125 such realizations).

4   Closing Remarks

The methods used in this paper can be easily extended to to other finite element approximations of the empirical characteristic function. The analysis of FFT-based multivariate kernel density estimation and the relative performance to the Greengard-Strain algorithm ([1]) in the multivariate case is also of interest.

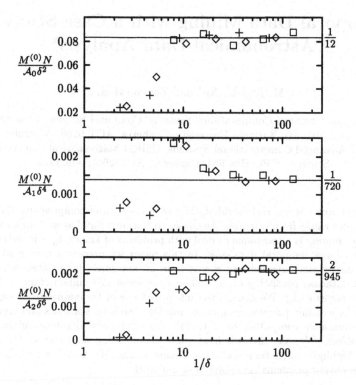

Fig. 1. Comparison of theory and simulation results for the constant factor in the $\delta \longrightarrow 0$ asymptotic convergence rate. The legends are standard gaussian $\Diamond$, gaussian mixture $+$, and uniform distribution $\square$.

References

1. L. Greengard and J. Strain, *The Fast Gauss Transform*, SIAM J. of Stat. Comput., 12 (1991), pp. 79-94.

2. P. J. Huber, *Projection Pursuit (With Discussion)*, Annals of Statistics, 9 (1985), pp. 435-475.

3. M. C. Jones and H. W. Lotwick, *On the errors involved in computing the empirical characteristic function*, J. Statist. Comput. Simul., 17 (1983), pp. 133-149.

4. M. C. Jones and H. W. Lotwick, *Remark AS R50: A Remark on Algorithm AS 176: Kernel density estimation using the Fast Fourier Transform*, Applied Statistics, 33 (1984), pp. 120-122.

5. B. W. Silverman, *Algorithm AS 176: Kernel density estimation using the Fast Fourier Transform*, Applied Statistics, 31 (1982), pp. 93-99.

6. B. W. Silverman, *Density Estimation for Statistics and Data Analysis*, Chapman and Hall, London (1986).

Temporal Data Mining with a Case Study of Astronomical Data Analysis*

Michael K. Ng[1] and Zhexue Huang[2]

[1] CRC for Advanced Computational Systems, Computer Sciences Laboratory,
The Australian National University, Canberra, ACT 0200, Australia
[2] CRC for Advanced Computational Systems, CSIRO Mathematical and Information
Sciences, GPO Box 664, Canberra, ACT 2601, Australia

Abstract. Many real world databases possess time components. Examples range from scientific databases to business databases. Temporal data mining is a technique to deal with problems of knowledge discovery from large temporal databases. In this paper we present a case study of discovering microlensing events from an extremely large astronomical database consisting of 40 million time series (20 million stars with two series each). We discuss the integrated use of techniques from signal processing, pattern recognition, machine learning, statistics and high performance computing to solve this large temporal database mining problem. We take the star light curve classification as a special focus to highlight some temporal data mining issues. We also discuss some computing problems encountered in our work.

1 Introduction

Across a wide variety of fields, data is being collected and accumulated at a dramatic pace. There is an urgent need for a new generation of computational techniques and tools to assist humans in extracting useful information (knowledge) from the rapidly growing volumes of data. These techniques and tools are the subject of the emerging field of *data mining*, that is, *knowledge discovery in databases* (KDD)[6].

Many real world databases possess time components. Temporal (having time components) databases [4] arise in science and engineering domains such as meteorological and hydrological databases, as well as in business such as stock exchange databases and the databases storing various kinds of business transactions. Temporal data mining deals with problems of knowledge discovery from these large temporal databases. One objective of temporal data mining is to find and characterize interesting sequential patterns from temporal data sets. For example, some sequential patterns of patients visiting doctors may indicate fraud committed in claiming benefit from health insurance companies. Although demands for such a technique from many application areas are high, the technique itself is still in its infancy.

---

* The research is supported by The Cooperative Research Centre for Advanced Computational Systems (ACSys) in Australia.

In this paper we present a temporal data mining case study. The objective is to discover microlensing events from an extremely large astronomical database consisting of 40 millions time series (20 million stars with two series each). Our current focus is on classification of star light curves from a sample data set. We discuss the integrated use of techniques from signal processing, pattern recognition, machine learning, statistics and high performance computing to solve this large temporal database mining problem. In particular the techniques used in data preprocessing and unsupervised classification of star light curves are presented. We also discuss some computing problems encountered in our work.

2   MACHO Database

The term MACHO stands for massive compact halo objects which astronomers hypothesise may constitute the dark matter in our own and other spiral galaxies. The objective of the ongoing MACHO project is to search for evidence of MACHO through detecting the gravitational microlensing events (Alcock et al [1] and Aubourg et al [2]).

To detect the gravitational microlensing events the light of about 20 million stars in the Large and Small Magellantic Cloud as well as the Galactic Bulge is measured with a specially designed CCD camera every clear night at the Mt Stromlo and Siding Spring Observatories (MSSSO) in Australia. The measuring work started in the autumn of 1992 and will continue until the year 2000.

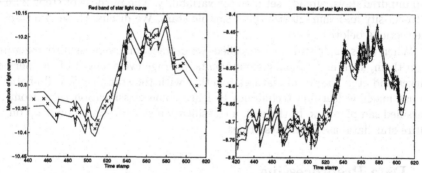

Fig. 1. The light curves of a star in the red spectral band (left) and blue spectral band (right). The crosses are the measurements of the magnitudes and the two bounding lines represent the lower and upper bounds of the magnitudes.

The star light data is collected in two spectral bands, red and blue. Two time series (light curves) containing the red and blue spectral band measurements have been collected for a period of four years for each star. An example of the light curves of a star is shown in Figure 1. The MACHO database has accumulated about 40 million star light curves which occupy 0.5 terabytes of storage space. The expansion of the database will continue until the year 2000.

Because the probability of occurrence of a microlensing event is very small $(1/10^6)$, we require analysis of millions of star light curves in order to find a microlensing event. The computational burden is so large that many traditional data analysis techniques are no longer practical and specific data mining technology has to be used.

3   Methodology for Star Light Curve Classification

We perform the MACHO data analysis in two stages, star light curve classification and microlensing signature search. We envisage that if the star light curves are appropriately classified, the star light curves carrying the microlensing signatures will concentrate in one or a few classes. After classification we will be able to focus our search on the subsets of the star light curves that possibly carry the microlensing signatures and ignore the other star light curve subsets.

The methodology we are using in the star light curve classification consists of the following steps:

1. sample star selection,
2. data preprocessing,
3. unsupervised star light curve classification,
4. feature extraction from star light curves,
5. learning classification rules from sample stars, and
6. building up a classification over the whole star database.

In the present stage a set of sample stars was selected by an astronomer from the large database. The sample data contains two hundred stars for a period of two hundred days, already selected for variability. The purpose of steps 2 and 3 is to classify and characterize these sample stars. We discuss these two steps in the sections below.

When the sample stars are classified, we extract some representative features from them, such as frequency components, average cycle length of the periodic curves and etc. The set of features together with the star classes will enable us to use machine learning techniques to learn some classification rules from the classified sample stars. In the end, a classification system can be built up on the entire star database from these rules.

4   Data Preprocessing

The raw data of the star light curves displayed in Figure 1 contains noise and missing values because of cloudy nights. The light curves also have different times for light measurements. The objective of data preprocessing is to

- remove the noise or otherwise enhance the original star profile,
- interpolate the missing values, and
- align the measurements to the same times.

Noise in a signal often transforms the process of recovering the light curve into an ill-posed problem in the sense that a small perturbation in the input may cause the estimate to vary drastically. In order to deal with noise, regularization is often used to constrain the solution to become unique and well-posed [5].

Assume the noisy light curve function $y(t)$ for the sample observations is given by $y(t_i) = g(t_i) + e(t_i)$ for $i = 1, \ldots, n$, where $g(t_i)$ are samples of an ideal noiseless light curve function $g(t)$ in an interval $[a, b]$ and $e(t_i)$ are sampled i.i.d. noise. The curve $g(t)$ is estimated by $f(t)$ obtained by

$$\min_{f \in H^m} \left\{ \frac{1}{n} \sum_{i=1}^n \left(\frac{f(t_i) - y(t_i)}{w_i} \right)^2 + \lambda \int_a^b \left(\frac{d^m f(t)}{dt^m} \right)^2 dt \right\},$$

where w_i is the weight for point t_i and $H^m = \{f : f(t) \in C^{m-1}[a, b], f^{(m)} \in L_2[a, b]\}$ is the Sobolev Hilbert space containing the solution. The parameter λ is chosen to control the tradeoff between the regularity of the solution and the infidelity to the data. Such a regularization is equivalent to the formulation of smoothing splines [3,5]. Computing this smoothing spline fitted to n distinct, not necessarily equally spaced data points requires the order of $m^2 n$ operations.

The advantages of using polynomial splines to represent the light curves are:

• Polynomial splines have good regularity properties. Among all interpolants of a given degree of smoothness, they are those that oscillate the least [3].

• Polynomial splines have a simple explicit form that makes them easy to manipulate. Operations such as differentiation and integration can be performed in a straightforward manner.

5   Unsupervised Classification of Star Light Curves

The classification is based on similarity of the star light curves. Because the profiles of the star light curves are very diverse, we use a multiple classification scheme with different dissimilarity measures. We first classify the star light curves into two general classes, periodic and non-periodic. Then we treat the periodic star light curves in the frequency domain whereas the non-periodic curves in the time domain. Finally, we partition the star light curves into subsets using a hierarchical clustering method [7].

5.1   Dissimilarity Measures

Let $f_s^{(r)}(t)$ and $f_s^{(b)}(t)$ be the scaled (between 0 and 1) smoothing splines of the red and the blue spectral bands of the light curves of star s. Proper amplitude scaling is necessary before determining if the light curves are similar.

Time Domain The dissimilarity measure in the time domain uses the L_1-norm distance between two light curves because this distance measure is less sensitive to outliers than the L_2-norm distance measure. The distance measure between two stars s_1 and s_2 in the red spectral band is defined by

$$d_r(s_1, s_2) \equiv \min_{-T \leq \tau \leq T} \left\{ \frac{\int_0^{\min\{T_{s_2}, T_{s_1} - \tau\}} \left| f_{s_1}^{(r)}(t + \tau) - f_{s_2}^{(r)}(t) \right| dt}{\min\{T_{s_2}, T_{s_1} - \tau\}}, \right.$$

$$\left. \frac{\int_{k=0}^{\min\{T_{s_1}, T_{s_2} + \tau\}} \left| f_{s_1}^{(r)}(t) - f_{s_2}^{(r)}(t - \tau) \right| dt}{\min\{T_{s_1}, T_{s_2} + \tau\}} \right\}$$

where $-T_{s_2} < T < T_{s_1}$ and τ is the shift of one curve against the other. The distance measure for the blue spectral band is defined the same.

Frequency Domain To define the dissimilarity measure in the frequency domain we first apply the Fourier transformation to the star light curves as it proved effective for classification purposes [10]. The Fourier transformation generates a set of complex numbers, i.e., the Fourier descriptors that describe the shapes of the light curves in the frequency domain. We select the first 50 Fourier descriptors to discriminate the light curves. Let

$$\left\{ f_{s_1}^{(r)}(T_{s_1} + \Delta k) \right\}_{k=0}^{n_{s_1}}, \text{ and } \left\{ f_{s_2}^{(r)}(T_{s_2} + \Delta k) \right\}_{k=0}^{n_{s_2}}$$

be the sampling values of the magnitudes of the stars s_1 and s_2, and let $\{F_{s_2}^{(r)}(\cdot)\}_{k=0}^{n_{s_1}}$ and $\{F_{s_2}^{(r)}(\cdot)\}_{k=0}^{n_{s_2}}$ be their Fourier descriptors. The distance measure in the red spectral band in the frequency domain is defined by

$$D_r(s_1, s_2) \equiv \sum_{k=0}^{m} \left| |F_{s_1}^{(r)}(k)| - |F_{s_2}^{(r)}(k)| \right|$$

where $0 \le m \le \lfloor \min\{n_{s_1}, n_{s_2}\}/2 \rfloor$. The distance measure in the blue spectral band is defined in a similar manner.

5.2   Classification Scheme

The multiple classification scheme consists of the following major steps:

Step 1: We first compute the average magnitude value of each light curve and the average values of the lower and upper bounds of each light curve using the given error estimate. We then count the number of changes of signs (cs) of the star profile with respect to its average lower and upper bounds. After this, we consider a star profile is non-periodic if $cs \le 1$ and periodic if $cs > 1$, and classify stars into two classes according to the following rules:

- **C1** either the red or the blue spectral bands are non-periodic;
- **C2** both the red and blue spectral bands are periodic;

Step 2: For **C1**, we construct the "dissimilarity" matrix from the distance measures $d_r(\cdot, \cdot) + d_b(\cdot, \cdot)$. For **C2**, we construct the "dissimilarity" matrix from the distance measures $D_r(\cdot, \cdot) + D_b(\cdot, \cdot)$;

Step 3: We use a hierarchical clustering algorithm to further classify the star profiles for both **C1** and **C2**.

Given a set of preprocessed star light curves a classification tree can be constructed using this classification scheme. The example of a classification tree is shown in Figure 2. Each leaf (a box) of the tree represents a subset of the star light curves that are similar according to the dissimilarity measures. The graphs in the box show the typical profiles of the star light curves in the subset. Using a classification tree we are able to visually characterize the star light curves in the subsets.

Creating a sound classification system for the star database requires a large sample star set. However, constructing a classification tree from a large star set is computationally difficult. Presenting a large classification tree that is comprehensible and characterizing large subsets of star light curves are also problematic. The visualization technique to tackle the presentation and characterization problems is under development[8]. In the next section we discuss a few computational issues encountered in this work.

6   Computational Issues

- The hierarchical clustering method used in unsupervised classification requires $O(N^2)$ time (N is the number of stars)[9]. This cost is prohibitive for large star sets. Several approaches to parallelizing this method should be considered and developed to take advantage of the computing power of multiple processor systems. For instance, parallel access of both the star profile and the dissimilarity matrix poses an interesting and challenging task. The construction and maintenance of the nearest neighbour chain, as well as the carrying out of agglomeration both offer possibilities for parallelization.

- Another direction of future work is to map the time series of the star profiles to some k-dimensional space ($k \ll$ the number of times in the star profiles) such that dissimilarities are preserved. There are two benefits from this mapping: (i) efficient retrieval, (ii) revealing potential clusters in a more efficient manner and (iii) a tool for visualization.

Acknowledgements

We thank Dr T. Axelrod at MSSSO for providing us the sample data and Drs. G. Williams and M. Hegland and Mr P. Milne for their comments.

References

1. C. Alcock et al., *Possible Gravitational Microlensing of a Star in the Large Magellantic Cloud*, Nature, 365 (1993), pp. 621–623.
2. E. Aubourg et al., *Evidence for Gravitational Microlensing by Dark Objects in the Galactic Halo*, Nature, 365 (1993), pp. 623–625.
3. C. de Boor, *A Practical Guide to Splines*, Appl. Math. Sci., 27, Springer-Verlag, 1978.
4. J. Clifford and A. Tuzhilin, *Recent Advances in Temporal Databases*, Springer-Verlag, 1995.
5. P. Craven and G. Wahba, *Smoothing Noisy Data with Spline functions*, Numer. Math. 31 (1979), pp. 377–403.
6. U. M. Fayyad, G. Piatetsky-Shapiro, P. Smyth and R. Uthurusamy, eds, *Advances in Knowledge Discovery and Data Mining*, AAAI Press / The MIT Press, 1996.
7. A. Jain and R. Dubes, *Algorithms for Clustering Data*, Prentice Hall, 1988.
8. T. Lin and Z. Huang, *A Visual Environment for Interactive Partitioning of Large and Complex Data Sets in Data Mining*, submitted to INTERACT'97, Sydney, Australia, 1997.
9. F. Murtagh, *Complexities of Hierarchic Clustering Algorithms: State of the Art*, Comput. Stat. Quar. 1 (1984), pp. 101–113.
10. E. Persoon and K. Fu, *Shape Discrimination using Fourier Descriptors*, IEEE Trans. Syst. Man. and Cybern. 7 (1977), pp. 170–179.

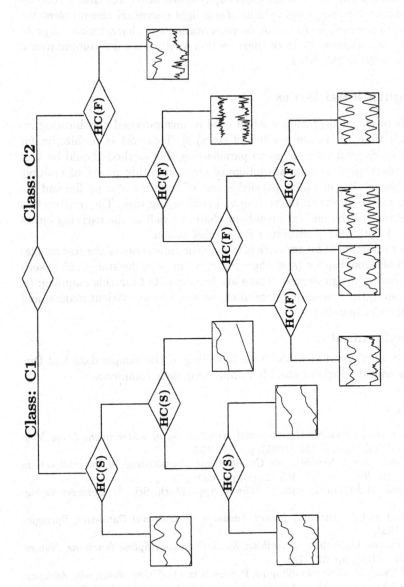

Fig. 2. The example of a classification tree. The upper curves in the boxes are the star profiles in the red spectral band and the lower curves are the star profiles in the blue spectral band. The subtree was created from time domain and the subtree of Class C2 was created from frequency domain.

A Note on Adaptive Restarting Procedure for Pseudo Residual Algorithms

Takashi Nodera and Takatoshi Inadu

Keio University, 3-14-1 Hiyoshi, Kohoku,
Yokohama 223, Japan

Abstract. This paper is concerned with PRES(pseudo-residual) algorithm one of the iterative algorithms for solving a large and sparse nonsymmetric linear systems of equations. The main variants of PRES algorithms are exact, restarted, truncated, and combined versions. In this paper, we propose the new combined algorithm, which is called the adaptive combined PRES algorithm, to monitoring the convergence behavior of residual. This adaptive procedure is very attractive because it only has the restarting process. Finally, we present some numerical experiments to illustrate the above fact using the parallel computer AP1000.

1 Introduction

In recent years, there has been a great deal of improvement in the iterative methods for solving a large sparse linear systems of equations

$$Ax = b \tag{1}$$

where A is nonsingular matrix with order n, and b is a vector with n components. The quality of the iterates $\{x_k\}$ produced by a method is often judged by the behavior of norm of residual $\|r_k\|$, where $r_k = b - Ax_k$. Namely, it is generally useful that the norm of residual converges smoothly to zero.

When the coefficient matrix A is symmetric positive definite, it is well known that the conjugate gradient (CG) algorithm is an effective iterative procedure for solving the linear systems of equations (1). However, in the case, when the matrix A is nonsymmetric, it is substantially more difficult to solve efficiently by means of iterative techniques. For instance, the CG algorithm is not able to be generalized to nonsymmetric case without a serious loss of some of its useful properties [1,2]. This difficulty has led to the development of a wide variety of generalized conjugate gradient (GCG) algorithms having varying degrees of success [2]. However, as reported by lots of researchers none of these generalizations come out as a clear winner, so that users face a difficult to choice when trying to select the "best" method to solve a specific problem.

The pseudo residual (PRES) algorithms are the substantial variant of iterative solver of GCG algorithms. Because the PRES algorithms converge very quickly under certain condition among all the GCG algorithms. But in some

1. choose an arbitrary x_0, compute $r_0 = Ax_0 - b$, where Z is an auxiliary symmetric, and positive definite matrix.
2. **for** $k = 0, 1, 2, \cdots$

$$d_k = P_k r_k , \quad (P_k: \text{right hand preconditioner})$$

$$\alpha_{i,k} = -\frac{r_{k+1-i}^T Z A d_k}{r_{k+1-i}^T Z r_{k+1-i}}, \quad \text{for } i = 1, \cdots, \sigma_k,$$

$$\bar{r}_{k+1} = A d_k + \sum_{i=1}^{\sigma_k} \alpha_{i,k} r_{k+1-i} , \quad (\bar{r}_{k+1} : \text{pseudo-residual})$$

$$\phi_k = 1 \bigg/ \sum_{i=1}^{\sigma_k} \alpha_{i,k} ,$$

$$r_{k+1} = \phi_k \bar{r}_{k+1} ,$$

$$x_{k+1} = \phi_k (d_k + \sum_{i=1}^{\sigma_k} \alpha_{i,k} x_{k+1-i}),$$

if $\|r_{k+1}\|_Z$ converges, escape the loop.
endfor

Fig. 1. Algorithm of PRES Method

case, the norm of residuals resulting from the PRES algorithms may oscillate heavily and then unstable converge.

In this paper, we present an adaptive restarting procedure on the PRES methods, particularly the combined method that is better able to deal with an faster convergence. The out line of the paper is arranged as follows. In §2, we first briefly describe the PRES algorithms and then we present our main algorithm of the adaptive restarting procedure in §3. At last, we present some numerical examples and compare the efficiency of several PRES algorithms in §4, and we give concluding remarks in §5.

2   The Pseudo Residual (PRES) Algorithm

In this section, we briefly describe the PRES algorithms that are closely related to our algorithm. The PRES algorithms are the subset of the GCG algorithms, where the pseudo-residual is minimized according to the Z-norm ($\|y\|_Z = \sqrt{y^T Z y}$, for any vector $y \in R$). If we permit the matrix Z to be dependent on the iterative step k, the GCG algorithms that minimize the true residuals are able to be transformed to PRES algorithms.

In Figure 1, we give the entire description of the PRES algorithms for solving the linear systems of equations (1).

Table 1. Variants of the PRES Methods

Version	Algorithm	σ_k	Z	P_k
exact	ORTHORES-E	$k+1$	I	I
restarted	ORTHORES-R(σ_{res})	$(k \bmod \sigma_{\text{res}}) + 1$	I	I
truncated	ORTHORES-T($\sigma_{\max}$)	$\min(k+1, \sigma_{\max})$	I	I
combined	ORTHORES-C($\sigma_{\max}, \sigma_{\text{res}}$)	—	I	I
—	ATPRES(CGNE)	2	I	A^T

In PRES algorithms, generally Z is chosen to be the unite matrix I. If $Z = I$, then the Euclidean norm is minimized. The matrix P_k is the right hand preconditioner. If $P_k = I$, the algorithm is considered as the basic method. If $P_k \neq I$, the algorithm is considered as the preconditioned method with the preconditioning matrix P_k.

In Table 1, we describe the several well known variants of methods that are all PRES algorithms. The ORTHORES-E is exact version, i.e. all preceding residuals are used for the computation of the new residuals. The ORTHORES-R(σ_{res}) is the restarted version, i.e. a restart is made periodically after σ_{res} steps, and the ORTHORES-T($\sigma_{\max}$) is the truncated version, i.e. only $\sigma_{\max}$ preceding residuals are used for the computation of the new residuals. The ORTHORES-C($\sigma_{\max}, \sigma_{\text{res}}$) is the combined version, i.e. the truncated version ORTHORES-T($\sigma_{\max}$) is restarted. The detailed derivation of these algorithms can be given in Bruaset [2].

Unfortunately, in the exact version, the storage requirements and the computational work increase in the iteration steps. The restarted and truncated version also have the unstable convergence in some practical problems. The combined version can not breakdown, it is stable, because these procedures based on the restarting process of truncated version [7]. Nevertheless, the combined version that restarts periodically represents bad convergence behavior in some case. The possibility of the restarting techniques may be exploited to develop efficient algorithm and make better robustness.

3 Adaptive Restarting Procedures

Restarting of the PRES algorithms are normally needed to reduce storage requirements. However, lots of restartings slow down the convergence of the algorithm, because they require some cost of recomputing new residuals. Therefore, we have designed an adaptive procedure with automatic restarting for PRES algorithms. We start with truncated version of PRES algorithm, such as ORTHORES-T($\sigma_{\max}$), if it does not converge "completely well", we compute new residuals $r_0 = Ax_k - b$ and then restart the algorithm.

Before providing a suitable framework for the presentation of the adaptive restarting procedure, we give some kind of the relation of the norm of residual $\|r_{k+1}\|_2 / \|r_0\|_2$ and the coefficient ϕ_k in the PRES algorithm, by using the

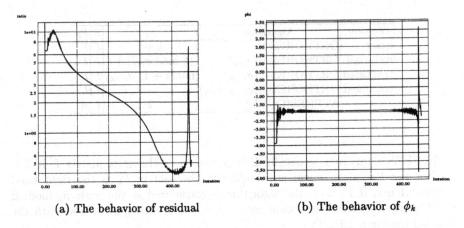

(a) The behavior of residual   (b) The behavior of ϕ_k

Fig. 2. Relation between the norm of residual ($\|r_{k+1}\| / \|r_0\|$) and the coefficient ϕ_k (Problem: $a = 30$, $b = 5$, ORTHORES-T(5)).

heuristic approach. For example, in our numerical problem ($a = 30$, $b = 5$) which are discussed in the next section, its convergence history of the residual $\|r_{k+1}\|_2/\|r_0\|_2$ and the coefficient ϕ_k can be plotted in Figure 2(a) and 2(b) respectively. From these figures, we are able to observe that the pointed peak of residual (the significant divergence of residual) can be occurred, when the ϕ_k turns on a positive real number at the 451 iteration step. In this point of view, we state the following conjecture.

[**Conjecture 3.1**] *In PRES algorithm, if ϕ_k turns on a positive real number, the norm of residual $\|r_{k+1}\|/\|r_0\|$ might be expected to be extremly large.*

This conjecture represents that the convergence behavior of PRES algorithm is determined by checking the value of ϕ_k. Namely, if ϕ_k turns on a positive real number, we might expect that the behavior of residual norm stands for a little bit ill convergence in some sense. It does not improve iteration process with respect to the convergence efficiently. Therefore, we have to perform the restart to correcting the ill-behaved residuals into the better situation. The detailed derivation of this conjecture can be given in Inadu and Nodera [8].

From the above conjecture and some stabilizing techniques of the coefficient ϕ_k, we propose the adaptive restarting procedures combined with the ORTHORES-T($\sigma_{\max}$) algorithm in Figure 3.

The outline of our proposed adaptive restarting algorithm is as follows. This procedure starts with ORTHORES-T($\sigma_{\max}$), and controls at the restarting point after checking the above mentioned restarting criterion and the following stabilizing criterion, which is based on the coefficients ϕ_k in the ORTHORES-T($\sigma_{\max}$).

The stabilizing criterion is as follows.

$$\frac{V(\phi_{k+1-i})}{E(\phi_{k+1-i})} \leq \epsilon_{\text{stab}} \tag{2}$$

```
1. set initial residual ||r_min|| = ||r_0||
2. do ORTHORES-T(σ_max) for σ_max iterations
   (Truncated Method Part)
3. minres_updated = FALSE
   for i = 0, ···, σ_max − 1
       if(||r_min|| > ||r_{k+1−i}||) ||r_min|| = ||r_{k+1−i}|| and
                minres_updated = TRUE
     endif
   endfor
4. if(minres_updated)
     go to (2)
   else
     go to (5)
   endif
5. for i = 1, ···, σ_max
       if(φ_{k+1−i} > 0)
       restart and go to (2)
       endif
   endfor
6. compute average and variance of φ_{k+1−i} (i = 1, ···, σ_max)
7. if(variance/average^2 < ε_stab) restart and go to (2)
   else
     go to (2)
   endif
```

Fig. 3. Algorithm of ORTHORES-AC(σ_{max})

where V is variance and E is arithmetic mean. The ϵ_{stab} is the tolerance for the stabilizing criterion. If the equation (2) is satisfied, the restart of ORTHORES-T(σ_{max}) is performed. It is done in the hope that this restart might be improved the very ill-behaved ORTHORES-T(σ_{max}) residual in the better direction. This parameter ϵ_{stab} is set up from the experimental results of numerical computation, normally $\epsilon_{stab} = 10^{-3}$.

We have named this procedure ORTHORES-AC(σ_{max}) in [8]. This algorithm is a typical "numerical engineering" approach that makes use of experience.

4   Numerical Experiments

For the test purposes we consider the problem arising from the centered difference discretization of the boundary value problem of the form

$$u_{xx} + u_{yy} + au_x + bu_y = f(x, y) \tag{3}$$
$$u(x, y)|_{\partial\Omega} = 0$$

Table 2. Results of experiments for various coefficients of equation (3): the computational time (TIME[sec]), the number of iterations (ITS), and the number of restarts (RST).

COEF.		ORTHORES-T(5)		ORTHORES-C(5,50)			ORTHORES-AC(5)		
a	b	TIME	ITS	TIME	ITS	RST	TIME	ITS	RST
3	5	—	diverge	—	—*	60	276	2500	4
3	50	—	diverge	124.5	1140	22	119.2	1085	11
3	500	59.6	537	53.4	487	9	56.0	511	7
3	5000	54.0	487	64.8	591	11	62.1	570.	13
30	5	—	diverge	175.8	1611	32	157.6	1433	9
30	50	158.0	1423	129.0	1177	23	117.0	1071	20
300	5	64.7	584	57.0	520	10	60.0	547	6
300	500	61.0	550	60.5	552	11	59.0	557	33
3000	5	50.2	453	54.8	499	9	52.3	482	13
3000	5000	78.8	711	79.8	726	14	88.8	814	16

*Norm of residual $\|r_{3000}\|/\|r_0\| = 10^{-10}$ after 3000 iterations.

on square region $[0,1] \times [0,1]$, and a and b are real coefficients. The right hand side $f(x,y)$ is chosen so that the solution is

$$u(x,y) = x(1-x)y(1-y)$$

The grid consists of square of 256 internal mesh points in each direction leading to a matrix of size $n = 65536$. The initial vector is chosen $x_0 = 0$. The convergence criterion used for stopping the iterative process is either when the residual satisfies $\|r_{k+1}\|/\|r_0\| \leq 10^{-12}$ or when the iteration limit is reached at 3000. These tests are run on AP1000 with 64 processors in double precision. The AP1000 is distributed memory MIMD multiprocessor machine which is produced by Fujitsu.

We are essentially concerned with the algorithm ORTHORES-AC(σ_{max}) with using a tolerance of $\epsilon_{stab} = 10^{-3}$ for the restarting criterion. For comparison, we also discuss the ORTHORES-T(σ_{max}), ORTHORES-R(σ_{res}) and ORTHORES-C(σ_{max}, σ_{res}).

As a results of lots of the tests with very large sparse matrix problems, the truncated parameter $\sigma_{max} = 5$ and the restarted parameter $\sigma_{res} = 50$ have been optimized resulting in a very efficient algorithm with limited storage requirements.

In Table 2, we show the numerical experiments for this problem with the various coefficients of a and b. Run for which convergence was not possible in 3000 iterations are labeled by (—) in the column of computational time. We make the following observations about these runs. For this problem, the ORTHORES-T(5) does not converge in a number of cases. However, ORTHORES-C(5,50) and ORTHORES-AC(5) are fairly good convergence in almost every cases. In particular, ORTHORES-AC(5) worked quite well in most cases, and also has the stable convergence and robustness.

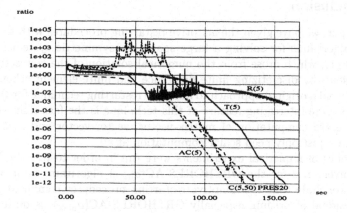

Fig. 4. The behavior of residuals vs. computational time ($a = 30$, $b = 50$).

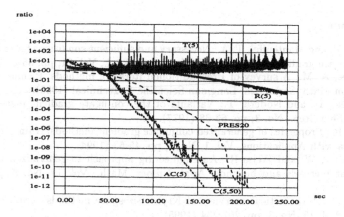

Fig. 5. The behavior of residuals vs. computational time ($a = 30$, $b = 5$).

Figure 4 shows representative plots of the convergence behavior of residuals $\|r_{k+1}\|/\|r_0\|$ vs. computational time (sec) for the case of coefficients $a = 30$ and $b = 50$. Figure 5 also shows the convergence behavior of residuals $\|r_{k+1}\|/\|r_0\|$ vs. computational time (sec) for the case of $a = 30$ and $b = 5$. In these figures, the PRES20 [4] is the ORTHORES-C(5,20) with applying the smoothing technique [5, p. 2]. From these figures, we can see that the norm of residuals of ORTHORES, just like ORTHORES-T(5), may oscillate heavily. These figures also show that the ORTHORES-AC(5) algorithm keeps the residuals size better behaved than the other algorithms over the course of the run.

5 Conclusion

In this paper, we have given the adaptive restarting procedures with the pseudo residual algorithms for solving a large sparse nonsymmetric systems of linear equations (1), which arose from the numerical solution of the convective diffusion problems by discretizing finite difference and finite element schemes. Our study involved a original approach to adaptive restarting technique for the PRES algorithms. One interesting feature of this technique is the fact that we do not need any extra computation of matrix-vector, vector-vector and scalar-vector products, we just only need scalar computation at all.

A number of numerical experiments were run in order to test the behavior of the convergence for the ORTHORES-AC(σ_{max}) algorithm. From these results, we conclude that the adaptive restarting procedures combined with the pseudo residual algorithm, especially ORTHORES-AC(σ_{max}), is quite capable and robust for the solution of this kind of nonsymmetric problem. More detailed numerical results of this algorithm are given in Inadu and Nodera [8].

Further research is necessary to better theoretical understanding of the adaptive restarting procedures which combined with the pseudo residual algorithms.

References

1. Faber, V. and Manteuffel, T. A.: Necessary and sufficient conditions for existence of a conjugate gradient method, SIAM J. Numer. Anal., Vol. 21, pp. 352-362 (1984).
2. Bruaset, A. M.: A survey of Preconditioned iterative methods, Pitman Research Notes in Math. Series 328, Longman Scientific & Technical (1995).
3. Takahasi, H. and Nodera, T.: Variants of the conjugate gradient method, Keio Math. Sem. Rep., No. 3, pp. 63-68 (1977).
4. Weiss, R.: Properties of generalized conjugate gradient methods, Numerical Linear Algebra with Applications, Vol. 1, No. 1, pp. 45-63 (1994).
5. Schönauer, W. and Weiss R.: An engineering approach to generalized conjugate gradient methods and beyonds, Appl. Numer. Math., Vol. 19, No. 3, pp. 27-34 (1995).
6. Weiss, R.: A theoretical overview of Krylov subspace methods, Applied Numer. Math., Vol. 19, No. 3, pp. 207-234 (1995).
7. Nodera, T. and Inadu, T.: The convergence acceleration of pseudo residual method using a restarted procedure, Trans. of IPSJ (in Japanese), Vol. 37, No. 6, pp. 1237-1240 (1996).
8. Inadu, T. and Nodera, T.: An adaptive restarting procedure for pseudo residual methods, Trans. of IPSJ (in Japanese), Vol. 37, No. 9, pp. 1637-1645 (1996).

Fixed Point Theorems and Error Bounds for Solutions of Equations

Georg Rex

Institute of Mathematics, University of Leipzig
Augustusplatz 10-11, D-04109 Leipzig, Germany

Abstract. We consider a system of equations $h(x) = 0$, where h is Gateaux-differentiable on an open convex set.

A computation of a solution of $h(x) = 0$ by numerical algorithms on digital computers in finite accuracy usually gives a good approximation x^0 of the exact solution x^*, but no verified error bounds.

On the basis of Brouwer's fixed point theorem, we give a new theorem and a new algorithm for a computation of error bounds for the computed x^0 and verify that within the error bounds the exact solution x^* exists and is possibly unique; especially if h is a linear system of equations.

1 Introduction

Let $h : D \subseteq \mathbf{R}^n \to \mathbf{R}^n, D$ open set, h Gateaux-differentiable on D with

$$h(x) = 0 \tag{1}$$

a system of equations. The computation of an exact solution $x^* \in D$ executed on a digital computer in finite accuracy usually gives a good approximation $x^0 \in D$ of x^*, but in general no verified error bounds.

One aim of this paper is the computation of a verified error bound $u \in \mathbf{K}^n := \{x \in \mathbf{R}^n \mid 0 \le x_i, 1 \le i \le n, x \ne 0\}$, $\mathbf{K}^n$ defined the *cone* of $\mathbf{R}^n$, such that $|x_i^* - x_i^0| \le u_i, 1 \le i \le n$ or equivalent

$$x^* \in x^0 \pm u =: X^0. \tag{2}$$

On the other hand, we will prove the existence and possibly the uniqueness of $x^* \in X^0$ despite the presence of conversion, rounding and cancellation errors on a digital computer.

A fundamental property of interval analysis is that the bounding of the range of a codable function is possible without any auxiliary knowledge about the function such as, for instance, Lipschitz continuity.

Because of this fact we choose, if (1) is nonlinear, a neighbourhood $X := x^0 \pm w, w \in \mathbf{K}_+^n := \{x \in \mathbf{K}^n \mid 0 < x\}$ and compute an interval enclosure $H'(X)$ of the Jacobian $h'(x^0)$. This step is not necessary if (1) is linear.

We show in section 2 by using Brouwer's fixed point theorem that such a u from (2) is the solution of

$$\langle H'(X)\rangle\, u = |h(x^0)| \tag{3}$$

if $H'(X)$ H-matrix and $u \le w$, where $\langle H'(X)\rangle \in \mathbf{R}^{n,n}$ is the comparison matrix of $H'(X)$, see section 1.1 and special (13) and (14).

We give in section 3 an algorithm which consists of two basic steps:

> **bstep 1:** prove $H'(X)$ H-matrix
>
> **bstep 2:** compute an $v \in \mathbf{K}^n$ with: $\qquad\qquad$ (4)
>
> $\qquad u \le v \le w$ if (1) nonlinear
>
> $\qquad u \le v$ $\quad$ if (1) linear .

If (4) is successful then we have verified that (1) has one and only one solution $x^* \in x^0 \pm v$.

The strict condition $H'(X)$ H-matrix can be satisfied in general by a precondition matrix $C \in \mathbf{R}^{n,n}$ such that we have to work with

$$\langle C\,H'(X)\rangle\, u = |C\,h(x^0)| \tag{5}$$

instead of (3).

Here we propose: choose C such that $C\,H'(X)$ is similar to a triangular matrix. One from (4) different algorithm, based on (5), where C is an approximation of the midpoint inverse, can be found in [8]. Modifications of the algorithm (4) were suggested in [11], [12].

A tool for a computation of error bounds is interval analysis. Here we point out that for the computation of the real boundary matrix $\langle C\,H'(X)\rangle$ and the real boundary vector $|C\,h(x^0)|$ we need only the interval tool. The basic steps (4) in our algorithm can be computed by directed rounding since we are working in $\mathbf{K}^n$.

1.1 Definitions, Notations and Preliminaries

In this section we briefly survey connections between interval analysis and the theory of nonnegative matrices which are needed in our context.

The symbol $\mathbf{IR} := \{X \mid X = [\underline{X},\overline{X}],\ \underline{X},\overline{X} \in \mathbf{R}\}$ denotes the set of all closed real intervals and following equivalent notations for an $X \in \mathbf{IR}$ are used, such that

$$X = [\underline{X},\overline{X}] = mid(X) \pm rad(X) \tag{6}$$

with the definitions: $mid(X) := (\underline{X} + \overline{X})/2$ for the *midpoint* of X and $rad(X) := (\overline{X} - \underline{X})/2$ for the *radius* of X.

An $x \in \mathbf{R}$ and a *point interval* $[x,x] \in \mathbf{IR}$ are identified by each other; in this sense intervals are an extension of real numbers.

The *absolute value* or *magnitude* of $X \in \mathbf{IR}$ can be computed by

$$|X| = |mid(X)| + rad(X). \tag{7}$$

However, we also need the *mignitude* of X computed by

$$\langle X \rangle = |mid(X)| - rad(X) \quad \text{if} \quad 0 \notin X; \quad \text{otherwise is} \quad \langle X \rangle = 0 . \quad (8)$$

The following equivalence

$$x \in Y \quad \Longleftrightarrow \quad |x - mid(Y)| \leq rad(Y) \quad (9)$$

shows that the radius $rad(Y)$ is a *measure* for the absolute accuracy of $mid(Y)$ as an approximation of the unknown number $x \in Y$. This is an important observation in our context, see (2) and *(s3)* in Theorem 1.

Interval operations are not restricted to arithmetic operations; they can be also extended of functions, called *interval functions*.

Let $f : D \subseteq \mathbf{R} \to \mathbf{R}$, $I(D) := \{X \in \mathbf{IR} \mid X \subseteq D\}$ then the interval function $F : I(D) \subseteq \mathbf{IR} \to \mathbf{IR}$ is an *interval enclosure* if

$$F(X) \supseteq f(X) := \{f(x) \mid x \in X \in I(D)\} . \quad (10)$$

This F gives an estimation for the range of f over X. For a computation of an enclosure of the range of a function over an interval we can make use of the monotonicity of the function over this interval. But also nonmonotonic functions can be executed over intervals by using a power series expansion with an estimation of the remainder term. More details of this subject and software packages can be found , for instance, in [2], [8],[1],[5],[6],[7].

For solving our problem we extend $\mathbf{IR}$ to $\mathbf{IR}^n$ componentwise, where $\mathbf{IR}^n$ denotes the *set of all n-dimensional column interval vectors*, and to $\mathbf{IR}^{n,n}$ elementwise, where $\mathbf{IR}^{n,n}$ denotes the *set of all (n,n)-dimensional interval matrices*. Furthermore the relations $>, \geq, <, \leq, \in, \subseteq, \supseteq, \supset$ and (6) - (9) are to be used component- and elementwise, respectively.

A matrix $\mathbf{A} \in \mathbf{IR}^{n,n}$ is called *regular* if each $A \in \mathbf{IR}^{n,n}$ with $A \in \mathbf{A}$ is nonsingular. Furthermore we define by $\mathbf{A}^{-1} := [inf \{A^{-1} \mid A \in \mathbf{A}\}, sup \{A^{-1} \mid A \in \mathbf{A}\}]$ the *inverse of a regular interval matrix* $\mathbf{A}$, and because of $M := \{A^{-1} \mid A \in \mathbf{A}\}$ is a bounded set, *inf* and *sup* exist componentwise. The hull of M is the tightest interval matrix enclosing M.

We shall now give some properties and definitions which are important for our considerations, where $\rho(\cdot)$ denotes the *spectral radius* of a real matrix.

Let $A \in \mathbf{IR}^{n,n}$, $C \in \mathbf{R}^{n,n}$ then: $\quad C A = C \, mid(A) \pm |C| \, rad(A)$. $\quad (11)$

For the following statements (12) - (16) let $A \in \mathbf{IR}^{n,n}$ then

$$A := D(A) - N(A) \quad (12)$$

is called *Jacobi splitting*, where $D(A)$ is the main diagonal and $N(A)$ is the off-diagonal part of A. The elementwise application of (7) and (8) on (12)

$$\langle A \rangle := \langle D(A) \rangle - |N(A)| \quad (13)$$

defined the *comparison matrix* of A and furthermore is

$$A \text{ H-matrix} :\Longleftrightarrow \exists u \in \mathbf{K}^n_+ \ (0 < \langle A \rangle \, u) \ \Longleftrightarrow \ \rho(\langle D(A) \rangle^{-1} |N(A)|) < 1. \quad (14)$$

$$\text{Let } A \text{ H-matrix} \quad \Longrightarrow \quad A \text{ is regular} \quad \text{and} \quad |A^{-1}| \leq \langle A \rangle^{-1}. \quad (15)$$

These given definitions, computation rules and propositions can be found in [8] and special for nonnegative matrices see e.g. [3],[15], and [9].

2 Basic Principle

In this section we are given more details with respect to equation (3).

Theorem 1. *Assume that*
(a1): $h : D \subseteq \mathbf{R}^n \to \mathbf{R}^n, D$ *open and* $x^0 \in D$ *be an approximation of*
     $x^* \in D$ *an exact solution of* $h(x) = 0$,
(a2): h *be Gateaux-differentiable on an open convex set* $D^0 \supset X := x^0 \pm w$,
     $w \in \mathbf{K}^n_+$ *sufficiently small*,
(a3): $H'(X) \in \mathrm{I\!R}^{n,n}$ *be an interval enclosure of the Jacobian* $h'(x^0)$,
(a4): $H'(X)$ *H-matrix*,
(a5): *the solution* u *of* $\langle H'(X) \rangle \, u = |h(x^0)|$ *satisfy* $u \leq w$,

then
(s1): $H'(X)$ *regular*,
(s2): x^* *is a unique solution with* $h(x^*) = 0$,
(s3): $x^* \in x^0 \pm u \subseteq X$.

Proof. Assume *(a1)-(a5)* are satisfied, then *(s1)* follows from *(a4)* and (15). Because of *(a2)*, *(a3)*, $q := x - x^0$, $x \in X$ and the transposed nabla operator ∇^T is

$$H'(X) \supseteq h'(x^0 \pm w) \ni Dh(x^0, q) := \begin{pmatrix} \nabla^T h_1(x^0 + t_1 q) \\ \vdots \\ \nabla^T h_n(x^0 + t_n q) \end{pmatrix}$$

for each fixed $t \in \mathbf{K}^n$ with $t_1, \cdots, t_n \in [0, 1]$,(in general the t_i will all be distinct) and $\forall |q| < w$ satisfied. The following sequence of inequalities is fulfiled for a fixed t and $\forall |q| \leq w$,

$$w \geq u = \langle H'(X) \rangle^{-1} |h(x^0)| \geq |H'(X)^{-1}| |h(x^0)| \geq |Dh(x^0, q)^{-1} h(x^0)|$$

this can be seen by using of *(a4)*, (15), *(a5)*, and *(a3)*. We define $W := [-w, w]$ and consider $G(q) := -Dh(x^0, q)^{-1} h(x^0)$. It is obvious that G is continous and since $-w \leq G(q) \leq w$ is $G : W \to W$. The application of Brouwer's fixed point theorem shows that, for at least a $t =: t^*$, for G a fixed point q^* exists with $|q^*| \leq w$, that is for $q^* := x^* - x^0$ and by application of the *generalized mean value theorem*, (see e.g. [9]) is

$$0 = h(x^0) + Dh(x^0, (x^* - x^0)) (x^* - x^0) = h(x^*).$$

Hence an $x^* \in X$ with $h(x^*) = 0$ exists.

We now show that x^* is unique by contradiction: Assume there is for this t^* or for a $t^{**} \neq t^*$ a fixed point q^{**} with $q^{**} \neq q^*$. Then an $x^{**} \in X, x^{**} \neq x^*, h(x^{**}) = 0$, exists in the same way as for x^*.

Since h is Gateaux-differentiable and because of the generalized mean value theorem there exists an $s^* \in \mathbf{K}^n$ with $s_1^*, \cdots s_n^* \in [0,1]$ such that

$$h(x^{**}) - h(x^*) = Dh(x^*, (x^{**} - x^*)) (x^{**} - x^*) = 0$$

that is $Dh(x^*, q^*)$ is singular. This contradicts *(a4)* in respect of (15). Therefore $x^* = x^{**}$ and *(s2)* is shown. The statement *(s3)* can be immediately seen from $|q^*| = |x^* - x^0| \leq u \leq w$. $\square$

We now show that the applicability of this theorem can be extended by preconditioning. Because of this we consider interval triangular matrices.

Lemma 2. *Let $A \in \mathrm{IR}^{n,n}$ be an upper (or lower) triangular matrix then:*

$$A \text{ regular} \iff 0 < \langle D(A) \rangle \iff \rho(\langle D(A) \rangle^{-1} |N(A)|) = 0 \iff A \text{ } H\text{-matrix}.$$

Proof. Because of the special structure of A follows that all real triangular matrices from A must be non-singular. Therefore $0 \notin A_{ii}$ $\forall i$. Note that also by the special structure $\rho(\langle D(A) \rangle^{-1} |N(A)|) = 0$ and applies to (14). $\square$

Let $C \in \mathbf{R}^{n,n}$ be a precondition matrix then (11) implies

$$C H' := C H'(X) = C \, mid(H'(X)) \pm |C| \, rad(H'(X)) \tag{16}$$

and Lemma 2 motivates the assumption *(a4)* of Theorem 1 to force through a suitable C. The best choice for C is $C = (mid(H'(X))^{-1}$, see, for instance, [13],[14]. But even an approximate computation of this C is often not desirable. On the other hand a C is often getatable from a matrix factorization for the computation of x^0, for instance, variants of $LU-$ or $QR-$factorizations, see [4], that is $C = L^{-1}$ or $C = Q^T$, respectively.

The comparison matrix (13) applied to (16) implies (17) with

$$D := \langle D(CH') \rangle, \quad N := |N(CH')|, \quad N = N_L + N_U,$$

where N_L the lower off-diagonal part and N_U the upper off-diagonal part of N denote, respectively,

$$\langle C H' \rangle = D - N = D - N_L - N_U. \tag{17}$$

The first splitting of (17) is known as Jacobi-splitting, see also (12), and the second is known as *Gauss-Seidel-splitting*. In the following these are denoted by *JS* and *GSS*, respectively. The precondition equation

$$\langle C H' \rangle u = b \quad \text{with} \quad b := |C h(x^0)|$$

implies a fixed point equation

$$u = P u + d \tag{18}$$

with

$$P := D^{-1}N, \quad d := D^{-1}b \qquad \text{for } JS$$

and

$$P := (D - N_L)^{-1}, \quad d := (D - N_L)^{-1}b \qquad \text{for } GSS.$$

With regard to a constuction of an algorithm we consider

$$u^{k+1} := P\,u^k + d, \qquad k = 0, 1, \ldots \tag{19}$$

instead of (18) and have the following fundamental facts:

Lemma 3. $\langle C\,H' \rangle\ H-matrix$ *if and only if the sequence (19) converges to* $u = \langle C\,H' \rangle^{-1}b,$ *for each* $u^0 \in \mathbf{K}^n,$

and

if $\langle C\,H' \rangle$ *H-matrix then:* $\rho((D - N_L)^{-1}N_U) \leq \rho(D^{-1}N) < 1,$

that is the convergence of (19) is for the GSS faster than for the JS.

Proof. Statement one see ([3],p.171,(3.1) Lemma) and use (14). Statement two see ([3], p. 187,(2.55)Corollary). $\square$

But because of the interval enclosure is $0 \leq N_L$ and we are not sure if $C\,H'$ is an H-matrix. To prove this, a compact algorithm based on Lemma 2 and (14) is given in next section.

3   Algorithm

In the following we use the notation from section 2 and the symbol $\sharp(\cdot)$ is meant to: compute the expression in parenthesis by means of an interval tool.
If h is linear, then $h(x) := A\,x - g,\ x, g \in \mathbf{R}^n,\ A \in \mathbf{R}^{n,n};\ h'(x) = A\ \forall x \in D$ and therefore $H'(X) = A$. We give now a compact algorithm.

ALGORITHM
- Assume that $x^0 \in D$ was obtained as an approximated zero of h
  (see: *(a1), (a2)* of Theorem 1), and furthermore assume that, from the computation of $x^0,$ a precondition matrix C is getatable.
- If h nonlinear then: choose a $w \in \mathbf{K}^n_+$ and compute $X = \sharp(x^0 \pm w).$
- Compute $D - N := \langle \sharp(C\,H'(X)) \rangle$ and $0 < b := |\sharp(C\,h(x^0))|;$
  -if $(\exists i\,(0 = D_{ii}, 1 \leq i \leq n))$ then **Stop**.
- Choose (*JS* or *GSS*)
    - *JS:*   Compute $P := D^{-1}\,N,$ and $d := D^{-1}b$
    - *GSS:*   Compute $P_L := D^{-1}\,N_L,\ P_U := D^{-1}\,N_U,$ and $d := D^{-1}b$
- Choose a $u^0 \in \mathbf{K}^n_+;$ set $k := 0,$
    repeat
        $k := k + 1$
    JS: $u^1 := P\,u^0 + d,$ (respectively, *GSS* $u^1 := P_L\,u^1 + P_U\,u^0 + d$)
        -If h nonlinear then: [(if $u^1 < u^0$ and $u^0 < w$) then **Stop-**

-$((s1)$-$(s3)$ in Theorem 1 are true.)]
-If h linear then: [(if $u^1 < u^0$) then **Stop**-
  -$((s1)$-$(s3)$ in Theorem 1 are true.)]
$u^0 := u^1$
until $k \geq 10$

Remarks. - Note: $u^1 < u^0$ implies $C H'(X)$ H-matrix.
This can be immediately seen, then from $u^1 = Pu^0 + d < u^0$, $b > 0$ and
$d = D^{-1}b$, follow $Pu^0 < u^0$, and the
application of (14) gives the assertion.
- If $k > 10$ and $u^1 < u^0$ but $u^0 \not< w$ then choose a new X and start the
algorithm new.
- If $k > 10$ and $u^1 \not< u^0$ the algorithm fails, because of singularity or a bad
condition number.

A connection between a posteriori error estimations, error bounds based on
Banach's fixed point theorem and error bounds based on Brouwer's fixed point
theorem can be found in [10], and especially by using norms in [12].

References

1. Alefeld, G. , Frommer, A. , Lang, B.: Scientific Computing and Validiated Numerics.
   Akademie Verlag, Berlin 1996
2. Alefeld, G. , Herzberger, J.: Introduction to Interval Computations. Academic Press,
   New York 1983
3. Berman, A. , Plemmons, R. J.: Nonnegative Matrices in the Mathematical Sciences.
   Academic Press, New York 1979
4. Golub, G. H. , Van Loan, C. F.: Matrix Computations. The Johns Hopkins University Press, Baltimore 1989
5. Hammer, R. , Hocks, M. , Kulisch, U. , Ratz, D.: Numerical Toolbox for Verified
   Computing I. Springer Verlag, Berlin 1993
6. Klatte, R. , Kulisch, U. , Neaga, M. , Ratz, D. , Ullrich, C.: PASCAL-XSC. Springer
   Verlag, Berlin 1991
7. Knueppel, O.: PROFIL / BIAS - A Fast Interval Library. Computing **53** (1994)
   pp.277–288
8. Neumaier, A.: Interval methods for systems of equations. Cambridge University
   Press, Cambridge 1990
9. Ortega, J. M. , Rheinboldt, W. C. : Iterative Solutions of Nonlinear Equations in
   Several Variables. Academic Press, New York 1970
10. Rex, G.: Zur Loesungseinschliessung linearer Gleichungssysteme. Wissenschaftl.
    Zeitsch. TH Leipzig **15** (1991) 6, pp.441–447
11. Rex, G.: Zur Einschliessung der Loesung eines linearen Gleichungssystems. Zeitsch.
    fuer angew. Math. und Mech. (ZAMM) **72** (1993) 7/8, T829–831
12. Rex, G. : Parameterabhaengige Loesungseinschliessungen linearer Gleichungssys-
    teme. Zeitsch. fuer angew. Math. und Mech. (ZAMM) **74** (1994) 6, T683–685
13. Rex, G. , Rohn, J.: A Note on Checking Regularity of Interval Matrices. Linear
    and Multilinear Algebra **39** (1995) pp.259–262
14. Rohn, J. , Rex, G.: Enclosing Solutions of Linear Equations. SIAM J. on Numer.
    Anal. (1997),(to appear)
15. Varga, R. S.: Matrix Iterative Analysis. Prentice-Hall, Englewood Cliffs 1962

Multiscale Edge Detection of Images from Its Radon Transform [*]

Tsi-min Shih[1], Zhi-xun Su[2], and Foo-tim Chau[3]

[1] Department of Applied Mathematics, The Hong Kong Polytechnic University, Kowloon, Hong Kong.
[2] Institute of Mathematical Sciences, Dalian University of Technology, Dalian 116024, P.R.China.
[3] Department of Applied Biology & Chemical Technology, The Hong Kong Polytechnic University, Kowloon, Hong Kong.

Abstract. This paper discusses wavelet multiscale edge detection for tomographic images. Methods to compute the wavelet transforms of the image directly from its Radon transform is presented, and then the maxima modulus method can be used in the tomographic image reconstruction process. Corresponding fast algorithms are provided.

1 Introduction

Tomographic imaging deals with reconstructing a cross-sectional image of an object from its projections. For nondiffracting tomography, a projection in a certain direction can be taken as a line integral of the image function. This is called the Radon transform and defined by (1) in section 2. A typical reconstruction method is the Convolution Back Projection(CBP) method.

Because the edge points of an image usually characterize important information, the edge detection and reconstruction techniques are very useful in image compression and denoising, and there have been many wavelet approaches which have proven to be powerful. Mallat et al([9]) introduced a maxima modulus method for edge detection and also a method to reconstruct the signal or image from its local maxima modulus of the wavelet transform. Although Meyer proved that the local maxima modulus of a wavelet transform does not characterize a function uniquely, a numerical algorithm developed in [9] is able to reconstruct a close approximation to the original data. It is also proved that this edge detection method is suitable for the noisy image processing([10]). Therefore this is an efficient method for image compression and denoising. For tomographic images, these techniques are also efficient.

To compute the wavelet transforms of a tomographic image directly from its Radon transform and then apply various wavelet methods is the goal of this paper. According to the Wavelet-Vaguelette Decomposition(W.V.D) method introduced by Donoho ([4]), $Wf(x,y)$, the wavelet transform of the image $f(x,y)$

[*] The work is supported by The Hong Kong Polytechnic University Research Grant No. 351/051 and NNSF of China.

can be computed directly from its Radon transform Rf by $< f, \psi_\lambda > = [Rf, \gamma_\lambda]$, where $\{\gamma_\lambda\}$ forms a near-orthogonal basis and is called Vaguelettes which are dilations and translations of a single "mother" Vaguelette. However, $Wf(x,y)$ can also be computed by other more general and fast methods. In fact, we can make use of fast wavelet transform together with the Convolution Back Projection reconstruction method to get a fast algorithm to compute $Wf(x,y)$ directly from Rf. Thus wavelet multiscale edge detection methods can be used in tomographic image processing efficiently. As a result, edge detection, together with denoising and compression of tomographic images can be employed in the reconstruction process.

In this paper, multiscale edge detection algorithms for tomographic images are studied. The complexity of the algorithm is of the same order as that of the Convolution Back Projection algorithm, which is the standard fast reconstruction method.

2   Radon transform and tomographic image reconstruction

Let $f(x,y) \in L^1(R^2) \cap L^2(R^2)$ denotes a tomographic image. For a given angle $\theta \in [0,\pi]$ and $s \in R$, the line integral

$$p(s,\theta) = Rf(s,\theta) = \int_{-\infty}^{\infty} f(s\cos\theta - t\sin\theta, s\sin\theta + t\cos\theta)dt \qquad (1)$$

is called the projection or raysum of $f(x,y)$ at (s,θ). The function $Rf(s,\theta)$ is known as the Radon transform of $f(x,y)$. Tomographic imaging aims to reconstruct $f(x,y)$ from a set of data consisting of finitely many measurements $\{p(s_n,\theta_m)\}$.

The most important property of the Radon transform is the following Projection-Slice Theorem:

Theorem 1. *Given $f(x,y) \in L^1(R^2) \cap L^2(R^2)$, $p(s,\theta)$ defined in (1), let*

$$\hat{p}_\theta(w) = \int_{-\infty}^{\infty} p(s,\theta)e^{-i2\pi ws} ds \qquad (2)$$

be the one-dimensional Fourier transform of $p(s,\theta)$ with respect to the parameter s for any given θ, and

$$F_2(f)(u,v) = \int_{-\infty}^{\infty}\int_{-\infty}^{\infty} f(x,y)e^{-i2\pi(ux+vy)} dx\,dy \qquad (3)$$

be the two-dimensional Fourier transform of $f(x,y)$. Then

$$\hat{p}_\theta(w) = F_2(f)(w\cos\theta, w\sin\theta). \qquad (4)$$

Due to the Projection-Slice Theorem and the inverse Fourier transform, $f(x, y)$ can be reconstructed by the following formula:

$$f(x, y) = \int_0^\pi \int_{-\infty}^\infty \hat{p}_\theta(w) e^{i2\pi w(x\cos\theta + y\sin\theta)} |w| dw d\theta$$

$$= B[C[p(s, \theta)]], \tag{5}$$

where $s = x\cos\theta + y\sin\theta$,

$$B(g) = \int_0^\pi g(\theta) d\theta \tag{6}$$

and

$$C[p(s, \theta)] = \int_{-\infty}^\infty \hat{p}_\theta(w) e^{i2\pi w s} |w| dw \tag{7}$$

are called the back projection operator and the convolution operator respectively.

Also, it is well known that

$$B(Rf)(x, y) = \int_0^\pi Rf(s, \theta) d\theta = f(x, y) * \frac{1}{\sqrt{x^2 + y^2}},$$

which implies

$$F_2 B(Rf)(w) = |w|^{-1} (F_2 f)(w), \tag{8}$$

where (*) denotes the 2-D convolution and F_2 denotes the 2-D Fourier transform. Let F_2^{-1} denotes the 2-D inverse Fourier transform, then

$$f(x, y) = F_2^{-1} |w| F_2 B(Rf)(x, y) \tag{9}$$

The above inversion technique is known as the Convolution Back Projection(CBP). In practice, discrete approximations of (5) and (9) will be needed for reconstructions of discretized images.

3   Wavelet transforms and multiscale edge detection of images

3.1   Wavelet transforms of images

Consider first a particular class of 1-D wavelets. Let $\phi_0(x)$ be a smoothing function, $\psi(x)$ be a wavelet and $\chi(x)$ be a reconstructing wavelet satisfying

$$\hat{\phi}_0(2w) = e^{-isw} H(w) \hat{\phi}_0(w),$$

$$\hat{\psi}(2w) = e^{-isw} G(w) \hat{\phi}_0(w),$$

and

$$\hat{\chi}(2w) = e^{isw} K(w) \hat{\phi}_0(w),$$

where $H(w), G(w)$, and $K(w)$ are functions with period 2π and $|H(w)|^2 + G(w)K(w) = 1$. For image processing, 2-D wavelets should be introduced. To minimize the amount of operations, the 2-D wavelets $\psi^1(x,y)$ and $\psi^2(x,y)$ can be chosen to be $\psi^1(x,y) = \psi(x)2\phi_0(2y) = \frac{\partial\theta(x,y)}{\partial x}$ and $\psi^2(x,y) = 2\phi_0(2x)\psi(y) = \frac{\partial\theta(x,y)}{\partial y}$, where $\theta(x,y)$ is a smooth function. The smoothing function $\phi(x,y)$ can be defined as $\phi(x,y) = \phi_0(x)\phi_0(y)$. The wavelet transform of a function $f(x,y) \in L^2(R^2)$ at the scale 2^j has two components defined by

$$W_{2^j}^1 f(x,y) = f * \psi_{2^j}^1(x,y) \text{ and } W_{2^j}^2 f(x,y) = f * \psi_{2^j}^2(x,y), \qquad (10)$$

where $\psi_{2^j}^1(x,y) = \frac{1}{2^j}\psi^1(\frac{x}{2^j}, \frac{y}{2^j})$ and $\psi_{2^j}^2(x,y) = \frac{1}{2^j}\psi^2(\frac{x}{2^j}, \frac{y}{2^j})$, $j \in Z$. Let $\chi_{2^j}^1 = \chi^1(2^j x, 2^j y)$ and $\chi_{2^j}^2 = \chi^2(2^j x, 2^j y)$ be the 2-D reconstructing wavelets whose Fourier transforms satisfy

$$\widehat{\chi}^1(2w_x, 2w_y) = e^{isw_x}K(w_x)L(w_y)\widehat{\phi}(w_x)\widehat{\phi}(w_y),$$
$$\widehat{\chi}^2(2w_x, 2w_y) = e^{isw_y}K(w_y)L(w_x)\widehat{\phi}(w_x)\widehat{\phi}(w_y),$$

where $L(w)$ is also a function with period 2π and satisfies $L(w) = (1+|H(w)|^2)/2$. Then $f(x,y)$ can be reconstructed from its wavelet transforms as follows

$$f(x,y) = \sum_{j=-\infty}^{+\infty} [W_{2^j}^1 f * \chi_{2^j}^1(x,y) + W_{2^j}^2 f * \chi_{2^j}^2(x,y)]. \qquad (11)$$

Define in the following a smoothing operator S_{2^j} as:

$$S_{2^j}f(x,y) = f * \phi_{2^j}(x,y) \text{ with } \phi_{2^j}(x,y) = \frac{1}{2^j}\phi(\frac{x}{2^j}, \frac{y}{2^j}). \qquad (12)$$

Then the 2-D dyadic wavelet transform of $f(x,y)$ can be defined as follows

$$\left\{ S_{2^J}f, (W_{2^j}^1 f)_{1\leq j\leq J}, (W_{2^j}^2 f)_{1\leq j\leq J} \right\}. \qquad (13)$$

In practice, an image is measured at finite resolution and can be taken as a finite energy 2-D discrete signal $\{d_{n,m}\}$. Then there exists a function $g(x,y) \in L^2(R^2)$ such that

$$S_1 g(n,m) = d_{n,m}, (n,m) \in Z^2. \qquad (14)$$

Let $H_j(w) = H(2^j w), G_j(w) = G(2^j w), K_j(w) = K(2^j w), L_j(w) = L(2^j w)$ and $\widetilde{H}_j(w) = \overline{H_j(w)}$. Fast algorithms for wavelet decomposition and reconstruction of $S_1 g(n,m)$ will be formulated as follows respectively

$$\begin{cases} W_{2^{j+1}}^1 g = \frac{1}{\lambda_j}S_{2^j}g * (G_j, D) \\ W_{2^{j+1}}^2 g = \frac{1}{\lambda_j}S_{2^j}g * (D, G_j) \ , 0 \leq j \leq J - 1 \\ S_{2^{j+1}}g = S_{2^j}g * (H_j, H_j) \end{cases} \qquad (15)$$

and

$$S_{2^{j-1}}g = \lambda_j W_{2^j}^1 g * (K_{j-1}, L_{j-1}) + \lambda_j W_{2^j}^2 g * (L_{j-1}, K_{j-1})$$
$$+ S_{2^j}g * (\widetilde{H}_{j-1}, \widetilde{H}_{j-1}), 1 \leq j \leq J, \qquad (16)$$

where λ_j is a real number, D is the Dirac filter, i.e. $D(w) = \begin{cases} 1, & \text{if } w = 0 \\ 0, & \text{if } w \neq 0 \end{cases}$, and $g * (A, B)$ denotes the separable convolution of rows and columns of g with the 1-D filters A and B respectively. The computation complexity of both algorithms are $O(N^2 \log(N))$.

3.2   Multiscale edge detection of images

For an image function $f(x, y)$, the modulus of the gradient vector is defined as

$$M_{2^j} f(x, y) = \sqrt{\|W_{2^j}^1 f(x, y)\|^2 + \|W_{2^j}^2 f(x, y)\|^2}. \tag{17}$$

Edge detection is to find the edge points where $M_{2^j} f(x, y)$ has a local maxima in the direction $W_{2^j}^1 f(x, y) + i W_{2^j}^2 f(x, y)$. This edge detection method is suitable to deal with noisy images([10]).

To reconstruct an image from its wavelet transform data at the edge points, the whole set of data must be computed first. Mallat et al([9]) gave an interpolation method for the approximation. The interpolation function is

$$\begin{cases} W_{2^j}^1 h(x, y) = \alpha_x e^{2^j x} + \beta_x e^{-2^j x}, \text{ for fixed } y \text{ and } x \in [x_0, x_1], \\ W_{2^j}^2 h(x, y) = \alpha_y e^{2^j y} + \beta_y e^{-2^j y}, \text{ for fixed } x \text{ and } y \in [y_0, y_1], \end{cases} \tag{18}$$

where $\alpha_x, \beta_x, \alpha_y$ and β_y satisfy

$$\begin{cases} \alpha_x e^{2^j x_0} + \beta_x e^{-2^j x_0} = W_{2^j}^1 f(x_0, y) \\ \alpha_x e^{2^j x_1} + \beta_x e^{-2^j x_1} = W_{2^j}^1 f(x_1, y) \end{cases} \text{ and } \begin{cases} \alpha_y e^{2^j y_0} + \beta_y e^{-2^j y_0} = W_{2^j}^2 f(x, y_0) \\ \alpha_y e^{2^j y_1} + \beta_y e^{-2^j y_1} = W_{2^j}^2 f(x, y_1). \end{cases}$$

To use these techniques in tomographic image processing, we will discuss how to compute the wavelet transform of a tomographic image from its Radon transform in the next section.

4   Wavelet transforms of tomographic images

Given a function $f(x, y)$, other than Donoho's Wavelet-Vaguelette decomposition([4]), one can also compute the wavelet transforms from its Radon transform Rf directly.

Theorem 2. *Given a filter $\psi(x, y) \in L^2(R^2)$ and an image function $f(x, y) \in L^2(R^2)$, then we have*

$$f * \psi(x, y) = F_2^{-1} |w| h(w) F_2 B(Rf)(x, y), \tag{19}$$

and

$$f * \psi_j(x, y) = F_2^{-1} |w| h_j(w) F_2 B(Rf)(x, y), j \in Z, \tag{20}$$

where Rf denotes the Radon transform of $f(x,y)$, B is the back projection operator defined as (6), F_2 and F_2^{-1} denote the 2-D Fourier and its inverse transform respectively, $\psi_j(x,y) = \psi(2^j x, 2^j y)$ and

$$h(w) = (\overline{F_2\psi})(-w_x, -w_y), \text{ and } h_j(w) = (\overline{F_2\psi_j})(-w) = \frac{1}{2^{2j}} h(\frac{w_x}{2^j}, \frac{w_y}{2^j}).$$

Proof. Since

$$f * \psi_j(x,y) = \int_{R^2} \hat{f}(w)\overline{\hat{\psi}_{(j,x,y)}(w)}dw$$

$$= \int_{R^2} \hat{f}(w)\overline{\hat{\psi}_j(-w)}e^{i2\pi(xw_x + yw_y)}dw$$

$$= F_2^{-1}\left(\hat{f}(w)\overline{\hat{\psi}_j(-w)}\right)(x,y), \tag{21}$$

by (9),

$$f * \psi_j(x,y) = F_2^{-1}|w|(\frac{1}{2^{2j}}(\overline{F_2\psi})(-\frac{w_x}{2^j}, -\frac{w_y}{2^j}))F_2 B(Rf)(x,y). \tag{22}$$

The formula (19) is a special case of (20).

If the filter $\psi_j(x,y)$ is replaced by $\psi_{2^j}^1(x,y), \psi_{2^j}^2(x,y)$ and $\phi_{2^j}(x,y)$, then $W_{2^j}^1 f, W_{2^j}^2 f$ and $S_{2^j} f$ can be computed by (20) respectively.

For a function $g(x,y) \in L^2(R^2)$, if the image $f(n,m)_{(n,m)\in Z^2}$ is considered as $S_1 g(n,m)$, $W_{2^j}^1 g, W_{2^j}^2 g$ and $S_{2^j} g$ can be computed by (15) and (20) by taking ψ_j as $H_j(x)H_j(y), G_j(x)D(y)$ and $D(x)G_j(y)$ respectively.

5   Reconstruction and edge detection algorithms for tomographic images

Given a set of detected projection data $\{p(s_n, \theta_m), 1 \le n \le N_s, 1 \le m \le N_a\}$, an algorithm for computing the wavelet transform of the tomographic image $f(n,m) = S_1 g(n,m), 1 \le n, m \le N_t$ is recommended as follows

Algorithm 1

1. Compute the back projection $B(Rf)$ and get an $N \times N$ matrix;
2. Compute $F_2 B(Rf)$ using 2-D FFT;
3. For each level j,
   (a) multiply the filter $|\cdot|h_j(\cdot)$, by taking ψ_j as $G_j(x)D(y), D(x)G_j(y)$ and $H_j(x)H_j(y)$ respectively
   (b) compute $W_{2^j}^1 g, W_{2^j}^2 g$ and $S_{2^j} g$ by 2-D IFFT;
4. Employ (16) to reconstruct $S_1 g(n,m) = f(n,m), 1 \le n, m \le N_t$.

In the algorithm, the back projection step which costs $O(N_a N_s^2)$ operations is employed only one time, and the filter step is done approximately $3\log_2(N)$ times. Thus the reconstruction and edge detection for tomographic images by algorithm 1 together with (15) and (16) has the same order of complexity as the typical Convolution Back Projection method.

The following algorithm is for edge detection in tomographic processing.

Algorithm 2

1, 2 and 3 are the same as those in algorithm 1;
4. For each level j,
  (a) compute $M_{2^j}g$ by (17),
  (b) find the edge points (x_i^j, y_i^j) at which $M_{2^j}g$ has local maxima in the direction $W_{2^j}^1 g + i W_{2^j}^2 g$.

The algorithms have been applied on the typical example Shepp-Logan Phantom([6]) and the numerical results are not presented here due to the limited space.

6 Conclusions

The wavelet multiscale edge detection method is suitable for tomographic image processing. The computation complexity is of the same order as the CBP method. Using algorithm 1, other wavelet methods in image processing can also be applied to tomographic images.

References

1. Chui, C. K., An introduction to wavelets. Academic Press (1992)
2. Daubechies, I.: Ten lectures on wavelets. SIAM, Philadelphia (1992)
3. Deans, S.R.: The Radon transform and some of its applications, John Wiley & Sons, Inc. (1983)
4. Donoho, D. L.: Nonlinear solution of linear inverse problems by wavelet-vaguelette decomposition. Technical Report No. **403**, Dept. of Statistics, Stanford University (1992)
5. Donoho, D. L., Johnstone, I. M., Kerkyacharian, G. and Picard, D.: Wavelet shrinkage: asymptopia?. J. Royal Stat. Soc., Series B **57** (1995) 301–369
6. Kak, A. C. and Slaney, M.: Principles of computerized tomographic imaging, IEEE press, New York (1988)
7. Kolaczyk, E. D.: A wavelet shrinkage approach to tomographic image reconstruction. Technical Report No. **403**, Dept. of Statistics, The University of Chicago (1995)
8. Lin, B.: Wavelet phase filter for denoising in tomographic image reconstruction(Ph. D. Thesis). Illinois Institute of Technology (1994)
9. Mallat, S. and Zhong, S.: Characterization of signals from multiscale edges, IEEE trans. Patt. Anal. Machine Intell. **14** (1992) 710–732
10. Mallat, S. and Hwang, W. L.: Singularity detection and processing with wavelets. IEEE trans. Inform. Theory **38** (1992) 617–645
11. Meyer, Y.: Wavelets and operators. Cambridge studies in advanced mathematics 37, Cambridge University Press (1992)
12. Meyer, Y.(translated by Ryan, R.) Wavelets Algorithms and Applications. SIAM, Philadelphia (1993)
13. Milanfar, P.: Geometric estimation and reconstruction from tomographic data(Ph. D. Thesis). Massachusetts Institute of Technology (1993)

Parallel Block Elimination Algorithm for Generalized Eigen-problem and It's Applications on Crystal Electronic Structure *

Jiachang Sun[1], Jianxin Deng[1], Jianwen Cao[1], Dingsheng Wang[2], and Li Jun[3]

[1] R & D Center for Parallel Software, Institute of Software,
Chinese Academy of Sciences. Beijing P.O. Box 2704 (47)
[2] Institute of Physics, Chinese Academy of Sciences, Beijing 100080.
[3] Institute of Applied Physics, Beijing University of Science and Technology,
Beijing 100083.
E-mail of Jiachang Sun: sun@par25t.ict.ac.cn;
E-mail of Jianxin Deng: deng@par25t.ict.ac.cn;
E-mail of Jianwen Cao : cao@par25t.ict.ac.cn.

Abstract. A class of parallel block elimination algorithms for solving large scale generalized eigen decomposition was proposed in this paper. The main idea of this new algorithm is called "Black Box", which is coding length independent. The parallel algorithm has been implemented on the distributed parallel system DAWNING-1000 made in China and it has been applied to analyse the crystal electronic structure of LBO, a nonlinear opitical crystal invented by Chinese scientists. The order of corresponding generalized eigen-decomposition problem is 1572. The iterative convergence, error analysis, parallel efficiency and numerical analysis of the new algorithm have been given.

Keywords: Parallel algorithm, Eigen-decomposition, Crystal, Electronic

1 Introduction

Many computing tasks arisen from computational physics, chemistry and biology are related to solve so-called generalized eigen-decomposition problems. [1], [2]. There are a lot of existing computing methods which may be used in this field. However, it still is a challenging problem to solve very large scale generalized eigen-decomposition problem, especially on distributed parallel systems. Our aim is to present a main framework for solving this task efficiently both in manual and computing effort.

The main idea of our proposed algorithm is to adopt the most efficient sequential solver, such as QR algorithm, to do block elimination as a "Black Box", which is coding length independent, then to do block data transportation for next elimination iterations. This procedure can be taken as a kind of parallel block Jacobi methods.

* Work supported by the National Research Center for Intelligent Computing Systems of China.

2 Parallel Block Iterative Elimination Procedure

Consider the following generalized eigen-decomposition problem

$$AX = BXD$$

where both A and B are Hermitian matrix with order n, moreover B is positive definite. Matrix X and D are unknown resulted eigenvectors and diagonal eigenvalues, respectively.

To deal with the large scale matrix problem on multi-processor system, we denote a partition

$$A = (A_{ij}), B = (B_{ij}), i, j = 1, 2, ..., m.$$

For simplicity of description, we suppose all submatrices are quadratic with order n/m and it is integer.

2.1 Parallel block iterative elimination

Step 1. Parallel block eliminate according to a given ordering. As an example, suppose the block number $m = 8$ and processor number $p = 4$, the following seven transformation groups are needed:

$$
\begin{aligned}
&1 \; (1,2) \; (3,4) \; (5,6) \; (7,8)\\
&2 \; (1,3) \; (2,4) \; (5,7) \; (6,8)\\
&3 \; (1,4) \; (2,3) \; (5,8) \; (6,7)\\
&4 \; (1,5) \; (2,6) \; (3,7) \; (4,8)\\
&5 \; (1,6) \; (2,5) \; (3,8) \; (4,7)\\
&6 \; (1,7) \; (2,8) \; (3,5) \; (4,6)\\
&7 \; (1,8) \; (2,7) \; (3,6) \; (4,5)
\end{aligned}
$$

where (i,j) represent the corresponding sub-problem solved by a "black box", the related matrices are A_{ii}, A_{ij}, A_{ji}, A_{jj} and B_{ii}, B_{ij}, B_{ji}, B_{jj}. This step reduces matrices A_{ij}, A_{ji} and B_{ij}, B_{ji} to be zero, and A_{ii}, A_{jj} to be diagonal simultaneously. Moreover, B_{ii} and B_{jj} become identity. The whole seven substeps are called a sweep on which each off-diagonal block has been iteratively elimination once. For example of transformation (1,2), the iterative procedure can be written as follows

$$
A^{(1)} = \begin{pmatrix} D_1 & 0 & \\ 0 & D_2 & A_{ij}^{(1)} \\ A_{ji}^{(1)} & & A_{ij} \end{pmatrix} = \begin{pmatrix} Q_1^H & 0 \\ 0 & I \end{pmatrix} \begin{pmatrix} A_{11} & A_{12} & \\ A_{21} & A_{22} & A_{ij} \\ A_{ji} & & A_{ij} \end{pmatrix} \begin{pmatrix} Q_1 & 0 \\ 0 & I \end{pmatrix}
$$

and

$$B^{(1)} = \begin{pmatrix} I_1 & 0 & \\ & & B_{ij}^{(1)} \\ 0 & I_2 & \\ B_{ji}^{(1)} & & B_{ij} \end{pmatrix} = \begin{pmatrix} Q_1^H & 0 \\ 0 & I \end{pmatrix} \begin{pmatrix} B_{11} & B_{12} & \\ & & B_{ij} \\ B_{21} & B_{22} & \\ B_{ji} & & B_{ij} \end{pmatrix} \begin{pmatrix} Q_1 & 0 \\ 0 & I \end{pmatrix}$$

where A_{ij}, A_{ji} and B_{ij}, B_{ji} represent the current values and

$$\begin{pmatrix} D_1 & 0 \\ 0 & D_2 \end{pmatrix} = Q_1^H \begin{pmatrix} A_{11} & A_{12} \\ A_{21} & A_{22} \end{pmatrix} Q_1, \begin{pmatrix} I_1 & 0 \\ 0 & I_2 \end{pmatrix} = Q_1^H \begin{pmatrix} B_{11} & B_{12} \\ B_{21} & B_{22} \end{pmatrix} Q_1,$$

$$A_{ij}^{(1)} = Q_1^H A_{ij}, B_{ji}^{(1)} = B_{ij}^H Q_1, Q^{(1)} = Q_0 Q_1$$

Step 2. Check the convergence. Compute off(A) and off(B):

$$off(A) = \sum_{i \neq j} \|A_{ij}\|_F^2 \quad off(B) = \sum_{i \neq j} \|B_{ij}\|_F^2$$

where $\|A\|^2 = \sum_{i,j} |a_{ij}|^2$.

If $off(A) \leq \delta$ and $off(B) \leq \delta$ hold for a given error δ, simultaneous, then the iteration procedure stops. Otherwise, go back to the step 1 to continue.

2.2 Convergence of Block Iterative Elimination

The convergence analysis of block Jacobi algorithm for matrix eigenvalue problem hasn't completely been done till now. It is very difficult to evaluate convergence of block Jacobi algorithm exactly. In this section, for our block iterative elimination method which uses ideas from block Jacobi algorithm, the convergence tendency is studied and the eigenvalue's appraisal is obtained. There are three theorems to describe them.

Theorem 1. Solving eigenvalue problem

$$AX = XD$$

which A is a Hermitian matrix using parallel block iterative elimination methods, the algorithm is asymptotic quadratic convergent.

Theorem 2. Assume $A^{(k)}$ is the matrix of k_{th} iterative step and written as $A^{(k)} = D + E$. At the time of $\|E\| \leq \delta$, the error of solving $AX = XD$ with parallel block iterative elimination method is as follows:

$$|d_i - \lambda_i| \leq \delta,$$

where $D = \text{diag}(d_1, d_2, ..., d_n)$.

Theorem 3. If there are the following forms of matrix A and B

$$A = \begin{pmatrix} D_1 & W \\ W^H & D_2 \end{pmatrix}, B = \begin{pmatrix} I & S \\ S^H & I \end{pmatrix}, W = \begin{pmatrix} W_1 & W_3 \\ W_4 & W_2 \end{pmatrix},$$

$$S = \begin{pmatrix} S_1 & S_3 \\ S_4 & S_2 \end{pmatrix}.$$

parallel block iterative elimination algorithm is convergent monotonically if and only if

$$\frac{Max(\sigma(S_1), \sigma(S_2))}{\sqrt{(1 - \sigma(S_1))(1 - \sigma(S_2))}} < \sigma(S)$$

where $\sigma(S)$ is the maximum singular value of matrix S .

3   Implementation on Dawning-1000

3.1   Introduction of Dawning-1000

Parallel processing system Dawning-1000 has been developed by the National Research Center for Intelligent Computing Systems (NCIC) of China in 1995 which represents one of the most advanced high-performance computer in China today. Its main technical features are uniquely-designed, stable and reliable wormhole communication chips, which extensively exploits the combination of asynchronous mode and synchronous mode, with higher message passing speed. Its main technical Targets are as follows:

Node number: total 36 nodes, among which there are 32 computing nodes (the CPU is i860xr), 2 service nodes and 2 I/O nodes;

Peak computing speed: 2.56 GFLOPS for single precision and 1.92 GFLOPS for double precision;

Memory volume: 1 GB(32 MB per node);

Harddisk volume: over 5 GB;

Exploiting wormhole 2D mesh communication network, with 4.8 GB/s total communication capacity;

The bandwidth between computing nodes and communication network is 2.8 Gb/s.

3.2   Algorithm Design based on MPP computer

Our algorithm is based on MPP computer using $p \times p$ processors(They are composed to be 2D meshy and named as Nodes(i,j) where i,j=0,1,...,p-1). Consider the following generalized eigenvalue problem:

$$A \times x = \lambda \times B \times x$$

Divide matrix A and B into $p \times p$ submatrices $(A_{(i,j)}, B_{(i,j)})$ and load them to Nodes(i,j) respectively. In order to use all memories of MPP computer effectively so as to compute large scale eigenvalue problem, our algorithm is designed as follows:

Step 1:   Nodes(i,i)(i=1,2,...,p) solve generalized eigenvalue problem which is parallelized:

$$A_{(i,i)} \times x_{(i,i)} = \lambda \times B_{(i,i)} \times x_{(i,i)}$$

Step 2:   Nodes(i,j)$(i \neq j)$ receive eigenvector $x_{(i,i)}$ and $x_{(j,j)}$ respectively, then implement matrix multiplication operation: $x^*_{(i,i)} \times A_{(i,j)} \times x_{(j,j)}$, where,$x^*_{(i,i)}$ is the conjugate transpose matrix of $x_{(i,i)}$.

Step 3:   Implement row and column block permutation of matrix A and B. The purpose is put submatrix (i,j)'s$(i \neq j)$ non-zero elements to submatrix (i,i) so as to call "black box" to eliminate them till zero. Each submatrix (i,j) is divided into 2×2 subblocks. Each permutation put one of Nodes(i,j)'s $(i \neq j)$ subblocks to diagonal Nodes(i,i)'s. When all subblocks are put to Nodes(i,i)'s and are eliminated by "black box" subroutine, we say one sweep has been done.

Step 4:   Check whether or not the result satisfies our request. If not, go to Step 1 to continue.

Generally, for $p \times p$ processors, 2p-1 parallel iterations construct one sweep. Numerical experiments show that, for generalized eigenvalue problem, about six sweeps are required in order to obtain good enough results. If matrix B is unit matrix, then only three sweeps is enough. The advantage of the algorithm is that it is easily to adopt well-established solver in one PE to obtain parallel programs, and the amount of programming is less. However, this algorithm costs more time to solve a problem than parallel QR algorithm.

The essence of parallel block elimination method concurrently reduces matrix A and B to diagonal form. Assume we have found matrix X, so that

$$X^*AX = \mathrm{diag}(d_1, d_2, ..., d_n)$$

$$X^*BX = \mathrm{diag}(1, 1, ..., 1)$$

In this case, we say that $d_1, d_2, ..., d_n$ is eigenvalues of the generalized eigenvalue problem, and $X = (X_1, X_2, ..., X_n)$ is eigenvectors of the problem. So, in above algorithm, from Step 1 to step 4, some non-diagonal elements of A and B is eliminated to zero. Iterations is going on until all non-diagonal elements of A and B is changed to zero, we say the algorithm converges and all eigenvalues and eigenvectors obtained.

This algorithm can be accelerated by a preconditioner. As we known, if B is a unit matrix, after three sweeps, we can get good enough results. However, if B is not a unit matrix, about six sweeps required. If we can find a preconditioner which reduces B to unit matrix within one sweep, then costs for two sweeps are saved. After the preconditioner is done, an eigenvalue problem is formed. the amount of calculations will much less than original problem. Theoretical analyses show that, with a preconditioner which reduces B to unit matrix within one sweep, the amount of calculations is only two-thirds of non-preconditioner algorithm. One of this preconditioner is $B^{1/2}$. Other preconditioner we have done is as follows (made for four processors):

(1) Apply "Black Box" to matrix B so that it is reduced to

$$\begin{pmatrix} I & S \\ S^* & I \end{pmatrix}$$

(2) Find the singular value decomposition of matrix S:

$$S = U^*\mathrm{diag}(\sigma_1, ...\sigma_s, 0, ...0)V$$

where both U and V are complex orthogonal matrix. In other words, finding U and V so that

$$\begin{pmatrix} U^\star & \\ & V^\star \end{pmatrix} \begin{pmatrix} I & S \\ S^\star & I \end{pmatrix} \begin{pmatrix} U & \\ & V \end{pmatrix} = \begin{pmatrix} I & diag(\sigma_1, ...) \\ diag(\sigma_1^\star, ...) & I \end{pmatrix}$$

(3) Implement row and column block permutation so that B is reduced to the matrix whose non-diagonal subblocks are zero.

(4) Implement block elimination method(Step 1 to Step 4, and recycle it until satisfied results are obtained).

4   Electronic structure of Optical Crystal LiB$_3$O$_5$

The optical crystal LiB$_3$O$_5$(LBO) has a complex orthorhombic cell with lattice constants a=8.47Å, b=5.13Å   and c=7.38Å.

Four fundamental ionic group $(B_3O_7)^{-5}$ are included in one unit cell. With the insertion of Li atom, the crystal has a network structure consisting of connecting $(B_3O_7)^{-5}$ groups.

There are five nonequivalent O and three nonequivalent B sites. Two of the B atoms, B(1) and B(3), form three-fold bonding with O, the other B atom, four-fold B(2), is bonded to its neighboring O atoms.

There are 36 atoms in one unit cell of the LBO crystal. The parallelized self-consistent Linearized Augmented Plane Wave method(LAPW)[7] with parallel eigen-problem solver discussed in section III are used to calculate the electronic structure. More than 44 LAPW's per atom are used to solve the semi-relativistic Kohn–Sham equation of the valence electrons.

The dimension of the eigen problems is thus about 1600.

The convergence of the self-consistent band calculation is assumed as the average root mean deviation between the input and output charge density is less than 5 me/$(a.u)^3$.

It is estimated that the time for solving the eigen problem is about 60% of the total time for physical iteration.

The calculated Valence-band (VB) and Conduction-band (CB) structures show that the LBO is a direct-gap material. The calculated band gap is 7.34eV at Γ, in quite good agreement with the measured gap value of 7.37eV. The total electronic density of states(DOS) and the atom-projected DOS are obtained by summation over the the data of 9 special k points in the BZ, and only the total DOS is depicted in Fig.2. The Fermi level has been shifted to zero in the figure. The DOS of VB and CB are located under or above zero energy respectively. The results show that the VB of the LBO crystal is mainly composed by the $2s$ and $2p$ orbitals of the O atoms. The $2s$ and $2p$ orbitals of the Li and B atoms are shifted upward into the CB because of their interaction with O atoms. It is worth mentioning that the valence orbitals of the B(2) atom, which have different bonding comparing with B(1) and B(3), are shifted upward so high that the projected DOS are almost zero within the given energy window. Possibly, this is an important reason resulting in the differences between the LBO and $\beta-BaB_2O_4$(BBO) crystals.

The charge density distribution of the LBO crystal in the plane containing three B atoms in the $(B_3O_7)^{-5}$ unit are shown in Fig.2, The bonding pattern between the B and O atoms can be seen clearly in the figure. Two B atoms, B(1) and B(3), forms the three-fold bonds with O atoms. Since the four nearest-neighbor O atoms bonded with B (2) are not in the same plane, only two of them can be seen. The other two are located above or under this plane respectively. These results are consistent with those obtained by Xu et al..[9] This shows the reliability of our parallel algorithm for solving matrix eigen problem and parallel LAPW code.

5   Conclusion

Solving large scale generalized eigen-decomposition is a difficult issue now. In this paper, we describe a new algorithm so-called parallel block elimination method which is based on the idea of "Black Box". The merit of this method is that it may take advantage of the currently most efficient sequential solver and is easy to be implemented. It can be taken as a kind of parallel block Jacobi methods and has all merits of Jacobi method. We analyse the convergence of this algorithm and describe its design based on distributed parallel systems. It has been applied to analyse the crystal electronics structure of LBO which is a nonlinear optical crystal. The problem is that this algorithm costs more time to solve a problem than parallel QR algorithm even though it is accelerated by a preconditioner. This algorithm need to be made better.

References

1. G.E.Forsythe and P.Henrici, The cyclic Jacobi method for computing the principle values of a complex matrix. Trans. Amer. Math. Soc. (1960), 1-23.
2. Gautam M. Shroff, A parallel algorithm for the eigenvalues and eigenvactors of a general complex matrix. Numer. Math. 58(1991), 779-805.
3. Deng Jian Xin, On parallel fast Jacobi algorithm for the eigen problem of real symmetric matrices,Journal of Computational Mathematics,7(1989),pp.412-417.
4. D.N.Parlett,The symmetric eigenvalue problem,1980 Prontice-Hall Inc.
5. W.f.Mascarenhas,On the convergence of the Jacobi method for arbitrary ordering, SIAM J.Matrix Anal.Appl. 16(1995),pp.1197-1209.
6. C.T. Chen et al, Inter. Rev. Phys. Chems. 2, 389(1989).
7. H.Krakauer and M.Posternak, and A. J. Freeman, Linearized augmented plane-wave method for the electronic band structure of thin films, Phys. Rev. B19, pp. 1706-1719(1979).
8. H. von Kronig and R. Hoppe, Zur Kenntnis von LiB_3O_5Z. Anorg. Allg. Chem, 439, 71-79(1978).
9. Yong-nan Xu and W.Y.Ching, Electronic structure and interatomic bonding of crystalline $\beta - BaB_2O_4$ with comparison to LIB_3O_5, Phys. Rev. B48, 17695-17701(1993);

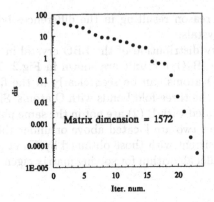

Fig. 1. Iterative Error Chart of Parallel Block Elimination Algorithm

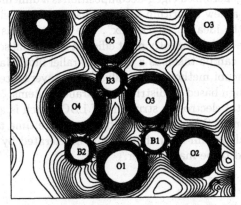

Fig. 2. The calculated total electronic density of states of the LBO crystal. The Fermi level has been shifted to zero in the figure.

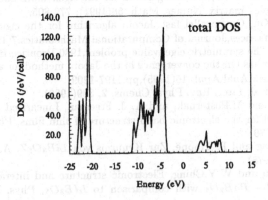

Fig. 3. The electronic charge density distribution of the LBO crystal in the plane containing three B atoms in the $(B_3O_7)^{-5}$ unit.

Adaptive Methods for Singular Perturbation Problems

Tao Tang[1] * and D. M. Sloan[2]

[1] Department of Mathematics, The Chinese University of Hong Kong, Hong Kong
[2] Department of Mathematics, The University of Strathclyde, Glasgow, Scotland.

1 Introduction

We consider numerical methods for the singular perturbation boundary value problem (BVP) given by

$$\begin{cases} -\epsilon u''(x) + p(x)u'(x) + q(x)u(x) = f(x), & x \in (a,b), \\ u(a) = \alpha, \qquad u(b) = \beta, \end{cases} \qquad (1)$$

where a, b, α, β are given constants, $\epsilon > 0$ denotes a fixed (small) constant. In many applications, (1) possesses boundary or interior layers, i.e. regions of rapid change in the solution near the endpoints or some interior points, with widths $o(1)$ as $\epsilon \to 0$. The problem (1) models many practical problems, such as linearized Navier-Stokes equations at high Reynolds number, the drift-diffusion equation of semiconductor device modeling, etc.

Many numerical methods have been proposed for the solutions of (1). Most of the methods require some of or all of the following information:

- **C1:** There are boundary and/or interior layers;
- **C2:** There are boundary layers only;
- **C3:** The positions and widths of the layers are known.

For some simple problems, the above information can be obtained by analyzing the given equations. The purpose of this paper is to investigate efficient methods which can handle more difficult problems. In Section 2, we discuss some methods for solving boundary layer problems. We begin with a quite well known method, i.e. the upwind finite difference with a Shishkin piecewise equidistance mesh (see, [7]). As will be shown below, this method requires the information (C2) and (C3). We then discuss a boundary layer resolving spectral method developed by Tang and Trummer [8]. The spectral method requires the information (C2) only and can resolve very thin boundary layers. In Section 3, we discuss an adaptive method developed by Mulholland et al. [5] that resolves interior layer problems.

We point out that the numerical methods considered below can be used to treat the relevant nonlinear two-point BVPs, e.g. see Example 2 in Section 2.

* Permanent address: Dept of Mathematics and Statistics, Simon Fraser University, Burnaby, B.C., Canada V5A 1S6. Supported by an NSERC Canada Grant.

2   Boundary layer problems

It is well known that (1) may possess boundary layers for some coefficient functions p, q and f. In particular, for the so-called Helmholtz type equation, $p(x) \equiv 0$ and $q(x) > 0$, the width of the boundary layer is $O(\sqrt{\epsilon})$. However, for general coefficient functions p, q and f, and especially for nonlinear two-point BVPs, the width of the boundary layer is unknown.

2.1 Upwind scheme with Shishkin mesh To understand the principal idea of the method, we consider the model problem:

$$-\epsilon u''(x) + p(x)u'(x) = f(x), \quad x \in (0,1); \quad u(0) = \alpha, \ u(1) = \beta, \quad (2)$$

where $p(x) > 0$ for all $x \in [0,1]$. The solution u has a boundary layer of width $O(\epsilon)$ of width $O(\epsilon)$ at $x = 1$. Let a mesh be $0 = x_0 < x_1 < \cdots < x_N = 1$. For each $j \geq 1$, we set $h_j = x_j - x_{j-1}$. Given a function $v = \{v_j\}$ defined on the mesh, define the following difference operators:

$$D_+ v_j = \frac{v_{j+1} - v_j}{h_{j+1}}, \quad D_- v_j = \frac{v_j - v_{j-1}}{h_j}, \quad D_+ D_- v_j = \frac{2(D_+ v_j - D_- v_j)}{h_{j+1} + h_j}. \quad (3)$$

Since ϵ is small, it is well known that on an equidistant mesh the use of central difference schemes for $u'(x)$ in (2) leads to nonphysical oscillations unless the mesh spacing is unsatisfactorily small. Upwind scheme, where D_- (or one of its variants) is used instead, is of the following form:

$$-\epsilon D_+ D_- v_j + p_j D_- v_j = f_j, \quad 1 \leq j \leq N - 1; \quad u_0 = \alpha, \ u_N = \beta. \quad (4)$$

A Shishkin piecewise equidistance mesh is given by setting

$$\tau = \min\{0.5, \epsilon \ln N\} \quad (5)$$

and by dividing each of the subintervals $[0, 1 - \tau]$ and $[1 - \tau, 1]$ into $N/2$ equal subintervals (see [7]). On the Shishkin mesh, the scheme (4) is almost first-order uniformly convergent with respect to ϵ, i.e. $\max_j |u(x_j) - u_j| \leq CN^{-1}\ln^2 N$, where the constant C is independent of ϵ and N. There have been extensive studies on the above mentioned method and mesh (see the recent books of Miller et al. [3] and Roos et al. [6]).

The above approach is useful for the simple problem like (2), for which the information (C2) and (C3) mentioned in Section (1) is provided. However, if (C3) is not given, i.e. if the order of the layer width is unknown, then Shishkin's method will not work. Consider the following example (**Example 1**):

$$-\epsilon u''(x) + (1 - x)u'(x) = f(x), \quad u(0) = \exp(-1/\sqrt{\epsilon}), \quad u(1) = \sin(1) + 1,$$

where $f(x) = (-1 + (1 - x)/\sqrt{\epsilon}) \exp(-(1 - x)/\sqrt{\epsilon}) + \epsilon \sin x + (1 - x) \cos x$. We solve the test problem with $\epsilon = 10^{-6}$ by using the Shishkin mesh (5). It is seen from Figure 1 that the boundary layer is not well resolved even with $N = 1000$. The reason for this is that we do not know the width of the boundary layer, which is not $O(\epsilon)$ for this test problem.

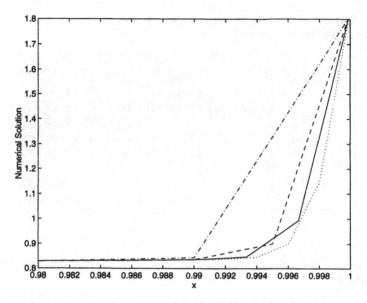

Fig. 1. The numerical solution for Example 1 with $N = 200$ (dashdot), 400 (dashed), 600 (solid) and 1000 (dotted).

2.2 Spectral method with coordinate stretching Without losing generality, we assume the solution interval is $(-1, 1)$. In [8] an effective procedure based on coordinate stretching and the Chebyshev pseudospectral method to resolve the boundary layers was introduced. In the following we briefly repeat the idea. We transform the singularly perturbed BVP (1) via the variable transformation $x \mapsto y(x)$ (or $x = x(y)$) into the new BVP

$$\epsilon v''(y) + P(y)v'(y) + Q(y)v(y) = F(y), \tag{6}$$

where v is the transplant of u, $v(y) = u(x(y))$. The transformed coefficients are

$$P(y) := \frac{p(x)}{y'(x)} + \epsilon \frac{y''(x)}{y'(x)^2}, \quad Q(y) := \frac{q(x)}{y'(x)^2}, \quad F(y) := \frac{f(x)}{y'(x)^2}, \tag{7}$$

where again $x = x(y)$. It is clear from (6)-(7) that for any variable transformation $x \mapsto y(x)$ the two quantities $1/y'(x)$ and $y''(x)/[y'(x)]^2$ are of interest and should be easy to calculate. We choose the transformation $x = x(y) := g_m(y)$, where

$$g_0(y) := y, \quad g_m(y) := \sin\left(\frac{\pi}{2} g_{m-1}(y)\right), \quad m \geq 1. \tag{8}$$

Then the two quantities $1/y'(x)$ and $y''(x)/[y'(x)]^2$ can be calculated easily by computer (a few lines by MATLAB; see Section 3 of [8]). Once the coefficient functions in (1) were supplied, less than 15 lines will be sufficient to form the transformed equation (6). Then we use the standard pseudospectral solver to solve (6).

Table 1. Maximum errors for Example 2 ('$\star$' indicates an error > 1 or convergence difficulties in the Newton process)

		$N = 32$	$N = 64$	$N = 128$	$N = 256$
	$m = 0$	$\star$	1.8144(-02)	3.6293(-04)	3.3776(-07)
$\epsilon = 10^{-3}$	$m = 1$	3.5818(-02)	4.5561(-04)	1.3573(-07)	3.4528(-14)
	$m = 2$	1.6063(-02)	3.6709(-04)	6.3134(-08)	4.9238(-14)
	$m = 1$	$\star$	$\star$	4.3554(-02)	1.3762(-03)
$\epsilon = 10^{-6}$	$m = 2$	$\star$	2.5004(-02)	2.4636(-03)	9.7656(-07)
	$m = 3$	$\star$	3.7734(-02)	7.1848(-04)	2.3793(-07)
	$m = 1$	$\star$	$\star$	$\star$	$\star$
$\epsilon = 10^{-9}$	$m = 2$	$\star$	$\star$	$\star$	6.1103(-03)
	$m = 3$	$\star$	$\star$	6.7784(-03)	1.0602(-04)

We consider a nonlinear example, namely the stationary Burgers' equation (**Example 2**):

$$\epsilon u''(x) + u(x)u'(x) = 0, \qquad x \in [-1, 1], \tag{9}$$

with boundary conditions chosen such that the function $u(x) = \tanh\{(x+1)/2\epsilon\}$ is an exact solution. This function has a boundary layer of width $O(\epsilon)$ at $x = -1$. The transformed equation with new variable $y = y(x)$ is simply

$$\epsilon v''(y) + \left[\frac{1}{y'(x)}v(y) + \epsilon\frac{y''(x)}{y'(x)^2}\right]v'(y) = 0, \qquad y \in [-1, 1]. \tag{10}$$

The solution is computed by Newton's method with $v \equiv 1$ as initial guess; for small values of the parameter ϵ a continuation procedure for ϵ to obtain better initial guesses is advisable. Table 1 lists the results for $\epsilon = 10^{-3}$, $\epsilon = 10^{-6}$ and $\epsilon = 10^{-9}$. Notice that we have a free parameter (positive integer) m in the transformation (8). This parameter plays an important role in the calculation: the value of m should be increased as the boundary layer width decreases. The procedure for obtaining reliable solutions is as following: fix a value of N, set $m = 1, 2, 3$ and plot the resulting solution, respectively. If the curves obtained by two different values of m are graphically indistinguishable (especially near the boundaries), then we regard the corresponding numerical solutions as good approximate solutions. If the three curves are different (especially near the boundaries), we increase the value of N. This procedure will end up with a satisfactory solution. In the above example, $N = 64$ solves Burgers' equation with $\epsilon = 10^{-3}$, $N = 128$ solves the problem with $\epsilon = 10^{-6}$, and $N = 512$ solves the problem with $\epsilon = 10^{-9}$.

Numerical experiments show that the above method is simple (the code is just a few lines longer than the standard spectral method code for (1)), robust (it can resolve very thin boundary layers), and does not require the information (C3). Theoretical error analysis for this method can be found in [2].

3   Interior layer problems

In this section we outline a method proposed in [5] that is applicable in situations where only information (C1) is applicable. The method makes use of adaptivity based on equidistribution. The idea in this approach is to place nodes in a manner that recognises solution variation: the node density is in some sense proportional to local computational error, or solution steepness. Nodes are placed by equidistributing a monitor function, M, over the computational domain, where M is commonly based on a scaled arc-length defined by

$$M(u(x), x) = \sqrt{1 + \alpha^2 (u_x)^2}. \tag{11}$$

Here α is a real parameter that determines the extent to which the solution gradient influences node placement. In its simplest form, nodes $a = x_0 < x_1 < \cdots < x_N = b$ are located using the condition

$$\int_{x_j}^{x_{j+1}} M(u(x), x)dx = \text{constant} = C, \quad j = 0, 1, \ldots, N - 1. \tag{12}$$

A discrete approximation of (12) readily yields the system of nonlinear algebraic equations

$$M_{j+\frac{1}{2}}(x_{j+1} - x_j) = M_{j-\frac{1}{2}}(x_j - x_{j-1}), \quad j = 1, 2, \ldots, N - 1, \tag{13}$$

where

$$M_{j+\frac{1}{2}} = \sqrt{1 + \alpha^2 \left(\frac{u_{j+1} - u_j}{x_{j+1} - x_j}\right)^2}$$

and u_j denotes an approximation to $u(x_j)$.

In practice, the system (13) is solved in conjunction with a discretisation of the boundary value problem that is under consideration. For example, if we wish to solve (2) with the nodal locations computed adaptively, we may solve (3.3) combined with

$$-\epsilon D_+ D_- u_j + p_j D_- u_j + q_j u_j = f_j, \quad j = 1, 2, \ldots, N - 1, \tag{14}$$

where D_+ and D_- are defined by (3). Equations (13) and (14) represent a set of $2(N - 1)$ equations for the unknowns $\{x_j\}_{j=1}^{N-1}$ and $\{u_j\}_{j=1}^{N-1}$. To obtain a useful grid, an element of smoothing is incorporated into (13), and details of the smoothing process are described in [5]. It is also convenient to solve the nonlinear system (13) and (14) using continuation in the parameters ϵ and α, and this is also described by Mulholland *et al.* [5].

The adaptive process that is outlined above effectively constructs a discrete map between the physical coordinate x and a computational coordinate ξ, where x and ξ are defined on $[-1, 1]$. The values $\{x_j\}_{j=0}^{N}$ given by the adaptive process

Table 2. Maximum errors in the computed solution of (18) with $\epsilon = 10^{-6}$, using $N = 32, \alpha = 4, m = 64, \delta = 32$ and $\gamma = 4$.

J	maximum error	J	maximum error
16	8.913(-3)	20	4.472(-3)
32	1.122(-4)	40	9.967(-6)
64	4.309(-8)	80	7.920(-9)
128	1.830(-12)	160	1.781(-13)

may be considered to be values related to evenly spaced nodes in the ξ coordinate. Thus,

$$x_j = x(\xi_j) = x\left(-1 + \frac{2j}{N}\right), \quad j = 0, 1, \cdots, N. \tag{15}$$

The information contained in this data set could be used in a prolongation process to construct a continuous map $x = x(\xi)$ relating ξ and x. Also, this map will mimic the features of the computed solution, so it will contain information related to boundary or interior layers that are present in the solution. The map is analogous to that described by (8), but the map presented in this section is more flexible since it is constructed without prior knowledge of the solution features. The adaptively constructed map has been used to great effect by Mulholland et al. [5] in forming a pseudospectral post-processing scheme for singular perturbation problems.

We construct a polynomial approximation to the map (15). First, the BVP is solved using (13) and (14) on a fairly coarse grid. This provides values of x at the evenly spaced nodes $\xi_j = -1 + 2j/N, j = 0, 1, \cdots, N$. To construct a polynomial approximation of degree m, the values $\{\bar{x}_j\}_{j=0}^{m}$ that approximate x at the Chebyshev nodes $\bar{\xi}_j = -\cos\frac{\pi j}{m}, j = 0, 1, \cdots, m$ are obtained by linear interpolation on the solution $\{x_j\}_{j=0}^{N}$ computed on the evenly-spaced nodes. The transformation $x = x(\xi)$ is approximated by

$$x(\xi) = \sum_{k=0}^{m} a_k T_k(\xi), \tag{16}$$

and the interpolatory conditions $x(\bar{\xi}_j) = \bar{x}_j, j = 0, 1, \ldots, m$ permit the evaluation of the coefficients $\{a_k\}_{k=0}^{m}$. The map (16) is then replaced by a smoothed map $x(\xi) = \sum \sigma_k a_k T_k(\xi)$, where σ_k is a filter function which may take the form $\sigma_k = \exp[-\delta(k/m)^\gamma]$, with δ and γ representing positive real numbers.

Once the smooth map (16) is available the BVP is transformed to the new independent variable ξ and solved by a standard pseudospectral method. This part of the solution process is similar to that described in Section 2.2. The dependent variable is approximated at the Chebyshev nodes $\xi_j = -\cos(\pi j/J), j = 0, 1, \cdots, J$.

Consider the problem (**Example 3**):

$$\epsilon u''(x) + 2xu'(x) = 0, \quad x \in (-1, 1); \quad u(\pm 1) = \pm 1. \tag{17}$$

This problem has a steep interior layer of width $O(\sqrt{\epsilon})$ at $x = 0$ (however, we do not use this information in the calculations!!). The equation obtained by transformation using the map $x = x(\xi)$ is

$$\epsilon v''(\xi) + [2x\, x_\xi - \epsilon\, x_{\xi\xi}/x_\xi]\, v'(\xi) = 0, \quad \xi \in (-1, 1); \quad v(\pm 1) = \pm 1, \qquad (18)$$

where $v(\xi) \equiv u(x(\xi))$. This problem was solved using the adaptive scheme with pseudospectral post-processing. Details of the smoothing and filtering parameters are given in Mulholland et al. [5]. Here we give only the parameter subset that has been referred to above. Table 2 gives results for the maximum pointwise error at various values of J. Note the rapid convergence as J increases.

4   Comments

We have reviewed several numerical methods for the approximate solution of singularly perturbed two-point boundary value problems in one dimension. Another method that is of interest here is that proposed by Huang and Sloan [1]. These authors outline a pseudospectral upwind method for boundary layer problems that converges spectrally outside the boundary layer. The method differs from the others considered here in the sense that it does not resolve the boundary layer. The method outlined in Section 3 has been applied to steady problems in two space dimensions and to unsteady problems in one space dimension. Results on uniform convergence of the numerical solution with respect to the perturbation parameter have been obtained for Shishkin grids [3,6]. Recently, results have been obtained on uniform convergence on grids that are produced by adaptivity based on equidistribution [4].

References

1. W-Z Huang and D. M. Sloan, *A new pseudospectral method with upwind features*, IMA J. Numer. Anal. **13** (1993), 413-430.
2. W. B. Liu and T. Tang, Error analysis for a new Galerkin spectral method, Research Report 94-04, Dept of Mathematics, Simon Fraser University, B.C., Canada.
3. J. J. H. Miller, E. O'Riordan and G. I. Shishkin, *Solution of Singularly Perturbed Problems with ϵ-uniform Numerical Methods – introduction to the theory of linear problems in one and two dimensions.* World Scientific, 1996.
4. Y. Qiu, D. M. Sloan and T. Tang, Convergence analysis for an adaptive finite difference solution of a variable coefficient singular perturbation problem, Strathclyde Mathematics Research Report 42, Univ of Strathclyde, 1996.
5. L. S. Mulholland, W-Z Huang and D. M. Sloan, *Pseudospectral solutions of near-singular problems using numerical coordinate transformations based on adaptivity.* To appear in SIAM J. Sci. Comput.
6. H.-G. Roos, M. Stynes, and L. Tobiska, *Numerical Methods for Singularly Perturbed Differential Equations.* Springer-verlag, Berlin, 1996.
7. G. I. Shishkin, *Grid approximation of singularly perturbed elliptic and parabolic equations.*, Second Doctoral Thesis, Keldysh Institute of Applied Mathematics, USSR Academy of Sciences, Moscow, 1990.
8. T. Tang and M. R. Trummer, *Boundary layer resolving pseudospectral methods for singular perturbation problems*, SIAM J. Sci. Comput., **17**, (1996), 430-438.

Image Reconstruction from Projections Based on Wavelet Decomposition *

Yin Jiahong

Applied Math. Dept., State University of Campinas, 13083-970 SP, Brasil

Abstract. In image reconstruction, the most commonly used methods are filtered backprojection algorithms (FBP), derived from different implementations of the inverse Radon transform. In this paper, we discuss FBP based on wavelet decomposition. Multiresolution image reconstruction can be performed using wavelet decomposition. Because of the fact that the projections are expanded in a given wavelet basis, we can calculate the interpolations needed for FBP in advance so that the computation can be reduced.

Key words: Image Reconstruction, Filtered Backprojection, Wavelet Decomposition.

1.Introduction

Wavelets have received much attention by mathematicians and engineers in the past few years. Recently, several researchers have used wavelets in image reconstruction [1]-[5]. The advantages of using wavelet decomposition is that a multiresolution image reconstruction can be given, which is desirable for many practical problems.

Bhatia and etc.[1] present a wavelet-based method for multiscale tomographic reconstruction. They represent projections in a wavelet basis and give the standard ramp-filter operator of filtered back-projection (FBP) reconstruction in the wavelet basis. The resulting multiscale representation of the ramp-filter matrix operator is approximately diagonal. This wavelet-based representation can be used to formulate a multiscale image reconstruction. They represent the object in the projection domain in a wavelet basis. This has the advantage that their multiscale basis representation of the object is closer to the measurement domain and their multiscale reconstruction has the same computational complexity as the FBP reconstruction method. They use a maximum a posteriori probability model to estimate the projection data from the noisy data, using a similar FBP method, at no additional cost. Reference [2] discussed the wavelet localization of the Radon transform. Using this, references [2] and [3] presented algorithms which significantly reduce radiation exposure in X-ray tomography when a local region of the body is to be imaged. The key to their algorithms

* This work was supported by FAPESP Grant 96/000837-0.

is that the space-frequency localization property of many wavelets is essentially presevered after a Hilbert transform. Sahiner and Yagle [4] use the wavelet transform to perform spatially-varying filtering by reducing the noise energy in the reconstructed image over regions where high-resolution features are not present. Wu [5] presents a Bayesian image-reconstruction approach for positron emission tomography using a wavelet decomposition.

In this paper, we will express the projections at a fixed view as a linear combination in the wavelet basis, which is similar to the idea in [1]. Therefore, this algorithm can provide the multiresolution reconstruction from projections at each l-level resolution ($l = 0, \ldots, L$). Then, we will calculate the filter (derivation and Hilbert transform) of the wavelet basis. In fact, they can be calculated in advance so that it is not necessary to interpolate them for the back-projection operation. In algebraic reconstruction algorithms, one expresses the object which is to be reconstructed as a combination of basis functions. For any given basis function, one can calculate its projections at any fixed view, so a system of equations which includes unknown coefficients arises. The problem of image reconstruction becomes the solution of this linear system of equations . The idea in this paper is that, since the projections have been expressed as a linear combination in a given wavelet basis, then we can deal with the wavelet basis operations in advance, which is similar to the idea in algebraic reconstruction methods.

This paper is organized as follows: section 2 describes the FBP method for image reconstruction, section 3 reviews the essentials of the wavelet representation, section 4 presents the new approach to a wavelet based FBP method and the conclusions and future work are presented in section 5.

2. FBP Reconstruction from Projection

The inverse Radon transform problem is to reconstruct an image $f(x, y)$ from its projection $p(\theta, t)$, where,

$$p(\theta, t) = \Re(f(x, y))$$
$$= \int_{-\infty}^{+\infty} \int_{-\infty}^{+\infty} f(x, y)\delta(t - x\cos\theta - y\sin\theta)dxdy \qquad (1)$$

is the Radon transform of $f(x, y)$ (or line integral of $f(x, y)$).

A common method for approximately obtaining $f(x, y)$ from $p(\theta, t)$ is filtered backprojection (FBP), in which, first, the projections at the fixed view (angle) θ are filtered to yield $Q(\theta, t)$. The filtering operation approximates a derivation operation and the Hilbert transform H of $p(\theta, t)$ with respect to the variable t. That is,

$$Q(\theta, t) = H(\frac{\partial}{\partial t}p(\theta, t)) = \frac{1}{\pi t} \star [\frac{\partial}{\partial t}p(\theta, t)] \qquad (2)$$

where $\star$ is the convolution operation. We denote the filter operation as F,then

$$Q(\theta, t) = F(p(\theta, t)) \qquad (3)$$

Second, the image $f(x, y)$ is then obtained by the backprojection of $Q(\theta, t)$. If we express the image in terms of polar coordinates (r, ϕ), for convenience, we will describe the image as the same function $f(r, \phi)$ and the image contains J pixels. Then, the value of the j-th pixel is

$$f(r_j, \phi_j) = \frac{1}{2\pi} \int_0^\pi Q(\theta, r_j \cos(\theta - \phi_j)) d\theta, 1 \le j \le J \qquad (4)$$

In practical problems, we have only projections at N_θ views and for a fixed view we have $(2N + 1)$ sampling values of projections. Let $\theta = m\Delta$, $t = nd$, where Δ is the angle sampling interval, d is the space sampling interval. So, $p(m\Delta, nd), 0 \le m \le N_\theta - 1, -N \le n \le N$ are given and from (2) we can calculate $Q(m\Delta, n'd), 0 \le m \le N_\theta - 1, -N \le n' \le N$, and we have

$$f(r_j, \phi_j) = \Delta \sum_{m=0}^{N_\theta - 1} Q(r_j \cos(m\Delta - \phi_j)) \qquad (5)$$

Interpolation is needed to estimate the values of $\{Q(m\Delta, r\cos(m\Delta - \phi))\}$ from $\{Q(m\Delta, n'd)\}$. If we use linear interpolation, we select n such that $nd \le r_j \cos(m\Delta - \phi_j) < (n+1)d$; then, $Q(m\Delta, r_j \cos(m\Delta - \phi_j))$ is estimated by

$$Q(m\Delta, r_j \cos(m\Delta - \phi_j)) = \frac{(n+1)d - r_j \cos(m\Delta - \phi_j)}{d} Q(m\Delta, nd) \\ + \frac{r_j \cos(m\Delta - \phi_j) - nd}{d} Q(m\Delta, (n+1)d) \qquad (6)$$

Because the contribution to $f(r, \phi)$ of the filtered projection data $Q(m\Delta, n'd)$, for any one view (i.e., m fixed, n varies), can be calculated from that view alone, and (5) need not be repeated for every $(r_j, \phi_j), j = 1, \ldots, J$ at which $f(r, \phi)$ is evaluated, the computer algorithm is as follows: first set

$$f_0(r_j, \phi_j) = 0, 1 \le j \le J \qquad (7)$$

for each value of $m, 0 \le m \le N_\theta - 1$, then the $(m+1)$-th picture can be produced from the m-th picture by a two-step process. First, calculate $\{Q(m\Delta, n'd), -N \le n' \le N\}$ from $\{p(m\Delta, nd), n \le n \le N\}$, then, let

$$f_{m+1}(r_j, \phi_j) = f_m(r_j, \phi_j) + \Delta Q(m\Delta, r_j \cos(m\Delta - \phi_j)), 1 \le j \le J \qquad (8)$$

Reference [3] discusses the implementation of FBP. The authors show that a typical high-resolution reconstruction using linear interpolation requires about N_θ^3 multiplications to implement (5), where, N_θ is the number of projection angles. Therefore, backprojection is usually the most computationally expensive part of the FBP algorithm by far. For example, to reconstruct a 512×512 image requires about 134×10^6 multiplications for all the backprojections, and only about 16×10^6 multiplications to filter the projections (using an FFT routine for real data).

3. The Wavelet Representation

The orthonormal basis of compactly supported wavelets of $L^2(R)$ is formed by the dilation and translation of a single function $\psi(t)$

$$\psi_k^l(t) = 2^{-\frac{l}{2}}\psi(2^{-l}t - k) \tag{9}$$

where $l, k \in Z$. The function $\psi(t)$ has a companion, the scaling function $\varphi(t)$,

$$\varphi_k^l(t) = 2^{-\frac{l}{2}}\varphi(2^{-l}t - k) \tag{10}$$

and these functions satisfy the following relations:

$$\varphi(t) = \sqrt{2} \sum_{k=0}^{K-1} h_k \varphi(2t - k) \tag{11}$$

$$\psi(t) = \sqrt{2} \sum_{k=0}^{K-1} g_k \varphi(2t - k) \tag{12}$$

where

$$g_k = (-1)^k h_{K-k-1}, k = 0, \ldots, K - 1 \tag{13}$$

and

$$\int_{-\infty}^{+\infty} \varphi(t)dt = 1 \tag{14}$$

In addition, the function ψ has M vanishing moments for some M,

$$\int_{-\infty}^{+\infty} \psi(t)t^m dt = 0, m = 0, \ldots, M - 1 \tag{15}$$

The number K of coefficients in (11) and (12) is related to the number of vanishing moments M, and for the wavelets in [6], $K = 2M$.

The wavelet basis induces a multiresolution analysis on $L^2(R)$, the decomposition of $L^2(R)$ into a chain of closed subspaces:

$$\ldots \subset V_2 \subset V_1 \subset V_0 \subset V_{-1} \subset V_{-2} \subset \ldots \tag{16}$$

such that

$$\bigcap_{l \in Z} V_l = \{0\}, \overline{\bigcup_{l \in Z} V_l} = L^2(R) \tag{17}$$

By defining W_l as the orthogonal complement of V_l in V_{l-1}

$$V_{l-1} = V_l \bigoplus W_l \tag{18}$$

the space $L^2(R)$ is represented as a direct sum

$$L^2(R) = \bigoplus_{l \in Z} W_l \qquad (19)$$

On each fixed l-level resolution, the wavelets $\psi_k^l(t), k \in Z$ form an orthonormal basis of W_l and the functions $\varphi_k^l(t), k \in Z$ form an orthonormal basis of V_l. for any function $p(t) \in L^2(R)$, we have the following wavelet decomposition

$$p(t) = \sum_{l,k} < p(t), \psi_k^l(t) > \psi_k^l(t) \qquad (20)$$

where $<,>$ denotes the standard inner product in $L^2(R)$.

4. The Algorithm Based on Wavelet Decomposition

We use FBP to reconstruct the image. First, we will express the projection at a fixed view as in a given wavelet basis, that is,

$$p(\theta, t) = \sum_{l,k} c_k^l(\theta) \psi_k^l(t) \qquad (21)$$

Then, filtering the data $p(\theta, t)$ as a function of t gives

$$Q(\theta, t) = F[p(\theta, t)] = F[\sum_{l,k} c_k^l \psi_k^l(t)] = \sum_{l,k} c_k^l(\theta) F[\psi_k^l(t)] \qquad (22)$$

Backprojecting, we get

$$f(r, \phi) \approx \Delta \sum_{m=0}^{N_\theta - 1} Q(m\Delta, r \cos(m\Delta - \phi))$$
$$= \Delta \sum_{m=0}^{N_\theta - 1} \sum_{l,k} c_k^l(m\Delta) F(\psi_k^l)(r \cos(m\Delta - \phi)) \qquad (23)$$

The FBP algorithm has the following steps:
1). The wavelet decomposition of the projection

$$p(\theta, t) = \sum_{l,k} c_k^l(\theta) \psi_k^l(t) \qquad (24)$$

2). Compute the filtered wavelet basis $F[\psi_k^l(t)]$
3). Backprojection :

$$f(r, \phi) \approx \Delta \sum_{m=0}^{N_\theta - 1} \sum_{l,k} c_k^l(m\Delta) F(\psi_k^l)(r \cos(m\Delta - \phi)) \qquad (25)$$

Now, we give a further explanation of the algorithm above.
• The computation of $F[\psi_k^l(t)]$. For a given wavelet basis, we can calculate the filter $\psi_k^l(t)$ in advance. Reference [2] shows that the support of $F[\psi_k^l(t)]$ will

have essentially the same support as $\psi_k^l(t)$, So, we will get the filter function of the wavelet basis on the corresponding support. Reference [1] shows that matrix WRW^T (in that paper) is nearly diagonal. This is equivalent to considering $F[\psi_k^l(t)] \approx d_k^l \psi_k^l(t)$. Therefore, we can get the filtered wavelet basis easily. From (25)

$$f(r, \phi) \approx \Delta \sum_{m=0}^{N_\theta - 1} \sum_{l,k} c_k^l(m\Delta)[d_k^l \psi_k^l(r\cos(m\Delta - \psi))] \qquad (26)$$

where $\{d_k^l\}$ are the diagonal elements of the matrix WRW^T.

•To obtain a near diagonalization of the matrix WRW^T, we need to choose the wavelet as the Daubechies wavelet with several vanishing moments (for example M=6). On the other hand, reference [6] gives the representation of some operators (derivation , Hilbert transform and Shift) compactly in wavelet basis by algebraic methods. We can use a similar approach to get the representation of the filter operator F in the wavelet basis.

• In general, it is necessary to interpolate the filtered projection for backprojection, and for different views, the filtered projections are different, We have to interpolate all the filtered projections. While here, for any view, $F[\psi_k^l(t)]$ is the same, we can use the same function values for different views. We can interpolate $F(\psi_k^l)(r\cos(m\Delta - \phi))$ from $\{F(\psi_k^l(nd)), -N_l \le n \le N_l\}$ in advance. The number N_l depends on the support of $\psi_k^l(t)$.

• In practical problems, we have only discrete projection sampling values. We denote the projection at l-level resolution as $p^l(\theta, t)$, then the representation of the projection at the coarsest resolution is

$$p^0(\theta, t) = \sum_k c_k^0 \varphi_k^0(t) \qquad (27)$$

the detail of the projection at the l-level resolution will be denoted as

$$d^l(\theta, t) = \sum_k c_k^l \psi_k^l(t) \qquad (28)$$

and the representation of the projection at the $(l+1)$-level resolution will be

$$p^{l+1}(\theta, t) = p^l(\theta, t) + d^l(\theta, t) \qquad (29)$$

$$p^L(\theta, t) = p(\theta, t) \qquad (30)$$

If we take only the projection $p^l(\theta, t)$ from $p(\theta, t)$, then

$$f(r, \phi) \approx \Delta \sum_{m=0}^{N_\theta - 1} \sum_{i=0}^{l} \sum_k c_k^l(m\Delta) F(\psi_k^l)(r\cos(m\Delta - \phi)) \qquad (31)$$

will be the reconstructed image at the l-level resolution.

Similarly, if we take $p(\theta, t) = d^l(\theta, t)$, we will get the image at the l-level resolution. And so on.

5. Conclusion

In this paper, we present an image-reconstruction algorithm based on wavelet decomposition. The advantages are that we can get the multiresolution image reconstruction, and we can calculate the filter of the given wavelet basis in advance so that we can reduce the computation. In a future paper, we will give a simulation example using the algorithm in this paper. We will also estimate the bound of the elements of the matrix WRW^T of reference [1].

References

1. M. Bhatia, W.C. Karl and A.S. Willsky, " A Wavelet-Based Method for Multiscale Tomographic Reconstruction.", IEEE, Trans. Medical Imaging. Vol. 15, No. 3. 1996.
2. T. Olson and J. DeStefano, " Wavelet Localization of the Radon Transform", IEEE Trans on Signal Processing, Vol. 42, No. 8, P 2055-2067, 1994
3. Alexander H. Delaney and Yoram Bresler,"Multiresolution Tomographic Reconstruction Using Wavelet", IEEE Trans on Image Processing, Vol. 4, No. 6, P799-813, 1995.
4. B. Sahiner and A.E. Yagle, "Image Reconstruction From Projections Under Wavelet Constraints", IEEE, Trans, Signal Proc. Vol. 41, No. 12, Dec, 1993.
5. Zhenyu Wu, "MAP Image Reconstruction Using Wavelet Decomposition", Proc. of XIIIth Int. Conf. on Info, IPMI 93, June, 1993.
6. G. Beylkin, " On the Representation of Operators in Bases of Compactly Supported Wavelets", SIAM J. Numer. Anal, Vol. 6, No. 6, P1716-1740, 1992
7. S. Mallat, " A Theory for Multiresolution Signal Decomposition: The Wavelet Representation", IEEE Trans Pattern Anal. Machine Intell. Vol. 11, No. 7, P674-693, 1989
8. G. T. Herman, Image Reconstruction from Projections. New York, Academic Press, 1980
9. A. C. Kak and M. Slaney, Principles of Computerized Tomographic Imaging, IEEE Press, 1988.

Adaptive Image Smoothing Respecting Feature Directions

Sifen Zhong

Statistics and Operations Research Program
CEOR, Princeton University
Princeton, NJ 08544, USA

Abstract. Accurate extraction of the image feature directions is essential to image smoothing and other image processing tasks. We show that the gradient based feature direction extraction method can be very erroneous. The gradient is too local, and it cannot detect oscillations. We have developed two new methods: the Hessian method, an approach using higher order differentiation, and the Gabor method, an approach using local spectral analysis.

1 Adaptive Smoothing

In order to smooth the images without damaging the image features such as edges, lines, and textures, the smoothing has to be adaptively controlled with two principles: (1) as the adaptivety respecting the feature strengths, the smoothing should be small where the feature is strong; (2) as the adaptivety respecting the feature directions, the smoothing should be in the direction along the feature, not across the feature. While both principles are crucial, this paper is mostly concerned with the adaptivety respecting feature directions.

A classic example of adaptive smoothing is the Perona-Malik equation [2]:

$$u_t = \mathbf{div}(\mathbf{g}(|\nabla u|)\nabla u),$$

or, written in the η-ξ form:

$$u_t = \mathbf{g}(|\nabla u|)\left(\left(1 + \frac{|\nabla u|\,\mathbf{g}'(|\nabla u|)}{\mathbf{g}(|\nabla u|)}\right)u_{\eta\eta} + u_{\xi\xi}\right),$$

where $u_{\eta\eta}$ and $u_{\xi\xi}$ respectively denote the second order directional derivatives in the direction of η, the gradient direction, and in the direction of ξ, the contour direction, which is the direction perpendicular to the gradient direction.

The Perona-Malik equation in the η-ξ form demonstrates the two principles of adaptive smoothing. The first factor $\mathbf{g}(|\nabla u|)$ is responsible for the adaptivety respecting the feature strength. For the adaptivety respecting the feature direction, the factor $(1 + |\nabla u|\,\mathbf{g}'(|\nabla u|)/\mathbf{g}(|\nabla u|))$ in the η direction should allow little smoothing where $|\nabla u|$ is large, while the factor in the ξ direction is always 1, giving maximal smoothing.

A general evolution equation for adaptive smoothing is of the form:

$$u_t = c\,(a\,u_{\eta\eta} + b\,u_{\xi\xi}).$$

The functions a, b, and c are not limited to functions of $|\nabla u|$. More importantly, and this is the main concern of this paper, the η-ξ pair is not limited to the gradient-contour pair, since the gradient-contour pair can be a poor estimation of the across-along feature direction pair.

2   Gradient Method

In the Perona-Malik equation, the gradient direction is considered to be the direction across the feature, and the contour direction is considered to be the direction along the feature. This gradient based method, although widely used, can be very erroneous.

Consider the images of three typical features:

$$\text{edge}(x,y) = \begin{cases} \alpha x & \text{if } |x| < \frac{1}{\alpha}, \\ \text{sign}(x) & \text{otherwise;} \end{cases}$$

$$\text{line}(x,y) = \begin{cases} 1 - |\alpha x| & \text{if } |x| < \frac{1}{\alpha}, \\ 0 & \text{otherwise;} \end{cases}$$

$$\text{wave}(x,y) = \sin(\alpha x).$$

The contour directions agree with the feature directions because there is absolutely no variation in the y direction. See Fig. 1.

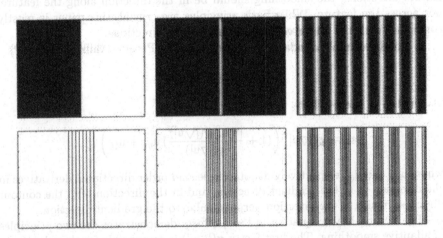

Fig. 1. top row: The images of three typical image features: edge, line, and wave. **bottom row:** The contours of the edge (zoom in), line (zoom in), and wave images.

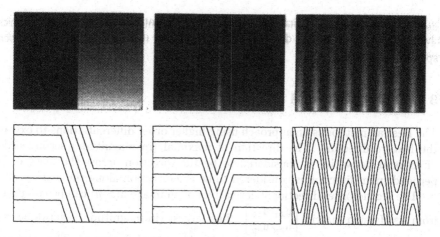

Fig. 2. top row: A plane with small slope is added to the edge, line, and wave images. **bottom row**: The contours of the edge (zoom in), line (zoom in), and wave images. The contour directions are perpendicular to the feature direction at many locations.

If a small variation is introduced, for example, by adding a plane with small slope in the y direction:

$$\mathtt{plane}(x, y) = \beta y,$$

then the contour directions will change while the feature directions remain the same, See Fig. 2.

In the edge image,

$$\widetilde{\mathtt{edge}}(x, y) = \mathtt{edge}(x, y) + \mathtt{plane}(x, y),$$

the edge is in direction $[0, 1]$, but the gradient is $[\alpha, \beta]$. Thus the contour is in direction $[-\beta/\alpha, 1]$. The contour direction does not agree with the edge direction unless $\beta = 0$ or $\alpha = \infty$.

The situation is much worse in the wave image:

$$\widetilde{\mathtt{wave}}(x, y) = \mathtt{wave}(x, y) + \mathtt{plane}(x, y).$$

The pattern is in direction $[0, 1]$, but the contour is in direction $[-\beta/(\alpha \cos(\alpha x)), 1]$. At many locations where $\cos(\alpha x)$ is zero, the contour directions are actually perpendicular to the pattern direction.

The situation is no better in the line image:

$$\widetilde{\mathtt{line}}(x, y) = \mathtt{line}(x, y) + \mathtt{plane}(x, y).$$

At the peaks of the line, the contour directions are perpendicular to the line direction.

A fundamental problem with the gradient method is that the gradient is too local to capture the direction information. In particular, the gradient cannot detect the oscillations, which are the most significant characteristics of features. The oscillations are better detected by higher order differentiations and by local spectral analysis.

3 Hessian Method

The Hessian method is an approach using higher order differentiations to extract the feature direction. The direction of maximal (in magnitude) second order directional derivative is considered to be the direction across the feature. Its perpendicular direction is considered to be the direction along the feature.

The second order directional derivatives of an image $u(x, y)$ can be computed from its Hessian matrix $\begin{bmatrix} u_{xx} & u_{xy} \\ u_{xy} & u_{yy} \end{bmatrix}$. The largest (in magnitude) eigenvalue, denoted by λ_η, is the maximal second order directional derivative. Thus the corresponding eigenvectors, denoted by v_η, is of the direction having the maximal second order directional derivative among all the directions. The other eigenvalue is denoted by λ_ξ. Its corresponding eigenvector, denoted by v_ξ, is perpendicular to v_η.

The general evolution equation using the Hessian method can be obtained by assigning the v_η-v_ξ pair to η-ξ pair. Because the eigenvalues are $u_{\eta\eta}$ and $u_{\xi\xi}$, the evolution equation can use the eigenvalues directly:

$$u_t = c\,(a\,\lambda_\eta + b\,\lambda_\xi).$$

The difference between the first order derivatives of the gradient method and the second order derivatives of the Hessian method is fundamental. The features are characterized by oscillations, which cannot be detected by the first order derivatives but can be captured by the second order derivatives. In particular, the lines and textures generate large second order derivatives but not large first order derivatives. The edges generate large second order derivatives at the sides and large first order derivatives at the centers. Overall, the second order derivatives are much more suitable for feature detection.

4 Gabor Method

The Gabor method is an approach using the local spectral analysis to extract feature direction. The direction of maximal spectral energy is considered to be the direction across the feature. Its perpendicular direction is considered to be the direction along the feature.

The spectral energy is obtained from the Gabor transform. The Gabor transform of an image $u(x, y)$ is the Fourier transform of the image being windowed by a Gaussian window $g(x, y)$ centered at each location:

$$\mathcal{F}(x, y; \omega, \theta) = \int u(r, s)g(r - x, s - y)\exp(-i\omega(r\cos(\theta) + s\sin(\theta)))\,dr\,ds.$$

Note that the Fourier variables $[\omega, \theta]$ are in a polar coordinate, ω being the absolute frequency and θ being the direction. At each location $[x, y]$, the Gabor transform $\mathcal{F}(\omega, \theta)$ is the response of the image value to the frequency ω in the direction θ.

The direction having maximal weighted accumulated spectral energy, denoted by θ_η, is considered to be the direction across the feature:

$$\theta_\eta = \arg \max_\theta \left\{ \int |\mathcal{F}(\omega, \theta)|^2 \, \mathbf{p}(\omega) \omega d\omega \right\}.$$

The weight function $\mathbf{p}(\omega)$ is for giving non-uniform scores to oscillations of different frequencies. For the same magnitude, the oscillations of higher frequencies should have higher score than that of lower frequencies. The discontinuities should have the highest score, higher than any oscillation. The weight function $\mathbf{p}(\omega) = \omega^3$ is used in our experiments, More study is needed in choosing the weight function.

The Gabor method is more general and more complete than the gradient method and the Hessian method. It utilizes information of all the frequencies in all the directions from a larger neighborhood, much more than the first and second order derivatives. However, this also makes the Gabor method more expensive to compute. A efficient implementation of the Gabor method is necessary in order to make it practically useful.

5   Experiment

The goal of our experiment is not to demonstrate the denoising properties of the smoothing, rather, it is to show the damaging effects of the smoothing reflecting different feature direction extraction methods.

We use the evolution equation of the Total Variation scheme [3]:

$$u_t = \frac{1}{(1 + |\nabla u|^2)^{\frac{1}{2}}} u_{\xi\xi},$$

with different methods to determine the direction of ξ and thus the value of $u_{\xi\xi}$. The image $u(x, y)$ is initialized to the input image $u_0(x, y)$ and is smoothed by running the evolution equation about 100 iterations until

$$\int |u(x, y) - u_0(x, y)|^2 \, dxdy = \sigma^2.$$

The value of σ is the same for all methods, thus the amount of the smoothing effects is the same for all cases. The smoothing has similar effects on noise, but quite different effects on features due to the different feature extraction methods.

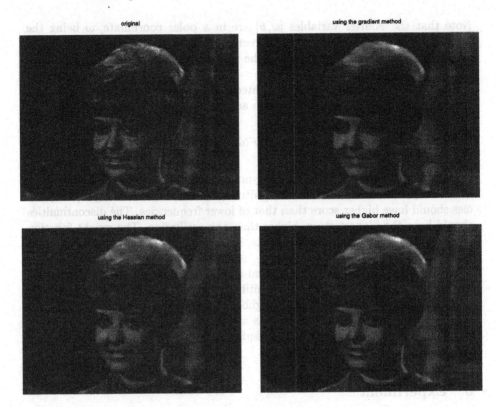

Fig. 3. top left: Image of 256 × 192 pixels. **top right**: Using the gradient method. **bottom left**: Using the Hessian method. **bottom right**: Using the Gabor method.

Fig. 3 shows the result of an image with the feature of waves. The hair on the forehead forms vertical wave textures. A horizontal highlight is across the hair. In the result of the gradient method, the smoothing mistakenly creates horizontal edges, responding to the large gradients due to the highlight but ignoring the oscillations that are more important to the hair. The Hessian method and the Gabor method capture the oscillations of the hair and thus handle the hair and highlight properly.

Fig. 4 shows the result of an image with the feature of lines. The fine lines are destroyed by the smoothing. The damage caused by the Hessian method and the Gabor method is significantly less than that by the gradient method.

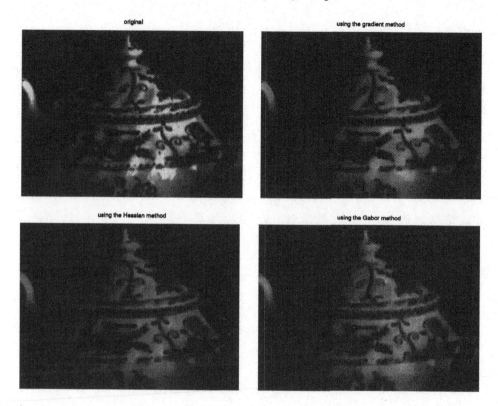

Fig. 4. top left: Image of 256 × 192 pixels. **top right**: Using the gradient method. **bottom left**: Using the Hessian method. **bottom right**: Using the Gabor method.

References

1. L. Alvarez, F. Guichar, P. Lions, and J. Morel, *Axioms and Fundamental Equation of Image Processing*, Archive for Rational Mechanics and Analysis, 123, pp. 199-257, 1993.
2. P. Perona and J. Malik, *Scale Space and Edge Detection Using Anisotropic Diffusion*, IEEE Transactions on Pattern Analysis and Machine Intelligence, 12, pp. 629-639, 1990.
3. L. Rudin, S. Osher, and E. Fatemi, *Nonlinear Total Variation Based Noise Removal Algorithms*, Physica D, 60, pp. 259-268, 1992.

Plate, top left: Image of 215 × 153 pixels, top right: Using the gradient method, bottom left: Using the Hessian method, bottom right: Using the Gabor method.

References

1. L. Alvarez, P. Guichar, F. Lions, and J. Morel. Axioms and Fundamental Equations of Image Processing. Archive for Rational Mechanics and Analysis, 123, pp. 199-257, 1993.

2. T. Poggio and J. Malik. Scale Space and Edge Detection Using Anisotropic Diffusion. IEEE Transactions on Pattern Analysis and Machine Intelligence, 12, pp. 629-639, 1990.

3. L. Rudin, S. Osher, and E. Fatemi. Nonlinear Total Variation Based Noise Removal Algorithms. Physica D, 60, pp. 259-268, 1992.

Author Index